Increases in computer power and technology have rapidly advanced the applications of numerical modeling in the environmental and earth sciences. The progress of numerical modeling in atmospheric, oceanic, and geophysical sciences was the topic of an international conference held by the National Autonomous University of Mexico (UNAM).

The review articles and research papers in this volume constitute a wide-ranging and up-to-date account of modeling environmental and earth processes through a variety of numerical simulations. The book is split into four parts. The first part covers General Circulation Models and global change. This is followed, in the second part, by chapters on the application of atmospheric modeling for time and space scales from the mesoscale to the single-cloud scale. The third part looks at methods of geophysical data assimilation, and the final part reviews the mathematical and computational methods that have potential applications in geophysics.

This book forms an excellent introduction and overview for graduate students as well as a critical update for researchers.

Numerical Simulations in the Environmental and Earth Sciences

Numerical Simulations in the Environmental and Earth Sciences

Proceedings of the Second UNAM-CRAY Supercomputing Conference

Edited by
Fernando García-García
Centro de Ciencias de la Atmósfera, Universidad Nacional Autónoma de México

Gerardo Cisneros
Silicon Graphics, S.A. de C.V. (Cray Research)

Agustín Fernández-Eguiarte
Instituto de Ciencias del Mar y Limnología, UNAM

Román Álvarez
Instituto de Geografía, UNAM

CAMBRIDGE
UNIVERSITY PRESS

PUBLISHED BY THE PRESS SYNDICATE OF THE UNIVERSITY OF CAMBRIDGE
The Pitt Building, Trumpington Street, Cambridge CB2 1RP, United Kingdom

CAMBRIDGE UNIVERSITY PRESS
The Edinburgh Building, Cambridge CB2 2RU, United Kingdom
40 West 20th Street, New York, NY 10011-4211, USA
10 Stamford Road, Oakleigh, Melbourne 3166, Australia

First published 1997

Printed in the United States of America

Typeset in LaTeX

Library of Congress Cataloging-in-Publication Data

UNAM-CRAY Supercomputing Conference (2nd : 1995 : National Autonomous
University of Mexico)
Numerical simulations in the environmental and earth sciences :
proceedings of the second UNAM-CRAY Supercomputing Conference /
edited by Fernando García-García ... [et al.]
p.   cm.
ISBN 0-521-58047-1
1. Environmental sciences – Mathematical models – Congresses.
2. Environmental sciences – Computer simulation – Congresses.
3. Earth sciences – Mathematical models – Congresses. 4. Earth
sciences – Computer simulation – Congresses. I. García-García,
Fernando, 1956–   II. Title.
GE45.M37U53 1997   96-39291
628–DC21   CIP

A catalog record of this book is available from the British Library.

ISBN 0 521 58047 1 hardback

Contents

I. General Circulation Models and Global Change

II. Dispersion and Mesoscale Modeling

III. Geophysical Data Assimilation

IV. Methods and Applications in Geophysics

Participants

Alberto Alonso y Coria, *DGSCA/UNAM, Mexico*
(alyco@servidor.unam.mx)

S. P. Arya, *North Carolina State University, USA*
(pal_arya@ncsu.edu)

Roberto Bonifaz, *IGg/UNAM, Mexico*
(bonifaz@indy2.igeograf.unam.mx)

Michael G. Brown, *Silicon Graphics, Inc., USA*
(mbrown@engr.sgi.com)

Abel Camacho-Galván, *IGf/UNAM, Mexico*
(abel@servidor.unam.mx)

David B. Carrington, *University of Nevada, Las Vegas, USA*
(swift@nge.nscee.edu)

Miguel Angel Castillo, *IGg/UNAM, Mexico*
(direc@igiris.igeograf.unam.mx)

Gerardo Cisneros-Stoianowski, *Cray Research de México, S.A. de C.V.,*
(gerardo@cray.com)

Alfredo Cortés, *IGg/UNAM, Mexico*
(direc@igiris.igeograf.unam.mx)

Felipe S. Cruz-Zárate, *DGSCA/UNAM, Mexico*
(felipe@labvis.unam.mx)

Marco de la Cruz-Heredia, *University of Toronto, Canada*
(marco@chinook.physics.utotonto.ca)

Terje O. Espelid, *University of Bergen, Norway*
(terje@ii.uib.no)

Agustín Fernández-Eguiarte, *ICMyL/UNAM, Mexico*
(eguiarte@mar.icmyl.unam.mx)

Rafael Fernández-Flores, *DGSCA/UNAM, Mexico*
(rafaelf@servidor.unam.mx)

Jorge Flores-Valdés, *CUCC/UNAM, Mexico*
(rafael@nadxeli.ifisicacu.unam.mx)

Artemio Gallegos, *ICMyL/UNAM, Mexico*
(gallegos@mar.icmyl.uanm.mx)

Elías Daniel Galván-Antonio, *FC/UNAM, Mexico*

Fernando García-García, *CCA/UNAM, Mexico*
(dire@mviica.atmosfcu.unam.mx)

Ian García-Olmedo, *DGSCA/UNAM, Mexico*
(igo@labvis.unam.mx)

José Gómez-Valdés, *CICESE, Mexico*
(jgomez@cicese.mx)

Jing Guo, *NASA/Goddard Space Flight Center, USA*
(guo@dao.gsfc.nasa.gov)

Catherine Gwilliam, *James Rennell Centre, NERC, UK*
(Catherine.S.Gwilliam@soc.soton.ac.uk)

Elizabeth Ann Hayes, *Cray Research, Inc., USA*
(hayes@mcnc.org)

Joaquín Hernández-Pérez, *DGSCA/UNAM, Mexico*
(jrhp@labvis.unam.mx)

Ismael Herrera-Revilla, *IGf/UNAM, Mexico*
(iherrera@tonatiuh.igeofcu.unam.mx)

Glenn Johnson, *Macquarie University, Australia*
(glenn@mpce.mq.edu.au)

Diana Liverman, *University of Arizona, USA*
(liverman@climate.geog.arizona.edu)

Tom Loveland, *U.S. Geological Survey, USA*
(loveland@edcsnw19.cr.usgs.gov)

Rong Lu, *University of California, Los Angeles, USA*
(rongl@atmos.ucla.edu)

Víctor Magaña, *CCA/UNAM, Mexico*
(victor@belenos.atmosfcu.unam.mx)

Gerard Meehl, *National Center for Atmospheric Research, USA*
(meehl@ncar.ucar.edu)

Gerardo M. Mejía-Velázquez, *ITESM-Monterrey, Mexico*
(gmejia@campus.mty.itesm.mx)

Carlos R. Mechoso, *University of California, Los Angeles, USA*
(mechoso@atmos.ucla.edu)

Víctor Manuel Mendoza-Castro, *CCA/UNAM, Mexico*

Anthony Meys, *Cray Research, Inc., USA*
(meys@cray.com)

Luis Rodolfo Meza-Peredo, *CCA/UNAM, Mexico*

Beatriz Oropeza-Villalobos, *IGf/UNAM, Mexico*
(beti@tonatiuh.igeofcu.unam.mx)

Elba Ortiz, *Instituto Mexicano del Petróleo, Mexico*
(elba@tzoalli.sgia.imp.mx)

Darrell W. Pepper, *University of Nevada, Las Vegas, USA*
(pepperu@nye.nscee.edu)

Héctor Perales-Valdivia, *DGSCA/UNAM, Mexico*
(hpv@labvis.unam.mx)

Enrique Pérez-García, *PUMA/UNAM, Mexico*
(enrique@servidor.unam.mx)

Ismael Pérez-García, *CCA/UNAM, Mexico*
(ismael@redvax1.dgsca.unam.mx)

George Philander, *Princeton University, USA*
(gphlder@splash.princeton.edu)

Participants

Mario Picazo-Soriano, *LABEIN Research Technological Center, Spain*
(mario@labein.es)

Eric Pitcher, *Cray Research, Inc., USA*
(eric.pitcher@cray.com)

Marc Pontaud, *CERFACS, France*
(pontaud@cerfacs.fr)

Graciela Raga, *CCA/UNAM, Mexico*
(raga@nefelos.atmosfcu.unam.mx)

Carlos Alberto Repelli, *FUNCEME, Brazil*
(repelli@zeus.funceme.br)

Julio Sheinbaum, *CICESE, Mexico*
(julios@cicese.mx)

Gustavo Silva-Guerrero, *ESPOL, Ecuador*
(gsilva@espol.edu.ec)

Nagendra Singh, *University of Alabama in Huntsville, USA*
(singhn@csparc.uah.edu)

Gustavo E. Sosa-Iglesias, *Instituto Mexicano del Petróleo, Mexico*
(gustavo@tzoalli.sgia.imp.mx)

Ricardo Todling, *NASA/Goddard Space Flight Center, USA*
(todling@dao.gsfc.nasa.gov)

Richard P. Turco, *University of California, Los Angeles, USA*
(turco@atmos.ucla.edu)

Diana N. Vázquez-Feregrino, *Servicio Meteorológico Nacional, Mexico*
(dvazquez@vmredipn.ipn.mx)

José A. Vergara, *University of Maryland, USA*
(jose@atmos.umd.edu)

Thomas T. Warner, *National Center for Atmospheric Research, USA*
(warner@monsoon.colorado.edu)

Jorge Zavala, *CICESE, Mexico*
(jzavala@cicese.mx)

YanChing Q. Zhang, *Lockheed Martin US/EPA Scientific Visualization Center, USA*
(yan@vislab.epa.gov)

Scientific Committee

Paul Densham
Department of Geography, University College London

Ismael Herrera
Instituto de Geofísica, UNAM

Diana M. Liverman
Department of Geography, Pennsylvania State University

Carlos R. Mechoso
Department of Atmospheric Sciences, UCLA

S. George Philander
Department of Geology, Princeton University

Julio Sheinbaum
Departamento de Oceanografía Física, CICESE

Ricardo Todling
Goddard Space Flight Center, NASA

Richard Turco
Department of Atmospheric Sciences, UCLA

Thomas T. Warner
National Center for Atmospheric Research

Organizing Committee

Román Álvarez-Béjar, Roberto Bonifaz-Alfonzo
Instituto de Geografía, UNAM

Eduardo Aguayo-Camargo, Agustín Fernández-Eguiarte
Instituto de Ciencias del Mar y Limnología, UNAM

Gerardo Cisneros-Stoianowski
Cray Research de México, S.A. de C.V.

Fernando García-García, Víctor Magaña-Rueda
Centro de Ciencias de la Atmósfera, UNAM

Víctor Guerra-Ortiz, Enrique Daltabuit-Godás, Alberto Alonso-y-Coria
Dirección General de Servicios de Cómputo Académico, UNAM

David Novelo-Casanova, Leticia Flores-Márquez, Marco Guzmán-Speziale
Instituto de Geofísica, UNAM

Octavio Rivero-Serrano, Simón González-Martínez, Enrique Pérez-García
Programa Universitario del Medio Ambiente, UNAM

Preface

Numerical modeling of atmospheric, oceanic and geophysical processes has always been a computationally demanding activity. The first supercomputers installed outside defense-oriented laboratories were devoted to numerical weather prediction and processing of geophysical data. Advances in the capability and performance of supercomputers have allowed researchers to tackle larger problems using fewer approximations, finer meshes and more realistic models. Newer, highly parallel computer architectures promise even larger performance gains, with a corresponding improvement in the physical insights and model accuracy to be gained.

The Second UNAM-CRAY Supercomputing Conference, **Numerical Simulations in the Environmental and Earth Sciences**, was held in the superb setting of *Universum*, a large science museum located in Ciudad Universitaria, the main campus of the National Autonomous University of Mexico (UNAM). The goal of this meeting was to bring together scientists working on a wide range of topics in the geosciences in order to exchange ideas and to stimulate new directions of research. A total of 59 participants from 10 different countries gathered together to discuss their most recent results, and the outcome of these fruitful three and a half days is reported in this book.

The meeting started with opening remarks by Dr. Gerardo Suárez-Reynoso, coordinator of scientific research at UNAM; Eric Pitcher, director of university marketing at Cray Research, Inc.; and Dr. José Sarukhán-Kermez, UNAM's rector. The conference consisted of a series of invited papers and submitted contributions; there was also a poster session. The presentations covered a wide range of topics that use supercomputers as a fundamental tool for research. Speakers organized their presentations as reviews of findings and discussion of challenges. Several papers described atmospheric and oceanic general circulation models, as well as coupled atmosphere–ocean models, and presented results of their investigations on El Niño/Southern Oscillation and climate change due to the increase in greenhouse gases. Other speakers addressed the advances and problems in the area of data assimilation for the atmosphere and the oceans. There was an invited presentation on mesoscale models of the atmosphere. Several speakers addressed the modeling of global and regional chemistry processes, with emphasis on air pollution. Advances in processing of satellite imagery were discussed. Another active area of research included in the program was seismology. A presentation focused on the importance for Latin America of modeling climate change. Submitted contributions complemented the invited presentations and extended consideration to other space and time scales. In the latter category, speakers discussed large eddy simulations, cloud models, atmospheric environmental management systems, and techniques to forecast marine productivity. Attention was also paid to computational methods used in large-scale problems. In the closing remarks of the Conference, Professor C. R. Mechoso (UCLA), representing the Conference's Scientific Committee, emphasized the importance to the Mexican as well as the international scientific community of the availability of supercomputer resources at UNAM. Those resources provide the support required by the development of successful research programs in areas of importance such as the environment. He also stated that continuing support was needed to maintain a state-of-the-art supercomputing facility at UNAM and encouraged earth and computer scientists to be prepared to work with the new computer architectures. The meeting motivated several discussions among participants on the establishment of future national and international collaborations.

These proceedings are organized in four parts, as follows: The first one includes presentations on the state of the art of both coupled and ocean and atmosphere General Circulation Models and some of their different applications to environmental problems.

In the second part a collection of papers is presented on applications of atmospheric modeling for a variety of time and space scales, from the mesoscale to the single-cloud scale. The third part covers specific methods applied to geophysical data assimilation, whereas the fourth part addresses general mathematical and computational methods with potential applications to a wide range of geophysical problems. The review papers provide a well-balanced overview of all these topics, and the peer-reviewed contributed papers illustrate current activities in each field. The final result is a comprehensive compendium of the most recent trends in environmental and geophysical simulations, with specific applications, which will allow the reader to have a general overview of these important and diverse topics of the geosciences.

We would like to acknowledge the members of the Scientific Committee, who reviewed and selected abstracts for inclusion in the program and refereed the full papers; Drs. John D. Farrara, Susana Gómez, Rong Lu, Chung-Chun Ma, Víctor Magaña and Everett Nickerson, who also helped review the full papers; and the other members of the Organizing Committee, for their excellent work in the design and organization of the meeting. We also thank our budget managers Ing. Enrique Pérez, of PUMA-UNAM, and Pedro Rocha, of DGSCA-UNAM, for sparing us the usual bookkeeping headaches with their very efficient financial organization, and to Toña Zimerman for her wonderful artwork and poster design. Very special thanks to Dr. Jorge Flores for graciously allowing us to use the wonderful auditorium at *Universum*, and to his staff, especially Arq. Serafín Pérez; to our talented executive secretaries Graciela González–de Hita and Luz del Carmen Pérez-Huerta, for maintaining the spirit of the meeting with their shining smiles and very efficient work, and to our enthusiastic crew Felipe Cruz, Ivonne Diéguez, Ian García, Juan Carlos de León-Violante, Lidia Maldonado, Héctor Perales-Valdivia, Rocío Reyes, Luis Felipe Rivera and Guadalupe Zárraga for keeping the meeting trouble-free.

Last, but certainly not least, we are very grateful for the generous grants provided by Cray Research, Inc., and Coordinación de la Investigación Científica (UNAM), and the financial support given by the Dirección General de Servicios y Cómputo Académico (UNAM).

Fernando García-García, Gerardo Cisneros,
Agustín Fernández-Eguiarte, Román Álvarez
México D. F., México, 1997

PART I
General Circulation Models and Global Change

A General Circulation Model of the Atmosphere–Ocean System

By Carlos R. Mechoso

Department of Atmospheric Sciences, University of California Los Angeles
405 Hilgard Avenue, Los Angeles, CA 90095-1565 USA

This paper examines the methodology used in studies with general circulation models (GCMs) of the atmosphere and oceans, the sensitivity of these models' performances to the representation of physical processes, and the computational challenges encountered in running their large codes. Our discussion of the atmosphere–ocean coupled GCM focuses on the difficulties in modeling the atmosphere–ocean system due to the complex interactions and feedbacks between its components. We present an example that underscores the sensitivity of the system to relatively modest changes in radiation. We show that enhanced stratus clouds off the coast of Peru result in significant local and remote cooling of the ocean surface and argue that those impacts are primarily due to different mechanisms. We discuss the relationships between surface heat fluxes and sea surface temperatures. The paper also includes a presentation of computational issues that arise in running the coupled GCM codes in a heterogeneous, distributed computer environment.

1. Introduction

To understand, and eventually predict, variations in climate is a particularly challenging task because of the complex interactions and feedbacks between different components of the system. General circulation models (GCMs) are computer codes that solve discretized expressions of the equations governing fluid motion, including parameterizations of physical processes at subgrid scales (e.g., cumulus convection, turbulence). Atmospheric and oceanic GCMs (AGCMs and OGCMs, respectively) have been extensively used in support of studies on the dynamics of the atmosphere and oceans. Running each of these models requires the specification of boundary conditions: (1) sea surface temperature (SST) and sea-ice cover for the AGCM and (2) surface fluxes of heat, water, and momentum for the OGCM. A coupled atmosphere–ocean GCM (coupled GCM) consists of an AGCM providing the surface fluxes of heat, water, and momentum to an OGCM, which returns the SST. (The sea-ice cover can be either prescribed or provided by a sea-ice module interacting with the other model components.) Coupled GCMs, therefore, can be used to study interactions and feedbacks of different components of the atmospheric and oceanic circulations.

This paper examines the methodology used in studies with GCMs, the sensitivity of these models' performance to the representation of physical processes, and the computational challenges found in running their large codes. Examples of outstanding problems that may be addressed with coupled GCMs are given in other papers of this volume. S. G. Philander focuses on the El Niño–Southern Oscillation (ENSO) events, which have dramatic impacts on the global climate (see, e.g., Ropelewski and Halpert [1987]; Pisciottano et al. [1994]). The ability to simulate and eventually predict ENSO with a coupled GCM is rightly considered a significant success of the model. G. A. Meehl and W. W. Washington show that a warming of the tropical Pacific Ocean (due to increased carbon dioxide and associated greenhouse effect, for example) would result in a stronger warming in the east than in the west, resembling ENSO conditions. M. Pontaud, L. Terray,

E. Guilyardi, E. Sevault, D. B. Stephenson and O. Thual discuss computational (as well as scientific) aspects of modeling the coupled atmosphere–ocean system.

2. The AGCM

The UCLA AGCM's prognostic variables are the horizontal velocity, potential temperature, water vapor mixing ratio, ozone mixing ratio, surface pressure, planetary boundary layer (PBL) depth, ground temperature, and snow depth. Formulation of the diabatic processes include parameterizations of cumulus convection [Arakawa and Schubert, 1974], PBL processes [Suarez *et al.*, 1983], and a radiation calculation (Katayama [1972] for the shortwave; and Harshvardhan *et al.* [1985] for the longwave). The geographical distributions of surface albedo, ground wetness, and sea ice are interpolated from prescribed monthly means based on the observed climatology. Details on the AGCM are given in Mechoso *et al.* [1987] and references therein. The model has two major components: (i) AGCM/Dynamics, which computes the evolution of the fluid flow governed by the primitive equations written in finite differences, and (ii) AGCM/Physics, which computes the effect of processes not resolved by the model's grid on processes that are resolved by the grid.

AGCMs have been extensively used to study the effects of Earth's orography on the atmospheric circulation. The traditional approach is based on contrasting a "control" simulation using full orography with "hypothesis-testing" experiments in which some or all orographic components are altered. As in Held [1983], we can refer to the "orographic component" of the response as the difference between the results obtained with and without orography, and "thermal component" of the response as the fields obtained without orography. Mechoso [1981], for example, showed that the orographic component of the response dominates the quasi-stationary wave field in the Southern Hemisphere. Tokioka and Noda [1986] showed that the orographic and thermal components of the response have comparable magnitude in the upper troposphere of the Northern Hemisphere. Mechoso [1981] and Tokioka and Noda [1986] found that removal of the mountains results in a warming of the south polar regions. The latter authors found that the removal also results in a cooling of the north polar regions.

A subset of the AGCM studies on orographic effects has focused on Antarctica. Antarctica is asymmetric about the pole, with massive elevations in the Eastern Hemisphere. This asymmetry has been linked to the dominant contribution of the component with zonal wavenumber 1 to the monthly-mean geopotential height field in the Southern Hemisphere. Such a quasi-stationary wave (QS-wave 1) is largest at about 60°S year round and reaches a maximum amplitude during September and October in the upper troposphere and stratosphere. An analysis of energy propagation in the Southern Hemisphere, however, suggests that the quasi-stationary wave field at high latitudes around Antarctica is primarily forced from lower latitudes, most prominently from the Indian Ocean [Quintanar and Mechoso, 1995a].

To address these issues, Quintanar and Mechoso [1995b] used a version of the UCLA AGCM. Their control simulation consisted of a long integration in the perpetual October mode (i.e., the solar insolation is fixed at its value for mid-October, and the prescribed geographical distributions of surface conditions correspond to the October climatology). The hypothesis-testing experiments consisted of AGCM integrations with surface conditions that excluded either (i) the Antarctic elevations, (ii) all orographic elevations, or (iii) all orographic elevations and zonal asymmetries in surface conditions (sea ice and sea surface temperature) south of 45°S. In another highly idealized experiment, they displaced the Antarctic elevations 180° in longitude while keeping all other surface boundary conditions unchanged from those in the control simulation.

The results obtained with the AGCM showed that none of the artificial modifications made in the model's boundary conditions has a drastic impact on the simulated QS-wave 1 around Antarctica. The amplitudes of QS-wave 1 in the experiments without the Antarctic elevations are only about 20% to 30% smaller than in the control simulation. There were, of course, large differences at polar latitudes just above the continent. These results strongly support the hypothesis that QS-wave 1 around Antarctica is primarily forced by factors *other than* the Antarctic orography and associated thermal effects. The Antarctic elevations also affect QS-wave 1 at high latitudes, but to a lesser extent and almost in phase with other forcings. Other planetary waves are significantly influenced by the zonal asymmetries of Antarctica and other geographic features of the southern polar region. For example, the zonal asymmetries in SST and sea ice seem to be crucial in determining the amplitude and phase of QS-wave 2 south of 50°S.

3. The Coupled Atmosphere–Ocean GCM

The atmospheric component of our coupled GCM is the UCLA AGCM. Concerning the OGCM, we have mostly used the Modular Ocean Model (OGCM/MOM), which evolved from the primitive-equation, finite-difference model developed at NOAA/Princeton University Geophysical Fluid Dynamics Laboratory (GFDL) by K. Bryan and M. Cox [Bryan, 1969; Cox, 1984]. The GFDL model, which uses depth as the vertical coordinate, has two major components: (1) OGCM/Baroclinic, which determines the deviations from the vertically averaged velocity, temperature, and salinity fields, and (2) OGCM/Barotropic, which determines the vertically averaged distributions of those fields. The OGCM/MOM uses a rigid lid as upper boundary condition. A sea-ice model [Flato and Hibler, 1992] is available for coupling to the OGCM. Another current version of the GFDL OGCM is the Parallel Ocean Program (OGCM/POP; Smith *et al.* [1992]). This version removes the rigid-lid approximation and includes sea surface height as a prognostic variable.

In this paper we use the Tropical Pacific version of MOM, which covers the tropical Pacific basin between 130°E and 70°W, 28°S, and 50°N, with a resolution of 1° longitude by 1/3° latitude between 10°S and 10°N. Poleward of 10° the meridional grid size increases gradually to 2° at the northern and southern boundaries. The model has 27 levels in the vertical between the surface and the ocean bottom at a constant depth of 4149 m.

3.1. *Sensitivity to the Radiation Calculation*

The performance of the coupled GCM can be extremely sensitive to even relatively modest changes in radiative heating. As indicated, the current version of the AGCM includes the Harshvardhan *et al.* [1987] scheme for parameterizing longwave radiative transfer; the previous version of the AGCM included the Katayama [1972] scheme. These schemes – which will be referred to as the "K" and "H" schemes in this paper – differ in the methods of calculation, absorbers, and effect of clouds on radiative transfer. Although the differences in the calculations did not appear to be crucial, we obtained unexpectedly drastic differences in the climate simulated by the nine-layer, 5° longitude × 4° latitude version of the AGCM with the two schemes coupled to the Tropical Pacific version of MOM [Ma *et al.*, 1994].

The simulation with the K scheme developed a severe climate drift almost from the beginning, as warm water surged eastward, eventually to cover most of the central equa-

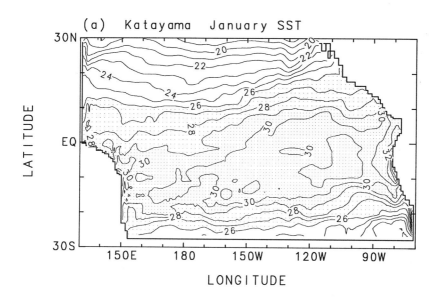

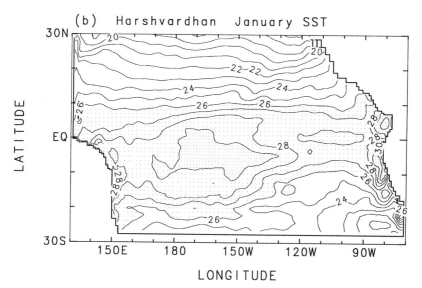

FIGURE 1. January sea surface temperature simulated by the coupled GCM using the longwave radiation scheme by (a) Katayama and (b) Harshvardhan *et al.*

torial Pacific Ocean (see Figure 1(a)). Consistently, regions of strong convective activity extended from the western to the central Pacific, and surface easterlies decreased in the central Pacific to be gradually replaced by surface westerlies. The experiment with the H scheme, on the other hand, simulated a realistic seasonal evolution of SST (see Figure 1(b)).

The difference shown in Figure 1 is *not* a simple manifestation of the difference in longwave radiative fluxes at the ocean surface. Our analyses of the results from the *uncoupled* AGCM demonstrated that, for identical distributions of SST, the net upward longwave flux at the surface is smaller in the tropics with the H scheme. This is consistent with the stronger downward flux due to the water vapor continuum effect included in the H scheme, but it is in apparent contradiction with the coupled GCM simulations, in which the H scheme resulted in colder SSTs. The net downward shortwave radiation is also smaller with the H scheme as a result of changes in cloudiness. The difference in sensible heat flux is very small in the tropics. The difference in latent heat flux, on the other hand, is much larger in magnitude than that corresponding to any other components of the heat flux. The sign of this difference is consistent with the behavior of the coupled GCM, in which the H scheme results in lower SSTs. The larger surface evaporation obtained with the H scheme is consistent with changes in other physical processes in the atmosphere. Interactions among radiation, hydrological, dynamical, and boundary-layer processes are essential in this difference, in which cumulus convection plays a key role.

3.2. *Influence of Stratus Clouds on Climate*

One of the most striking features of the climate over the eastern tropical Pacific is the large asymmetry of sea surface temperature (SST) about the equator throughout the seasonal cycle. There is no comparable asymmetry in the SST distribution simulated by almost all contemporary coupled GCMs [Mechoso *et al.*, 1995]. The simulated fields show warm biases extending thousands of kilometers off the coast of Peru in a zonal band just south of the equator. These SST errors go together with serious deficiencies in the simulated atmospheric circulation. The affected models produce a strong surface convergence over the anomalously warm waters just south of the equator, and their simulated seasonal cycle includes either a migration of the intertropical convergence zone (ITCZ) across the equator or a double ITCZ straddling the equator, depending on the model.

The extensive and persistent stratus cloud decks off the Peruvian coast are believed to play a key role in the coupled atmosphere–ocean processes that determine the SST throughout the eastern tropical Pacific. We used the coupled GCM to examine this notion [Ma *et al.*, 1996]. A control simulation, with an unrealistically low amount of simulated Peruvian stratus clouds, was compared to an idealized experiment with a prescribed stratus cloud deck persistently covering the ocean off the Peruvian coast (i.e., between 10°S and 30°S, and east of 90°W).

The coupled GCM results show great sensitivity to the amount of Peruvian stratus. The increase in stratus cloud quantity in the southeastern Pacific can lower SST by up to 5K by reducing shortwave radiation reaching the surface (see Figure 2). This cooling is sufficient to account for the regional warm bias in the simulations with the coupled GCM. The cooling also results in stronger and more realistic asymmetries – both north–south and east–west – in SST distribution. The enhanced north–south asymmetry prevents the ITCZ from crossing the equator during the northern spring. The surface wind stress and precipitation distributions also improve substantially. While the model's annual mean climatology is significantly improved in terms of overall pattern and strength of the low-level winds, this occurs partly at the expense of the SST seasonal cycle, whose amplitude in the eastern equatorial Pacific is reduced greatly in the experiment with enhanced stratus.

The results of the coupled GCM simulations support the notion that the Peruvian stratus clouds exert a strong influence on the climate in regions away from the coast of Peru. A substantial cooling even extends westward along the equator to the dateline. Different mechanisms are responsible for the remote effect in different regions, as shown schematically in Figure 3. The enhanced large-scale meridional and zonal SST gradients drive a thermally direct circulation that strengthens the Hadley/Walker circulation.

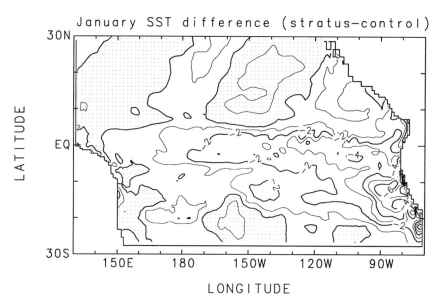

FIGURE 2. Difference in sea surface temperature in January between the experiment with stratus prescribed in the southeastern Pacific and the control simulation. Contour interval is 1 K; regions with positive values are stippled.

These changes in atmospheric circulation increase the surface wind speeds and thereby enhance evaporation over the eastern Pacific. This increased evaporation accounts for the cooling of SST in the proximity of the prescribed stratus region. At the equator, enhanced oceanic cold advection is largely responsible for the stronger cold tongue away from the coast in the stratus experiment.

Nevertheless, the lower SSTs obtained by increasing stratus off the coast of Peru do not seem to eradicate the coupled GCM's tendency to produce a zonally oriented belt of warm SSTs south of the equatorial cold tongue. There is still a region of relatively high SSTs near 5°S in the stratus experiment. Precipitation is greatly reduced, but there is still an unrealistic surface wind convergence. These results indicate that the coupled GCM simulations can benefit substantially from improved parameterization of the marine stratus clouds.

3.3. *Relationships between Surface Heat Fluxes and SST*

The coupled GCM performance can be highly sensitive to the successes and failures of the individual model components because of air–sea interaction and complicated feedbacks. One method to gain insight into this sensitivity is to compare the coupling fields obtained with the coupled GCM and those obtained in simulations with the uncoupled AGCM and OGCM using prescribed boundary conditions corresponding to either an observed climatology or a specified period of interest.

Figure 4 shows the errors in the surface heat flux simulated by the uncoupled AGCM and coupled GCM. [The verification dataset is Oberhuber's [1988] estimates from observational data mainly derived from the Comprehensive Ocean–Atmosphere Data Set (COADS).] The uncoupled AGCM overestimates the heat flux out of the ocean in the northern and southern subtropics, particularly along the coast of Mexico and Central America. It also overestimates the flux into the ocean along the coast of South Amer-

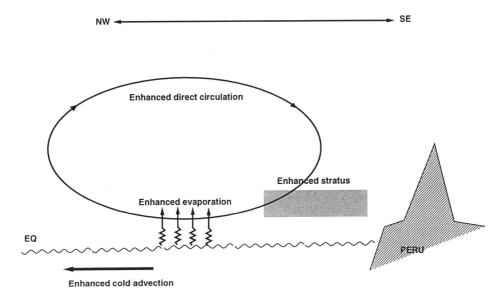

FIGURE 3. Schematic showing the mechanisms responsible for ocean cooling from increased
stratus off the Peruvian coast.

ica. The coupled GCM produces a similar error pattern, but with significantly weaker
magnitudes. An analysis of these fields shows that the principal contributor to the large
AGCM surface heat flux errors in the subtropics of the Northern Hemisphere is excessive
evaporation [Yu and Mechoso, 1996]. The large errors along the coast of South America, on the other hand, are primarily due to excessive net radiation flux into the ocean.
(These errors in net radiation flux are dominated by those in shortwave flux).

The coupled GCM, therefore, simulates heat fluxes that are generally closer to observational estimates than those simulated by the AGCM. Obviously, surface heat flux
errors associated with AGCM deficiencies in the parameterization of subgrid scale processes tend to be compensated in the coupled GCM through errors in SST. This is
broadly confirmed by a comparison between Figure 4(a) and Figure 4(c), which shows
the annual-mean SST errors produced by the coupled GCM. (In this paper, SST errors
of the coupled GCM are defined in reference to the dataset prescribed in the uncoupled
AGCM.) It is apparent that in regions off the equator where the uncoupled AGCM
produces excessive heat flux into (out of) the ocean, the coupled GCM produces SSTs
that have a warm bias (cold bias). This tendency toward compensation off the equator
between annual-mean surface heat flux and SST errors also holds for the corresponding
monthly-mean fields. In the monthly means, therefore, where the uncoupled AGCM
produces excessive heat flux into (out of) the ocean, the coupled GCM produces SSTs
that have a warm bias (cold bias). Off the equator, therefore, the surface heat flux is
the primary forcing to the seasonal cycle of SST, and the feedbacks between surface heat
flux and SST are negative and act locally.

At the equator, on the other hand, there is no clear tendency toward compensation
between surface heat flux and SST errors, particularly over the cold tongue in the eastern
Pacific. Here, the seasonal cycle of latent heat flux produced by the uncoupled AGCM is
primarily determined by the wind speed in the PBL, while that by the coupled GCM is

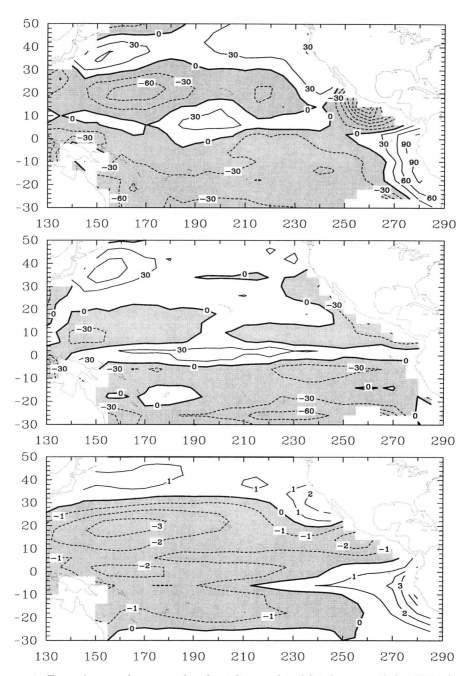

FIGURE 4. Errors in annual-mean surface heat flux produced by the uncoupled AGCM (upper panel) and coupled GCM (middle panel). The lower panel shows the errors in the annual-mean sea surface temperature produced by the coupled GCM. Contour intervals are 30 W/m^2 in the upper and middle panels, and 1 K in the lower panel. Negative values are shaded.

TABLE 1. Wall-clock time required to simulate 1 day (s) and performance (Gflop/s) with the 9-layer, 2.5° longitude × 2° latitude versions of the AGCM. The number in parentheses following the platform is the number of processors.

Platform	Total	AGCM/Dynamics	AGCM/Physics
CRAY C90 (1)	600/0.20	264/0.23	336/0.18
CRAY T3D (256)	115/1.0	93/0.65	22/2.7
IBM SP-2 (144)	94/1.3	72/0.84	22/2.7

primarily determined by the humidity difference between PBL air and the surface. This effect seems to be primarily due to nonlocal effects.

4. Computational Aspects

We have developed parallel versions of the AGCM and OGCM. When the model subdomains are distributed on several processors, AGCM/Physics scales very well since almost no communication is required between subdomains [Mechoso, 1995]. AGCM/Dynamics, on the other hand, scales less well since substantial amounts of interprocessor communication are required by the horizontal finite differencing scheme (see Table 1). The top performance obtained with the Tropical Pacific version of MOM (in which there are no islands) is 0.67 Gflop/s on 256 processors of an Intel Paragon.

To analyze the behavior of different codes in various computer environments, understand bottlenecks and scalability issues, and explore design tradeoffs, we have built a timing model that predicts the wall-clock running time as a function of code configuration, machine parameters, data layout, and algorithm. We are making significant progress toward building a framework for climate model simulations. In this context, we developed a system to archive model output in standard formats, and to catalogue descriptive information about the output in a database management system (DBMS) using a new data schema (BigSur). A prototype distributed system was tested, and estimates of its performance with and without remote data archival were obtained. The simulations with the coupled GCM resulting from all of these development efforts will have the following characteristics: (1) performance at tens of Gflop/s, (2) parallelized components so that the wall-clock time required to run the model is comparable to that of the slowest model component, (3) output consisting of metadata catalogued via a database management system using the BigSur data schema and data itself in standard formats, (4) code running in a single multiprocessor computer or distributed in a heterogeneous computer environment connected by high-speed networks.

5. Distribution of the Code

A unique feature of the UCLA coupled GCM is that its components are being configured to run both in a single supercomputer and distributed in several computers with eventually different architectures and connected via either local or wide-area networks with broad bandwidths [Mechoso, 1995]. An experimental run of the coupled GCM using two CRAY Y-MPs connected by a T1 link (NSFnet), with the AGCM at the National Center for Atmospheric Research (NCAR) in Boulder, Colorado, and the OGCM at the San Diego Supercomputer Center (SDSC) in San Diego, California, was performed in summer of 1992. To our knowledge, this is the first time that two supercomputers at different locations have been used for a coupled GCM integration. In this experiment, in which neither the computers nor the network was dedicated, the wall-clock time was dominated by communication time.

The availability of multiple nodes in the computer environment allows for the parallel execution of major model components. For example, for the AGCM on one processor of the CRAY C90 at SDSC coupled to the OGCM on 144 processors of the Intel Paragon at Caltech connected by the Gigabit Network we obtained a speedup of 1.483 in reference to the coupled GCM code on the C90. We can expect higher speedups if we use a scheme specifically designed for overlapping communications with computations in climate models [Mechoso et al., 1993]. In this case, for example, running the coupled GCM distributed such that the AGCM/Dynamics runs on one processor of a C90 and the AGCM/Physics coupled to the OGCM running each on 242 processors of a CRAY T3D will result in a speedup of 5.448 with respect to the C90 and 2.620 with respect to the T3D [Mechoso *et al.*, 1995].

We have developed and completed initial testing of remote coupling between model and DBMS. The AGCM, running on a DEC 3000/500 workstation at UCLA in Los Angeles, California, sent output to be archived at UCB in Berkeley, California, over the T3 (45 Mbits/s) Sequoia 2000 network that links several campuses of the University of California system. The data were registered with the Illustra DBMS.

6. Summary and Conclusions

We started our discussion of the UCLA coupled GCM by presenting an application of its atmospheric component. The AGCM was used to gain insight into the generation mechanisms of the quasi-stationary (monthly-mean) wave field around Antarctica. A comparison between integrations with realistic and idealized surface elevations strongly supported the hypothesis that the dominant component of the quasi-stationary wave field around Antarctica is primarily forced by factors *other than* the Antarctic orography and associated thermal effects. Studies with idealized models, which also include idealized forcings, have assigned a more important role to the Antarctic elevations. The intensity of these idealized forcings, however, depends on "tunable" coefficients. There is, therefore, some degree of ambiguity in the results since the values of the coefficients may be selected to produce realistic amplitudes in a feature of interest, while this feature may be strongly influenced by processes not included in the idealized model. The AGCM studies clarify that the response to Antarctica is one component, albeit not the most important, in the atmospheric circulation at the high latitudes of the Southern Hemisphere.

We presented an example that underscores the sensitivity of the coupled atmosphere–ocean system to relatively modest changes in radiation, which can easily be produced by relatively small changes in cloudiness. We showed that enhanced stratus clouds off the coast of Peru result in significant local cooling of the ocean, as expected from decreased solar radiation reaching the surface, and also in significant cooling over much of the eastern tropical Pacific south of the equator, and even along the equator well into the central Pacific. In this context, Peruvian stratus clouds are important in modulating the circulation of the tropical Pacific. The coupled GCM behavior simulations reveal that the remote ocean cooling in different regions can be due to different mechanisms. The SST cooling immediately to the west and north of the region with the prescribed stratus deck is primarily associated with increased evaporation as the southeasterly trade winds increase. The cooling along the equator in the central Pacific is mainly due to increased oceanic cold advection. Our simulations also indicate that, off the equator, surface heat

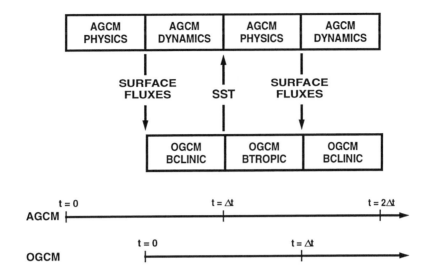

FIGURE 5. Schematic of the sequence of execution of the coupled AGCM/OGCM.

flux errors produced by AGCM deficiencies are compensated in the coupled GCM by SST errors. This tendency towards compensation of errors suggests that the surface heat flux is the primary forcing to the seasonal cycle of SST, and ocean dynamics plays a less important role. At the equator, on the other hand, the compensation between surface heat flux and SST is less clear as coupled atmosphere–ocean dynamics plays a more important role.

These selected examples emphasize the difficulties in modeling the coupled atmosphere–ocean system due to the complex interactions and feedbacks between its components. In view of these difficulties, it is not surprising to detect errors in the fields simulated by coupled GCMs. Interestingly, a comparison of results from several contemporary models indicated that, in spite of the considerable differences between the models, all had similar problems and deficiencies, especially in the southeastern tropical Pacific [Mechoso *et al.*, 1995a]. Here, the simulated cold tongue is often too strong, too narrow, overelongated, perennial, and premature in its onset; SSTs are overestimated off the coast of Peru; some models have a perennial double ITCZ associated with this, while in others, the ITCZ migrates across the equator with the seasons, and the semiannual harmonic is often overestimated. Preliminary results from a more detailed intercomparison of simulations performed with AGCMs at Princeton University and at UCLA coupled to the same OGCM (the Tropical Pacific version of MOM) indicate that, although the coupled models may have similar flaws, the reasons for those flaws are different. In the Princeton simulations, the absence of low-level stratus clouds seems to be the principal reason for that model's problems. In the UCLA simulations enhanced stratus clouds off the Peruvian coast largely alleviate the model's local warm bias in SST and produce a distribution of SST with more realistic interhemispheric asymmetries. On the other hand, the model's tendency to produce a zonal band of warm SSTs south of the cold tongue is not eradicated and the model's cold bias in the tropics is exacerbated. According to these results, the "double ITCZ" syndrome of the coupled GCM does not appear to be solely due to underpredicted stratus cloud cover and requires consideration of other processes in the model.

Coupled GCMs are making rapid progress toward a successful simulation of the coupled atmosphere–ocean system. Such progress requires the collaboration of meteorologists, oceanographers, and computer scientists. Their efforts are amply justified since coupled GCMs are fundamental tools for the study of climate and climate change.

Acknowledgments. J. D. Farrara, C.-C. Ma and J.-Y. Yu contributed to this paper with useful comments and by preparation of the figures. D. McDonald typed the manuscript. This work was supported by DOE/CHAMMP under Grant DE-FG03-91ER61214JA004. Model integrations were performed at the San Diego Supercomputer Center.

REFERENCES

Arakawa, A., and W. H. Schubert, Interaction of a cumulus cloud ensemble with the large-scale environment, Part I, *J. Atmos. Sci.*, **31**, 674–701, 1974.

Bryan, K., A numerical method for the study of the circulation of the world ocean, *J. Comp. Phys.*, **4**, 347–376, 1969.

Cox, M. D., A primitive equation three-dimensional model of the ocean, *GFDL Ocean Group Tech. Rep. No. 1*, 1984.

Flato, G. M., and W. D. Hibler III, Modeling pack ice as a cavitating fluid, *J. Phys. Ocean.*, **22**, 626–651, 1992.

Harshvardhan, R. D., D. A. Randall, and T. G. Corsetti, A fast radiation parameterization for general circulation models, *J. Geophys. Res.*, **92**, 1009–1016, 1987.

Held, I. M., Stationary and quasi-stationary eddies in the extratropical troposphere: Theory. *Large-Scale Dynamical Processes in the Atmosphere*, B. J. Hoskins and R. P. Pearce, Eds., Academic Press, 127–168, 1983.

Katayama, A., A simplified scheme for computing radiative transfer in the troposphere. *Numerical Simulation of Weather and Climate Tech. Rep. No. 6*, Dept. of Atmospheric Sciences, University of California, Los Angeles, 1972.

Ma, C.-C., C. R. Mechoso, A. Arakawa, and J. D. Farrara, Sensitivity of a coupled ocean–atmosphere model to physical parameterizations, *J. Climate*, **7**, 1883–1896, 1994.

Ma, C.-C., C. R. Mechoso, A. W. Robertson, and A. Arakawa, Peruvian stratus clouds and the tropical Pacific circulation: A coupled ocean–atmosphere GCM study, *J. Climate*, **9**, 1635–1645, 1996.

Mechoso, C. R., Topographic influences on the general circulation of the Southern Hemisphere: A numerical experiment, *Mon. Wea. Rev.*, **109**, 2131–2139, 1981.

Mechoso, C. R., High-performance computing and networking for climate research. *Lecture Notes in Computer Science 919: High-Performance Computing and Networking*, B. Hertzberger and G. Serazzi, Eds., Springer, 142–147, 1995.

Mechoso, C. R., A. Kitoh, S. Moorthi, and A. Arakawa, Numerical simulations of the atmospheric response to a sea surface temperature anomaly over the equatorial eastern Pacific Ocean, *Mon. Wea. Rev.*, **115**, 2936–2956, 1987.

Mechoso, C. R., C.-C. Ma, J. D. Farrara, J. A. Spahr, and R. W. Moore, Parallelization and distribution of a coupled atmosphere–ocean general circulation model, *Mon. Wea. Rev.*, **121**, 2062–2076, 1993.

Mechoso, C. R., A. W. Robertson, N. Barth, M. K. Davey, P. Delecluse, P. R. Gent, S. Ineson, B. Kirtman, M. Latif, H. Le Treut, T. Nagai, J. D. Neelin, S. G. H. Philander, J. Polcher, P. S. Schopf, T. Stockdale, M. J. Suarez, L. Terray, O. Terray, O. Thual, and J. J. Tribbia, The seasonal cycle over the tropical Pacific in general circulation models, *Mon. Wea. Rev.*, **123**, 2825–2838, 1995.

Meehl, G. A., Seasonal cycle forcing of El Niño–Southern Oscillation in a global coupled ocean–atmosphere GCM. *J. Climate*, **3**, 72–98, 1990.

Oberhuber, J. M., An atlas based on the COADS data set: The budgets of heat buoyancy and turbulent kinetic energy at the surface of the global ocean, *Max-Planck-Institut für Meteorologie Report No. 15*, Bundesstrasse 55, 2000 Hamburg 13, FRG, 1988.

Pisciottano, G., A. Diaz, G. Cazes, and C. R. Mechoso, El Niño–Southern Oscillation impacts on rainfall in Uruguay, *J. Climate*, **7**, 1286–1302, 1994.

Quintanar, A. I., and C. R. Mechoso, Quasi-stationary waves in the Southern Hemisphere, Part I: Observational data, *J. Climate*, **8**, 2659–2672, 1995a.

Quintanar, A. I., and C. R. Mechoso, Quasi-stationary waves in the Southern Hemisphere, Part II: Generation mechanisms, *J. Climate*, **8**, 2673–2690, 1995b.

Ropelewski, G., and M. S. Halpert, Global and regional scale precipitation patterns associated with El Niño–Southern Oscillation, *Mon. Wea. Rev.*, **115**, 2161–2165, 1987.

Smith, R. D., J. K. Dukowicz, and R. Malone, Parallel ocean general circulation modeling, *Physica D*, **60**, 38–61, 1992.

Suarez, M. J., A. Arakawa, and D. A. Randall, The parameterization of the planetary boundary layer in the UCLA general circulation model: Formulation and results, *Mon. Wea. Rev.*, **111**, 2224–2243, 1983.

Tokioka, T., and A. Noda, Effects of large-scale orography on the January atmospheric circulation: A numerical experiment, *J. Meteor. Soc. Japan*, **606**, 819–839, 1986.

Yu, J.-Y., and C. R. Mechoso, Relationships between surface heat flux and SST in the Tropical Pacific, Submitted to *Dyn. Atmos. Oceans*, 1996.

Coupled Ocean–Atmosphere Modeling: Computing and Scientific Aspects

By M. Pontaud, L. Terray, E. Guilyardi, E. Sevault,
D. B. Stephenson and O. Thual

Centre Européen de Recherche et de Formation Avancée en Calcul Scientifique
CERFACS, av. G. Coriolis, 31057 Toulouse-Cedex, France

Successful prediction of anthropogenic global climate change requires one to take into account the coupling phenomena between the ocean and the atmosphere. Coupled General Circulation Models are useful tools because they include a complete set of physical processes controlling the climate mean state and its seasonal and interannual variability. A coupling interface named OASIS has been developed in order to handle any type of General Circulation Model (GCM). One of its main features is the possibility of running distributed ocean–atmosphere simulations. The GCMs run on remote supercomputers and data are exchanged through a high-speed network. The second feature is both a database and analysis software called VAIRMER. It gives a quick, simple and unified way to analyze and visualize the large amount of data generated by this type of simulation. On the scientific side, analysis of the time-mean climatology and variability is presented for simulations with a tropical Pacific ocean and with all the three ocean basins.

1. Introduction

In order to simulate the main features of the global climate, and its variability, a climate model must interactively represent the atmosphere, ocean and cryosphere. With the advent of supercomputers, improved understanding of global climate processes and computationally efficient General Circulation Models (GCMs), global coupled models are now used for multidecadal climate integrations where the atmospheric and oceanic GCMs can be run synchronously.

In Section 2, the computing aspects are described: the OASIS coupler, an experiment in distributed computing between two remote CRAY supercomputers and the VAIRMER postprocessing software. The results of three different coupled simulations are presented in Section 3. These are the numerical results of the coupling of the following atmospheric and oceanic models.

The atmospheric component is the ARPEGE-Climat model [Déqué et al., 1994], derived from the spectral ARPEGE-IFS forecast model, developed jointly by Météo-France and the European Center for Medium-Range Weather Forecasts (ECMWF). A low horizontal resolution (triangular spectral truncation at wavenumber 21 – T21) and a high horizontal resolution (T42) have been used. The model has 30 vertical levels and a good resolution in the stratosphere, with prognostic ozone.

The oceanic components are the Pacific and global versions of the OPA primitive equations Oceanic GCM (OGCM) developed at the Laboratoire d'Océanographie Dynamique et de Climatologie (LODYC) [Delecluse et al., 1993]. The horizontal resolution in the tropics is $0.75° \times 0.3°$ (longitude $\times$ latitude) in the Pacific version, and $2° \times 0.5°$ in the global version. There are 28 vertical levels in the Pacific version and 31 in the global version. A turbulent kinetic energy scheme is used for the vertical diffusion [Blanke and Delecluse, 1993].

2. Computing Aspects

2.1. *The OASIS Coupler*

The OASIS coupler [Terray, 1994] is a set of Fortran routines which allows the coupling of oceanic and atmospheric GCMs on a supercomputer. OASIS is extremely modular as each routine has only one well defined task. It respects the GCM structure as very few changes have to be implemented in the two models being coupled. In particular, the specific options of each model (multitasking, input/output [I/O], number of central proccessing units [CPUs] used) are left untouched. Furthermore, the three programs are considered as different processes in the UNIX sense and so synchronization is done by using message passing with special files (named pipes). The coupler performs all the coupling stages sequentially, allowing an easy understanding of its structure. It offers multiple options regarding interpolation schemes, I/O formatting, and the type of models being coupled.

The current version of OASIS was extensively validated during 1994. It is now used routinely in the French climate community and recently at the ECMWF. Current efforts involve the implementation of the Coupling Library for Interfacing Models (CLIM) within the OASIS program. This library allows distributed coupled simulations as well as coupling of parallel models on massively parallel processor (MPP) machines. Version 2.0 will have these features as well as some changes in the code structure (use of the cpp C-preprocessor, portability to workstations, dynamic definition of the coupling fields).

2.2. *Distributed Computing*

Mechoso *et al.* [1993] have tested the distribution of a climate model on the basis of domain, task and I/O decomposition. Their results demonstrate that the distributed computing should facilitate performance of the long integrations required by climate studies. CLIM [Sevault, 1994; Sevault *et al.*, 1994a; Sevault *et al.*, 1994b] provides an easy way for coupling independent models running on various platforms under Parallel Virtual Machine (PVM). To build this tool, some specifications have been gathered:

• CLIM allows one to couple parallel codes even if the parallelism is expressed outside the temporal loop.

• Programs should remain independent UNIX processes with no hierarchical dependence.

• Reliability is of major importance. Therefore, multiple checks are performed before any communication operation. Time-out control and detailed trace file are provided as well as two levels of error codes.

• The global configuration of the coupled application is hidden as much as possible and has to be fixed only at execution time. Real-case simulations have been carried out: CLIM was used with OASIS to couple the atmospheric model ARPEGE to the oceanic model OPA.

The goal of the so-called CATHODe (Couplage ATmospHère Océan Distribué) project is to run the atmosphere and the ocean GCMs on two different supercomputers communicating through the network. A 2 year simulation has been performed in which ARPEGE ran on a CRAY C90 in Paris and the global version of OPA ran on a CRAY-2 in Toulouse (the two places are about 700 km apart). Both supercomputers were in shared mode. The two codes communicate through the 2 Mbit/s RENATER network. Each day of simulation, the two codes exchange their fluxes, which are interpolated from one grid to another by the coupler OASIS. A coupling procedure based on the PVM software was tested.

Our climate experiments are shared in "small" successive jobs which simulate 5 days. The elapsed time needed for one "5-day simulated" job is around 400 seconds with the distributed version of the coupling compared to the 300 seconds in the standard version (i.e., both GCMs running on the same computer). The increase of the elapsed time arises from the transfer of the coupling data through the network. On the other hand, the input waiting time for each job is around 7 seconds with the distributed version compared to the 2000 seconds with the standard version. With the standard version, the simulation runs on a low-priority batch queue (because of large memory requirements), while in the distributed version, each GCM can access queues with greater priority and number of running jobs. The distributed version induces a significant decrease in the duration of the climatic experiments and allows one to increase the GCM resolution for the same cost as the standard version. Nevertheless, performance depends very strongly on the speed and reliability of the network connecting the remote computers.

These achievements have confirmed that distributed computing is practical for big codes and opens the way for further applications in which the localization of the code environment is a critical factor.

2.3. *The VAIRMER Experiment Manager*

The aim of the experiment manager is to provide a set of tools for the launching, control and postprocessing of numerical climate experiments. The Global Change team of CERFACS, in charge of the design of the French community "coupler" OASIS, has developed an *integrated* package that can handle the analysis tools of a heterogeneous climate research community: VAIRMER [Guilyardi, 1994; Guilyardi *et al.*, 1994]. VAIRMER provides an efficient and homogeneous environment to store, analyze, compare and visualize experiments done with the French community General Circulation Models. It requires little time investment and no specific computer knowledge, and is simple and powerful. VAIRMER is an open system: its modular UNIX structure makes it easy for the user to implement new models and new functionalities. VAIRMER is now used routinely by the French climate community for storage, analysis and visualization purposes and a unified visualization format allows easy intercomparison of the results.

3. Scientific Aspects

3.1. *Strategy*

The ARPEGE–OPA coupled model is a first contribution of the French climate community to the simulation of global climate and its variability.

A first stage was testing with several forced mode experiments of both the atmosphere and ocean models. These were necessary to establish "reference" experiments and to establish the initial state for the coupled experiments.

The Atmospheric GCM (AGCM) ARPEGE was forced with the Center of Ocean-Land Atmosphere Interactions, University of Maryland – Climate Analysis Center (COLA–CAC), Washington, D.C., observations of sea surface temperature (SST) for the period 1979–1993 (the experiment was called BF5). This experiment has been used to validate the version 1.0 of ARPEGE-Climat.

The OGCM OPA was forced (in the experiment FG2, which spanned 15 years) by the daily averaged energy, momentum and water fluxes generated by experiment BF5. The model was run in *robust diagnostic* mode: Observed temperature and salinity were relaxed toward Levitus climatology below the thermocline, outside the tropics and away from the coast. The tropical Pacific SSTs obtained from this integration are close to observations (good seasonal cycle) and show a realistic interannual variability. Global poleward energy transports agree fairly well with observations.

The simple ice parameterization only involves a test on the SST. Ice cover is present when the SST is below the freezing temperature and fixed heat flux is imposed on the ocean: -2 W/m^2 in the Northern Hemisphere and -4 W/m^2 in the Southern Hemisphere [Maykut and Untersteiner, 1971; Stössel *et al.*, 1989]. More sophisticated schemes are currently being developed.

3.2. *Three Coupled Simulations*

The ARPEGE and OPA models have been coupled with the OASIS coupler developed at CERFACS. The coupled model is only forced by seasonally varying insolation. The three programs run in parallel and exchange coupling fields once per day, averaging out the diurnal cycle. SST and, for the global OGCM, sea ice extent are given to the AGCM, and surface fluxes of heat, momentum and fresh water are passed to the OGCM. The global coupled model uses neither artificial flux correction techniques at the air–sea interface nor relaxation towards climatology in the ocean. In the Pacific version, the temperature and salinity fields are constrained to climatological values in deep water away from the equator and from the coastline.

Three coupled simulations have been performed. First, the Pacific OGCM has been coupled with the T42 ARPEGE AGCM for 10 years (in the experiment CO1) [Terray *et al.*, 1995; Mechoso *et al.*, 1995]. In addition, the global OGCM has been coupled with the T42 ARPEGE AGCM for 25 years (the experiment CG3), and with the T21 ARPEGE AGCM for 50 years (the experiment CA1) [Guilyardi *et al.*, 1995].

3.3. *The Pacific Coupling*

After a short and weak initial warm shock, the Pacific coupled model shows no drift in SST and exhibits a regular seasonal cycle in the equatorial Pacific. The climatological mean state (Figure 1) is slightly warmer in the western Pacific and north of the equator, and significantly warmer along the coast of South America, but the seasonal cycle of equatorial SSTs captures the annual harmonic quite realistically (Figure 2 and also Figure 9.b in Terray *et al.* [1995]). The oceanic surface current pattern is accurately depicted, and the location and intensity of the Equatorial Undercurrent (EUC) are in good agreement with available data (Figure 3). Some deficiencies remain, including a weak zonal equatorial SST gradient, underestimated wind stress over the Pacific equatorial band, and an overly strong convergence zone extending south of the equator in northern winter and spring. The atmospheric surface layer over the eastern part of the tropical oceans simulated by the AGCM is too dry. Weak interannual variability is present in the equatorial SST with a maximum amplitude of 0.5°C.

Sensitivity experiments with cloud and convection parameterizations have been performed (L. Terray, personal communication – paper submitted to *J. Climate*). Decreasing the entrainment rate with height in the convective parameterization improves the representation of the convergence zones. In addition, not enough low clouds are generated, and the direct solar radiation is roughly 50 W/m^2 too high. When the low cloud parameterization is modified to be more sensitive to low values of humidity, the mean state simulated by the coupled model becomes too cool. It seems possible to fit the cloud parameterizations to simulate a realistic mean state and seasonal cycle in the tropics, but there is no assurance that the tropical improvement will maintain the quality in extratropical regions.

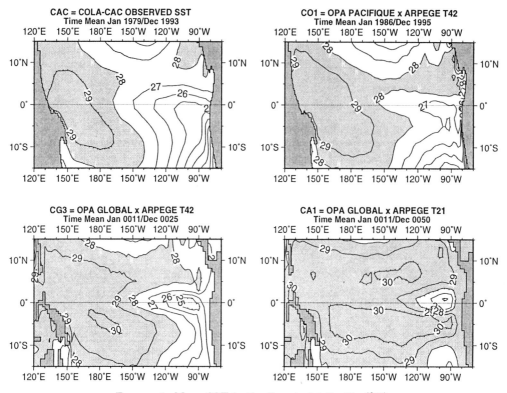

FIGURE 1. Mean SST in the Equatorial Pacific (°C).

3.4. *The Global Coupling*

The two global–ocean coupled simulations show initial tropical warming (within a few months) more pronounced with the T21 AGCM (+2°C) than with the T42 AGCM (+1°C). Both simulations lead to relatively stable mean states after about 10 years (Figures 1 and 3). The drift of the global mean oceanic temperature is between +0.3°C and +0.4°C per century. The warm pool is warmer than the observations and the 28°C isotherm is shifted eastward (T42: around 150°E; T21: around 120°E). The SST in the eastern Pacific Ocean is colder with the T42 AGCM (22°C) in the Northern Hemisphere summer, and reasonable with the T21 AGCM (23°C). The equatorial zonal SST gradient is increased. The simulated thermocline is quite tight and shows a marked latitudinal "W" structure. This positive result is from the turbulent kinetic energy scheme used for the vertical diffusion. The SST seasonal cycle in the tropical Pacific shows a spurious enhancement of the semiannual signal with the T42 AGCM and not with the T21 AGCM and the cold phase of the seasonal cycle is very short (Figure 2). The T42 coupling improves the summer monsoon simulation giving more realistic precipitation over Asia with less precipitation over Sumatra [Stephenson *et al.*, 1995]. In both simulations, there is interannual variability in the equatorial Pacific Ocean (T42: −3.5°C to +2.5°C; T21: −5°C to +5°C). Cold and warm events appear both in the eastern part of the equatorial Pacific basin and "basin-wide." Events can also appear simultaneously, and some coastal events are seen to propagate westward.

    In view of long range simulations, energy budgets and transports in the low atmospheric

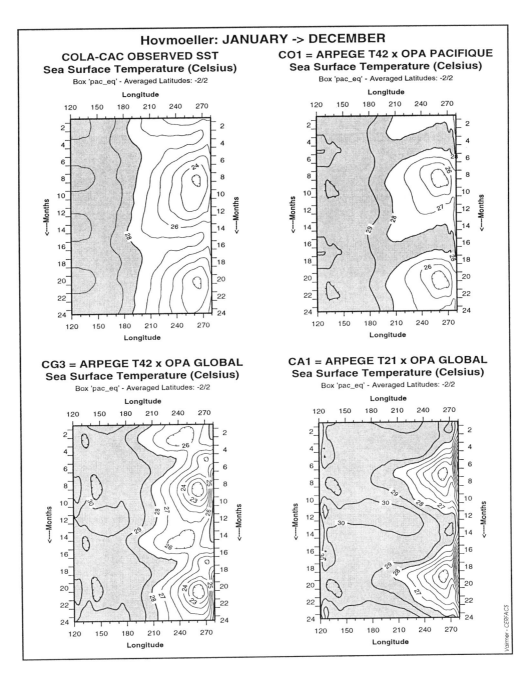

FIGURE 2. SST seasonal cycle.

resolution (T21) coupled model are assessed and prove to be within available estimates [Guilyardi and Madec, 1997]. The climate drift is on average moderate but is significant in some regions: initial tropical warming, lack of formation of intermediate water masses, and warming of the southern ocean. These regional drifts affect the mean global oceanic

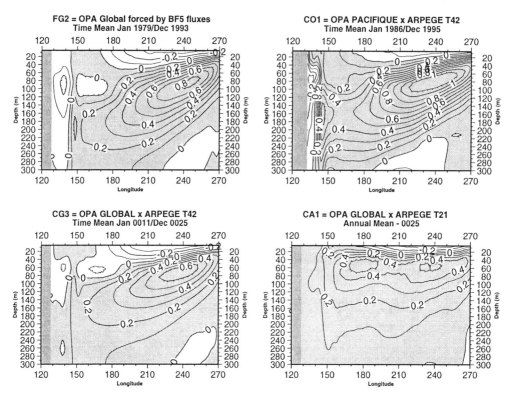

FIGURE 3. Vertical slice at the equator of mean zonal current $(\mathrm{m\cdot s^{-1}})$.

and atmospheric circulation. Despite these systematic biases, the model's robustness and stability are quite promising for future studies.

4. Conclusions

An initial step has been performed at CERFACS with the development of a set of tools for the control and the postprocessing of numerical climate simulations. The feasibility of distributed computing has been confirmed and allows new perspectives such as increasing of the GCM resolution.

The three different coupled simulations performed at CERFACS show different behavior. The global ocean volume-averaged temperature drift is not excessive despite the absence of flux correction. The cloud parameterization appears to be an important key to the control of the tropical Pacific mean state. Horizontal resolution of at least T42 in the atmospheric GCM seems to be necessary to obtain reasonable tropical ocean dynamics.

The SSTs simulated in the tropical Pacific with the Pacific OGCM and the Global OGCM, both coupled with the same T42 AGCM, have large differences which are still under investigation. The mean state simulated with the global ocean is significantly warmer than the one simulated with the Pacific Ocean only. It is also interesting to note that the SST and wind annual cycle simulated with the Pacific OGCM are very regular, while there is a significant interannual variability in the coupled simulation involving the global OGCM. An analysis based on "coupled instabilities theories" suggests that the increased zonal gradient of SST and a weakened mean upwelling along the equator are the sources of the high interannual variability in the global coupled simulations (Pontaud and Thual – paper submitted to *Q. J. R. Meteor. Soc.*).

Some interactions have been mentioned but the mechanisms of the climate still raise many challenging questions for the future.

Acknowledgments. The work on the distributed computing has benefited from the European Community contract "Anthropogenic Climate Change" of Environement Program number PL910387.

REFERENCES

Blanke, B., and P. Delecluse, Low frequency variability of the tropical Atlantic Ocean simulated by a general circulation model with mixed layer physics, *J. Phys. Ocean.*, **23**, 1363–1388, 1993.

Delecluse, P., G. Madec, M. Imbard, and C. Levy, OPA version 7: Ocean general circulation model reference manual, *Internal Report LODYC 93/05*, Paris, France, 1993.

Déqué, M., C. Dreveton, A. Braun, and D. Cariolle, The climate version of ARPEGE-IFS: A contribution to the French community climate modelling, *Clim. Dyn.*, **10**, 249–266, 1994.

Guilyardi, E., The VAIRMER experiment manager: User's guide and reference manual, Version 2.1.1, *Technical Report TR/GMGC/9404*, CERFACS, Toulouse, France, 1994.

Guilyardi, E., and G. Madec, Performance of the OPA/ARPEGE-T21 global coupled ocean–atmosphere model, *Clim. Dyn.*, **13**, 149–165, 1997.

Guilyardi, E., G. Madec, L. Terray, M. Déqué, M. Pontaud, M. Imbard, D. Stephenson, M. A. Filiberti, D. Cariolle, P. Delecluse, and O. Thual, Simulation couplée océan-atmosphère de la variabilité du climat, *C. R. Acad. Sci. Paris*, **320**, série IIa, 683–690, 1995.

Guilyardi, E., L. Terray, and O. Thual, The VAIRMER experiment manager. Part I. A description of the project, *Epicoa Letters 0222*, CERFACS, Toulouse, France, 1994.

Maykut, G. A. and N. Untersteiner, Some results from a time-dependent thermodynamic model of sea ice, *J. Geophys. Res.*, **76**, 1550–1575, 1971.

Mechoso, C. R., C.-C. Ma, J. D. Farrara, J. A. Spahr, and R. W. Moore, Parallelization and distribution of a coupled atmosphere–ocean general circulation model, *Mon. Wea. Rev.*, **121**, 2062–2076, 1993.

Mechoso, C. R., A. W. Robertson, N. Barth, M. K. Davey, P. Delecluse, P. R. Gent, S. Ineson, B. Kirtman, M. Latif, H. Le Treut, T. Nagai, J. D. Neelin, S. G. H. Philander, J. Polcher, P. S. Schopf, T. Stockdale, M. J. Suarez, L. Terray, O. Thual, and J. J. Tribbia, The seasonal cycle over the tropical Pacific in coupled ocean–atmosphere general circulation models, *Mon. Wea. Rev.*, **123**, 2825–2838, 1995.

Sevault, E., CLIM: Coupling library for interfacing models: User's guide and reference manual, Version 1.0, *Technical Report TR/GMGC/94*, CERFACS, Toulouse, France, 1994.

Sevault, E., P. Noyret, and L. Terray, Coupling library for interfacing models: User's guide and reference manual, Version 1.0, *Technical Report TR/GMGC/94*, CERFACS, Toulouse, France, 1994a.

Sevault, E., P. Noyret, L. Terray, and O. Thual, Distributed and Coupled Ocean–Atmosphere Modelling. *Proceedings of the VI ECMWF Workshop on the Use of Parallel Processors in Meteorology*, Reading, England, ECMWF, 1994b.

Stephenson, D. B., M. Pontaud, E. Guilyardi, L. Terray and O. Thual, The Asian monsoon in a globally coupled ocean atmosphere general circulation model. *Proceedings of the 1995 GEWEX/Game Conference*, Pattaya, March 1995.

Stössel, A., P. Lemke, and W. B. Owens, Coupled sea-ice-mixed layer simulations for the southern ocean, *MPI Report, 30*, Max-Planck-Institut für Meteorologie, D-2000 Hamburg, Germany, 1989.

Terray, L., The OASIS Coupler user guide, Version 1.0, *Technical Report TR/CMGC/94-05*, CERFACS, Toulouse, France, 1994.

Terray, L., O. Thual, S. Belamari, M. Déqué, P. Dandin, P. Delecluse, and C. Lévy, Climatology and interannual variability simulated by the ARPEGE-OPA coupled model, *Clim. Dyn.*, **11**, 487–505, 1995.

The OCCAM Global Ocean Model

By C. S. Gwilliam,[1,2] A. C. Coward,[1] B. A. de Cuevas,[1] D. J. Webb,[1]
E. Rourke,[1] S. R. Thompson[1] and K. Döös[1]

[1] Southampton Oceanography Centre, Empress Dock, Southampton, SO14 3ZH, UK

[2] Supported by a Cray Research Fellowship

Climate change is not affected by the atmosphere alone – the world's oceans also play an important role, hence the need for global ocean models. In the past ocean modeling has been restricted to coarse resolutions and to limited areas, for example the Antarctic. The arrival of powerful array processors is allowing global studies at resolutions high enough to resolve eddies in the ocean. This paper describes one such project, OCCAM, and some of the scientific results and problems from an initial 3 year run.

1. Introduction

The aim of the Ocean Circulation and Climate Advanced Modeling (OCCAM) project is to develop global ocean models which are suitable for climate studies. The OC-CAM code has been developed to investigate global ocean characteristics, for example, heat transport and global circulation, and to test improved schemes for modeling ocean physics. Examples of areas where work is being undertaken include improving the representations of bottom topography, deep convection and distribution of sea ice. The model will also be used to identify any problems that remain with respect to climate prediction.

In Section 2, the basic structure of the model is outlined. The model run to date and scientific problems that have been encountered are described in Section 3. Section 4 details some initial results, and Section 5 outlines areas of future work.

2. The OCCAM Model

The OCCAM code is a message passing version of the Modular Ocean Model–Array processor version (MOMA) [Webb, 1993], which is a development of the widely available Geophysical Fluid Dynamics Laboratory (GFDL) Modular Ocean Model [Pacanowski et al., 1990]. The basis for these models is the Bryan–Cox–Semtner ocean general circulation model [Bryan, 1969; Cox, 1984; Semtner, 1974], which uses time-dependent equations:

$$\frac{\partial \mathbf{u}}{\partial t} + (\mathbf{u} \cdot \nabla)\mathbf{u} + w\,\frac{\partial \mathbf{u}}{\partial z} + f \times \mathbf{u} = -\frac{1}{\rho_0}\nabla p + \mathbf{D}_u + \mathbf{F}_u \tag{2.1}$$

$$\frac{\partial S}{\partial t} + (\mathbf{u} \cdot \nabla)S + w\,\frac{\partial S}{\partial z} = \mathbf{D}_S + \mathbf{F}_S \tag{2.2}$$

$$\frac{\partial T}{\partial t} + (\mathbf{u} \cdot \nabla)T + w\,\frac{\partial T}{\partial z} = \mathbf{D}_T + \mathbf{F}_T \tag{2.3}$$

$$\rho = \rho(T,\,S,\,p) \tag{2.4}$$

$$\rho g = -\frac{\partial p}{\partial z} \tag{2.5}$$

$$\nabla \cdot \mathbf{u} + \frac{\partial w}{\partial z} = 0 \tag{2.6}$$

24

do main time loop	do main time loop
Calculate the baroclinic timestep	*Calculate the baroclinic timestep*
for the whole domain	*for the processor domain*
do j loop	do m=1, no. of points held
do i loop	i=ilist(m), j=jlist(m)
call baroclinic(i,j)	call baroclinic(i,j)
end i loop	end m loop
end j loop	*Exchange halo information for all levels*
Calculate the barotropic time-	*Calculate the barotropic timesteps*
steps for surface points	*for surface points of processor domain*
do barotropic timestep	do barotropic timestep
do j loop	do m=1, no. of points held
do i loop	i=ilist(m), j=jlist(m)
call barotropic(i,j)	call barotropic(i,j)
end i loop	end m loop
end j loop	*Exchange halo information for*
	surface points
end barotropic timestep	end barotropic timestep
end main time loop	end main time loop
(a) Sequential	(b) Parallel

FIGURE 1. Structure of the main timestepping routine (`step`) in the (a) sequential and (b) parallel codes.

Equations 2.1–2.3 describe the horizontal momentum, temperature and salinity changes; equations 2.4–2.6 describe the density, pressure gradient and incompressibility. The variables are $\mathbf{u}$, the horizontal velocity; w, the vertical velocity; S, the salinity; and T, the potential temperature. The pressure variable is p; the density is ρ. The Coriolis term, f, is

$$f = 2\Omega \sin(\theta)$$

where Ω is the rotation rate of the earth and θ is the latitude. $\mathbf{D}$ represents diffusion, and $\mathbf{F}$ the forcing terms.

The equations are discretized using a finite difference formulation and are stepped forward explicitly in time. However, very fast surface gravity waves in the ocean restrict the timestep that can be used, increasing the computational cost of solution. To overcome this, the preceding equations are split into their depth-averaged part, often called the barotropic equations, and the remainder, the baroclinic ones. This splitting allows a long timestep to be used for the baroclinic equations. The depth-averaged equations, whose solution includes the fast gravity waves, are solved by using a free surface method [Killworth *et al.*, 1989] where many barotropic timesteps are used for each baroclinic one. At the end of each baroclinic timestep, the barotropic velocities are added to the baroclinic ones.

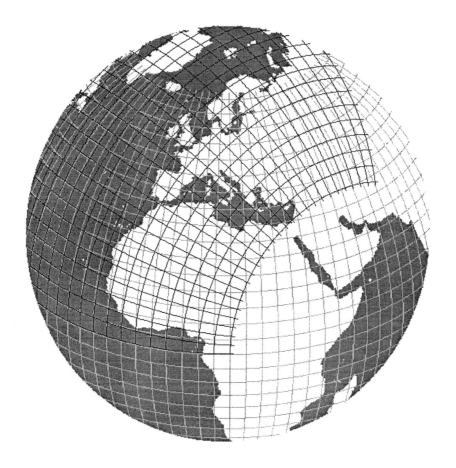

FIGURE 2. The two-grid model.

The main timestep calculation is carried out in subroutine `step`. A schematic is shown in Figure 1(a), where i and j are surface sea points. The baroclinic calculation is carried out over the column of points associated with each of these surface points. In the parallel implementation, each processor is assigned a region and makes a list of those points it holds which are sea points. The timestep calculation is carried out over these points with communication taking place as necessary (Figure 1(b)).

The OCCAM model is a global model based on a latitude–longitude grid. In order to avoid the problems that occur when the grid converges at the North Pole, a two-grid method [Coward *et al.*, 1994; Deleersnijder *et al.*, 1993; Eby and Holloway, 1994] has been introduced. A grid rotated through 90° is used for the North Atlantic, from the Equator to the Bering Straits, and the normal latitude–longitude grid is used everywhere else (Figure 2). The two grids meet orthogonally at the Equator. The calculation of the grids for a given resolution is done once off-line, so the only complications to the code are the need to communicate differently where the grids meet and the use of altered finite difference coefficients along the Equator. At the Bering Straits, the channel is very shallow and a stream tube model is used. For the current $\frac{1}{4}^\circ$ resolution only eight surface points and their associated columns of points are affected.

Another difference between the sequential and the parallel code is the treatment of input/output (I/O). The main purpose of OCCAM is analysis of model data, and regular output of the ocean state is required. For this reason one processor is set aside to deal with I/O. This processor writes out the required data sets in a contiguous manner for ease of postprocessing. These data sets are also used to restart the model; the I/O processor reads in the data at the start of a run. By using contiguous datasets, the number of processors and/or the division of work between the processors can be changed easily. The I/O processor also reads in the topography and processor maps at the start and surface forcing fields during the run. It also outputs diagnostics and two-dimensional (2-D) snapshots of the ocean.

3. The Initial Model Run

The model has been spun up for 3 years on a 256-node T3D. The resolution is $\frac{1}{4}^{\circ}$ at the Equator with 36 depth levels, requiring 29,911,680 points for grid 1 and 6,843,312 for grid 2. The baroclinic timestep is 15 minutes and the barotropic timestep is 9 seconds. The ocean temperature and salinity have been relaxed to the annual Levitus dataset on an annual timescale at depth and to the monthly Levitus datasets on a monthly timescale at the surface. The ocean surface has also been forced by using monthly European Centre for Medium Range Weather Forecasts (ECMWF) wind stresses. For these 3 years the Bering Straits have been closed and the Pacanowski and Philander [1981] vertical mixing scheme has been used from day 480.

During the initial 3 years various problems with the numerical aspects of the code have occurred. After a few days, what seemed to be standing waves, which were increasing in size, appeared off the Pacific coast of America. These were actually surface gravity waves which were moving their own wavelength every baroclinic timestep. To overcome these, a time-averaged scheme is used for the free surface calculation. For each baroclinic timestep, the free surface is stepped forward the equivalent of two baroclinic timesteps and the averages of the free surface height and surface velocity fields are used as the solution.

Another problem was plus–minus checkerboarding in areas where the barotropic flow was changing rapidly. This was reduced by adding a diffusion term to the free surface height. Two Laplacian operators were subtracted using a stencil of the form

$$\begin{bmatrix} 0 & X & 0 \\ X & X & X \\ 0 & X & 0 \end{bmatrix} - \begin{bmatrix} X & 0 & X \\ 0 & X & 0 \\ X & 0 & X \end{bmatrix}$$

where X denotes a contribution from a point and 0 denotes no contribution. These stencils were altered near landmasses to ensure the stability of the overall operator. This diffusion term filters out small waves and two-grid-point noise. To reduce computational cost it is carried out on the first free surface timestep of each baroclinic one.

The advection scheme used in the model is known to cause under- and overshooting of the tracer (salinity and temperature) values where large gradients in tracers and/or velocities occur. In the Kuroshio off Japan, sea surface temperatures of -5°C have been seen at the front between cold and warm water masses. Although these disappeared, improvements in the advection scheme need to be made for correct representation of such fronts. The QUICK scheme [Farrow and Stevens, 1995; Leonard, 1979], a second-order upwind method, for the advection terms in the tracer equations will be implemented from the end of year 4 of the model run.

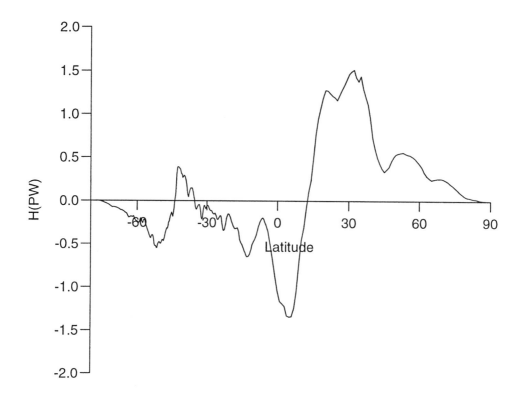

FIGURE 3. Heat transport from day 1020 of the OCCAM initial run

4. Initial Scientific Results

The analysis stage of the run has not yet been reached but investigation of some global properties and of events occurring on a more local scale has begun. Most of the following results are based on the model data from day 1020. At this stage the initial transients have died out and the model is starting to move toward a steady state. Thus the features observed are likely to be typical of those found later in the model run.

The instantaneous northward heat transport at day 1020 is shown in Figure 3. At 30°N, the model value of 1.5 PW compares with Bryden's ship based observation value of 2 PW [Bryden, 1993; Bryden *et al.*, 1991]. The values seen in the OCCAM model are larger than the mean values found in the FRAM model [Saunders *et al.*, 1993]. For example, the southward heat transport in the polar gyres near 50°S is 0.6 PW compared with 0.2 PW in FRAM. However, the seasonal variation for this value was found to be large in FRAM and it is possible that the final mean heat transports in OCCAM will be lower.

In the Southern Ocean, OCCAM produces a transport through the Drake Passage of 171 Sv. The jet along the north side of the channel contributes 125 Sv with the remainder coming from a broad current lying between 59°S and the southern boundary of the Strait. The northern jet is strongly barotropic, and 20 Sv of the transport is associated with the non–zero bottom velocity. The new total figure falls mid-way between earlier model estimates of around 190 Sv [Cox, 1975; Semtner and Chervin, 1988;

Semtner *et al.*, 1992; FRAM Group, 1991] and the observational estimates of around 140 Sv [Nowlin and Klinck, 1986]. The disagreement with the latter may be due to observations that missed part of the barotropic signal. However, the disagreement with the earlier model estimates is more intriguing. It may be due to differences in bottom topography, but the discrepancy needs to be investigated further.

In the North Atlantic the model shows a Gulf Stream transport of 10 Sv through the Florida Strait and an Antilles Current of 19 Sv. Of the total transport, 16 Sv returns as the eastern part of the subtropical circulation and the rest as a southward flow of deep water concentrated mainly along the flanks of the Bahamas Rise. In the Gulf of Mexico, the model shows the loop current producing a series of rings, as expected. In the 18 months up to day 1020, one of these is produced, on average, every 6 months, with a transport of 12 Sv.

More unexpectedly the model shows the Mindanao eddy producing a series of rings with a similar size and strength (12 Sv) in the Celebes Sea. However, the frequency of these is much greater, with a new one generated approximately every 50 days. After formation, these move westward at a speed of approximately 15 cm/s. The existence of the Mindanao current loop or eddy is well known [Wyrtki, 1961], but although it is known to be variable, we know of no previous work which has indicated that its behavior may be similar to that of the Gulf of Mexico loop current.

The model shows a total Indonesian throughflow of 15 Sv. Of this, 7 Sv passes into the Indian Ocean through the Lombock Channel, 6 Sv through the channel north of Timor, and the remainder through the section between Timor and Australia.

A series of rings is produced at the northern end of the Mozambique Channel. On average, a new one is produced every 40 days and drifts southward at an approximate speed of 10 cm/s. The model shows a southward transport through this channel of 24 Sv. Another current, with a strength of 15 Sv, is found running along the east coast of Madagascar. This turns westward at the southern end of the island, forming another series of eddies. Both sets of eddies become entrained in the Agulhas current, those from the Mozambique Channel forming significant disturbances which are carried southward along the core of the current.

5. Future Work

The model will continue to be spun up for a further 3 years but with the Bering Straits open. Also, a freshwater flux is being added to simulate the change in surface height due to precipitation and evaporation. After the end of the fourth model year the depth relaxation to Levitus will be removed and the surface relaxation will be weakened. ECMWF heat fluxes will be used as well as the wind fields for the surface forcing. After six years the model will be run without relaxation to Levitus. Instead, improved full surface forcing with global wind, precipitation, evaporation and heat fluxes will be used. These fluxes are being developed by a team at the James Rennell Centre, under Dr. P. Taylor.

Other improvements include the use of variable bottom topography from the end of year 4. This will improve the representation of sills, continental shelves and very deep ocean basins. The extension of QUICK to the momentum equations will be investigated, and a sea-ice model will be added. When these improvements have been made it is hoped that the resolution will be increased to $\frac{1}{6}^\circ$.

Acknowledgments. The model development work represented here has been carried out by the OCCAM group at the James Rennell Centre, consisting of D. J. Webb, B. A. de Cuevas, A. C. Coward and C. S. Gwilliam, in conjunction with M. E. O'Neill and R. J. Carruthers of Cray Research, U.K. Help with the results and graphics has been provided by S. R. Thompson, K. Döös and E. Rourke.

REFERENCES

Bryan, K., A Numerical method for the study of the circulation of the world ocean, *J. Comp. Phys.*, **4**, 347–376, 1969.

Bryden, H. L., Ocean heat transport across 24 N latitude, interactions between global climate subsystems: The legacy of Hann, *Geophysical Monograph, IUGG*, **15**, 65–75, 1993.

Bryden, H. L., D. H. Roemmich, and J. A. Church, Ocean heat transport across 24 N in the Pacific, *Deep-Sea Res.*, **38**(3), 297–324, 1991.

Coward, A. C., P. D. Killworth, and J. R. Blundell. Tests of a two-grid world ocean model, *J. Geophys. Res.*, **99**, 22725–22735, 1994.

Cox, M. D., A baroclinic numerical model of the world ocean: Preliminary results. *Numerical Models of Ocean Circulation*, R. O. Reid *et al.*, Eds. National Academy of Sciences, Washington, DC, 1975.

Cox, M. D., A primitive equation, 3-dimensional model of the ocean, Technical Report No. 1, GFDL Ocean Group, GFDL/NOAA, Princeton University, Princeton, NJ, 1984.

Deleersnijder, E., J.-P. Van Ypersele, and J.-M. Campin, An orthogonal, curvilinear coordinate system for a world ocean model, *Ocean Modelling*, **100**, 1993 (unpublished ms, available from authors).

Eby, M., and G. Holloway, Grid transform for incorporating the Arctic in a global ocean model, *Climate Dynamics*, **10**, 241–247, 1994.

Farrow, D., and D. Stevens, A new tracer advection scheme for Bryan and Cox type ocean general circulation models, *J. Phys. Ocean.*, **25**, 1731–1741, July, 1995.

The FRAM Group (Webb, D. J. *et al.*), An eddy-resolving model of the southern ocean, *EOS, Transactions, American Geophysical Union*, **72**(15), 169–174, 1991.

Killworth, P. D., D. Stainforth, D. J. Webb, and S. M. Patterson, A free surface Bryan–Cox–Semtner model, Report No. 270, Institute of Oceanographic Sciences, Deacon Laboratory, 1989 (available from authors).

Leonard, B. P., A stable and accurate convective modelling procedure based on quadratic upstream interpolation, *Computer Methods in Applied Mechanics and Engineering*, **19**, 59–98, 1979.

Nowlin, W. D., and J. M. Klinck, The physics of the Antarctic circumpolar current, *Reviews of Geophysics*, **23**(4), 469–491, 1986.

Pacanowski, R. C., K. Dixon, and A. Rosati, The GFDL modular ocean model users guide, Technical Report No. 2, GFDL Ocean Group, GFDL/NOAA, Princeton University, Princeton, NJ, 1990.

Pacanowski, R. C., and S. G. H. Philander, Parameterization of vertical mixing in numerical models of tropical oceans, *J. Phys. Ocean.*, **11**, 1443–1451, 1981.

Saunders, P. M., and S. R. Thompson, Transport, heat and freshwater fluxes within a diagnostic numerical model (FRAM), *J. Phys. Ocean.*, **23**, 452–464, 1993.

Semtner, A. J., An oceanic general circulation model with bottom topography, Technical Report No. 9, Dept. of Meteorology, UCLA, Los Angeles, CA, 1974.

Semtner, A. J., and R. M. Chervin, A simulation of the global ocean circulation with resolved eddies, *J. Geophys. Res.*, **93**, 15502–15522, 15767–15775, 1988.

Semtner, A. J., and R. M. Chervin, Ocean general circulation from a global eddy-resolving model, *J. Geophys. Res.*, **97**, 5493–5550, 1992.

Webb, D. J., An ocean model code for array processor computers, Internal Document No. 324, Institute of Oceanographic Sciences, Deacon Laboratory, 1993 (available from authors).

Wyrtki, K., Physical oceanography of the Southeast Asian waters, scientific results of maritime investigations of the South China Sea and the Gulf of Thailand 1959–1961. NAGA Report 2, Scripps Institute of Oceanography, La Jolla, CA, 1961.

Climatic Asymmetries Relative to the Equator

By S. G. H. Philander, T. Li and G. Lambert

Atmospheric and Oceanic Sciences Program
Princeton University, Princeton, NJ 08544, USA

Although solar radiation is symmetrical about the equator and has a maximum there, the coast of Panama is a tropical jungle with plentiful rainfall, whereas the coast of Peru, at the same latitude but south of the equator, is a barren desert. Such climatic asymmetry is present in both the eastern tropical Pacific and the Atlantic, regions with a surprisingly prominent annual (rather than semiannual) cycle at the equator even though the sun "crosses" that line twice a year. These curious features depend on the global distribution of continents, on the coastal geometries of western Africa and the Americas, and on unstable interactions between the ocean and atmosphere, interactions that involve low-level stratus clouds.

1. Introduction

In a world that is perfectly symmetrical about the equator, the climate could nonetheless be asymmetrical if symmetrical conditions were unstable to perturbations. But why would the processes that lead to asymmetries be more effective in some longitudes than others? Why are asymmetries in sea surface temperatures, rainfall and cloudiness most prominent in the eastern tropical Pacific and Atlantic Oceans? Figure 1(a) shows that the Northern rather than the Southern Hemisphere is favored with the warmest waters. Why is that hemisphere, which also has the heavier rains, favored over the other? The answers to these questions must involve the distribution and geometries of the continents, but first we need to investigate the processes that can convert symmetric conditions into asymmetric ones.

From an atmospheric point of view, the climatic asymmetries of the eastern tropical Pacific and Atlantic can be attributed to the asymmetry in sea surface temperatures. Atmospheric convection that involves rising air, cumulus towers and heavy rains occurs over the warm waters off Panama, not the very cold waters off Peru. The rising air is sustained by convergent low-level winds. From an oceanic point of view those winds cause the observed sea surface temperature pattern. The northward winds that prevail south of the convective region drive oceanic currents with a northward component. That component is small far from the equator because the Coriolis force deflects the wind-driven currents. As the equator is approached, and the Coriolis force becomes small, the northward component of the currents gains in speed and attains a maximum at the equator, where the Coriolis force vanishes. This means that the northward currents are divergent south of, and convergent north of, the equator. They therefore cause the upwelling of cold water and cause low sea surface temperatures south of the equator, and are responsible for downwelling and warm surface waters north of the equator [Philander, 1990; Neelin et al., 1994]. Hence the winds determine the surface temperature patterns. They also depend on those patterns, as explained earlier. This circular argument suggests that interactions between the ocean and atmosphere are at the heart of the matter. Similar interactions influence the ocean–atmosphere response to seasonal variations in solar radiation in the eastern equatorial Pacific and Atlantic but not the Indian Ocean. This can be inferred from the high correlations evident in Figure 1(c) and (d), which

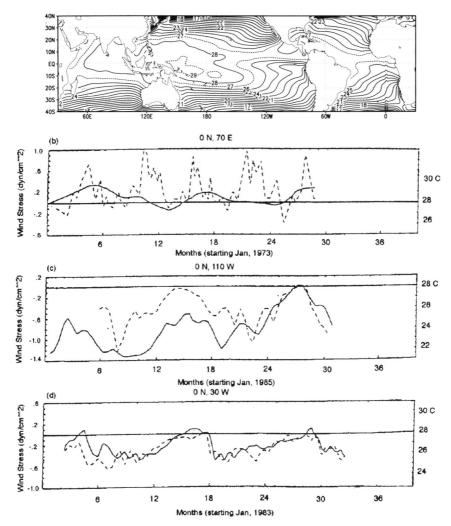

FIGURE 1. (a) Time-averaged sea surface temperatures for the period 1950 to 1979 [Halpert and Ropelewski, 1989]. The contour interval is 1°C. Dashed contours are 27°C and 29°C. In panels (b) through (d) the solid line is the sea surface temperature in degrees Celsius; the dashed line is the zonal component of the windstress (dynes/cm^2). Time, in units of a month, starts on January 1 in all three cases. The measurements were made on the Equator at (b) the island Gan in the Indian Ocean (70°W) by Knox [1976], starting in 1973; (c) at 110°W in the eastern Pacific by Halpern [1987], and McPhaden and McCarty [1992], starting in 1985; and (d) at 30°W in the Atlantic by Weingartner and Weisberg [1991], starting in 1983.

show seasonal variations in the sea surface temperature, and the zonal component of the wind, at two locations on the equator: at 110°W in the eastern Pacific and at 30°W in the Atlantic. Once again it appears that the winds both cause and are caused by the sea surface temperature variations. Intriguingly, this is the case where climatic asymmetries are large but is not the case at the island Gan in the Indian Ocean. There the correlation between the zonal wind and sea surface temperature is low, a semiannual harmonic is dominant, and the annual harmonic is minimal (Figure 1(b)). What is special about the eastern tropical Pacific and Atlantic for those to be the preferred regions of ocean–

atmosphere interactions? How do those interactions give rise to an annual harmonic at the equator?

The ocean–atmosphere interactions [Chang and Philander, 1994] that are relevant to the phenomena under consideration here have the following effect on a perturbation to symmetric conditions, a perturbation that displaces the warmest waters, initially at the equator, slightly northward (say). The southerly winds that converge onto the displaced warm waters (where the air rises into convective clouds) causes low sea surface temperatures south of the equator, warm surface waters north of the equator, for the reasons mentioned. The resultant sea surface temperature gradient creates pressure gradients in the lower atmosphere that intensify the northward winds [Lindzen and Nigam, 1987], which in turn strengthen the temperature gradient, and so on. In this positive feedback, divergent surface currents cause a decrease in sea surface temperatures [Chang and Philander, 1994]. The feedback is therefore most effective where the thermocline is shallow. (The thermocline is the layer of large vertical gradients that separates the warm surface waters from the cold water at depth.) The thermocline happens to be shallow in the eastern tropical Pacific and Atlantic Oceans because the Trade winds that prevail over the Atlantic and Pacific drive the warm surface waters westward and expose cold water to the surface in the east. Hadley explained why such winds would prevail in the tropics on a water-covered globe. (The conservation of angular momentum requires westerlies in the earth's middle latitudes, easterlies in the tropics.) The extent to which the presence of continents modifies the Trades is minimal over the tropical Atlantic and Pacific Oceans but is enormous in the Indian sector, where cross-equatorial monsoons are the dominant winds. (Depending on the season, they blow to or from the Indian subcontinent, whose seasonal temperature fluctuations are much more extreme than those of the adjacent ocean.) Because of the monsoons, the thermocline in the Indian Ocean is essentially uniformly deep along much of the equator. Hence the global distribution of continents, by determining where Trades and where monsoons prevail, determines where the equatorial thermocline is shallow – in the eastern tropical Pacific and Atlantic – and hence where air–sea interactions can create climatic asymmetries. Those interactions favor neither hemisphere. Why then are warmest waters north rather than south of the equator?

2. Asymmetries of the Time-Averaged State

To investigate the specific aspects of the continental geometry that cause climatic asymmetries we start with a General Circulation Model of the atmosphere that reproduces realistic surface winds if the observed sea surface temperature patterns are specified as a lower boundary condition [Manabe and Hahn, 1981]. In the following numerical experiments, in which the annual mean solar radiation is specified as the forcing function, the model calculates the winds when sea surface temperature is strictly a function of latitude and is symmetrical about the equator. (The specified sea surface temperatures correspond to the time-averaged temperatures observed along the dateline and in the Northern Hemisphere.) The continents are deformed in various ways as shown in Figure 2 and the land is assumed to be flat in all the calculations to be described here. In Figure 2(b) the shape of the continents is idealized so that coastlines are either lines of longitude or circles of latitude. In such a world, climatic asymmetries persist in the eastern tropical Atlantic because of the bulge of West Africa. Although the solar radiation is symmetrical about the equator, the West African land surface attains a temperature far higher than that of the ocean to the south. This contrast is similar to that which causes a land–sea breeze or the monsoons. In Figure 2(b) the winds over the eastern equatorial Atlantic are seen to aquire a component toward the bulge, a component that is absent from the winds over the eastern equatorial Pacific. In this model, with specified sea surface temperatures, the modification to the winds is modest, but it could be amplified considerably if the winds were allowed to influence the ocean. The northward winds can cause oceanic upwelling and cold surface waters, not only to the south of the equator as explained earlier, but especially along the southwestern coast of Africa, where they drive northward oceanic currents that the Coriolis force deflects offshore. That deflection induces coastal upwelling and low sea surface temperatures. A decrease in sea surface temperatures magnifies the land–sea contrast, which intensifies the northward winds, causing even lower sea surface temperatures, and so on.

In Figure 2(b), the Pacific Ocean has practically no asymmetry in the winds. We next explore whether the asymmetry that exists in reality could be caused by the greater land area of the Northern Hemisphere. To eliminate possible effects of the local coastal geometry, we have changed the Americas in such a way as to preserve the land area of each latitude while making the western coast coincide with a line of longitude. In such a world, shown in Figure 2(a), the winds over the eastern tropical Pacific do not appear to acquire any asymmetry that the various feedbacks mentioned earlier could amplify. In this model, at least, the climatic asymmetry of the eastern Pacific is not attributable to the greater land area of the Northern Hemisphere.

The final experiment explores how the inclination of the western coast of the Americas to lines of longitude affects the winds. Although a comparison of Figures 2(b) and (c) indicates that the winds over the tropical Pacific hardly change when the coast is inclined, that view is strictly an atmospheric one. From an oceanographic point of view there is a critical change when the coast is inclined: the northeast Trades to the north of the equator become essentially perpendicular to the coast while the southeast Trades to the south of the equator are parallel to the coast. It is well known that winds parallel to a coast drive an oceanic jet in that direction. Because of the Coriolis force, the flow veers offshore so that cold water from below rises to the surface. That can happen south of the equator, where the wind is parallel to the inclined coast but not to the north, where the wind is perpendicular to the coast. Thus the winds will cause the surface waters to be cold off the coasts of Ecuador and Peru, warm off the coast of Panama.

In principle, ocean–atmosphere interactions ought to amplify the modest asymmetries introduced by continents when sea surface temperatures are symmetrical about the equator. However, calculations with a coupled ocean–atmosphere model (which previously was used in realistic simulations of the Southern Oscillation and El Niño [Philander *et al.*, 1992]) indicate that the interactions mentioned earlier are not very effective amplifiers, not unless another feedback involving stratus clouds is taken into account. Decks of low-level stratus clouds cover the oceans off the coasts of Peru, California and Angola. Whereas the deep convective clouds over the warmest waters are associated with the release of substantial amounts of latent heat that drive atmospheric motion, the stratus clouds over the cold water are so shallow and thin that they merely increase the albedo of the earth. They are nonetheless important to the maintenance of the climatic asymmetries because they are involved in a crucial feedback process. The clouds depend on the vertical temperature gradient of the lower atmosphere and become more dense the colder the sea surface. Enhanced cloudiness shields the ocean from solar radiation, causes even lower sea surface temperatures, and thus leads to more cloudiness, and so on. Although our atmospheric model satisfies most of the conditions for the clouds to form – a temperature inversion in the lower atmosphere and subsiding air aloft – its vertical resolution is too coarse for the humidity to reach a critical value for stratus clouds in any one layer. An empirical formula for cloudiness in terms of the vertical temperature gradient and subsidence in the lower troposphere – derived on the basis of satellite and

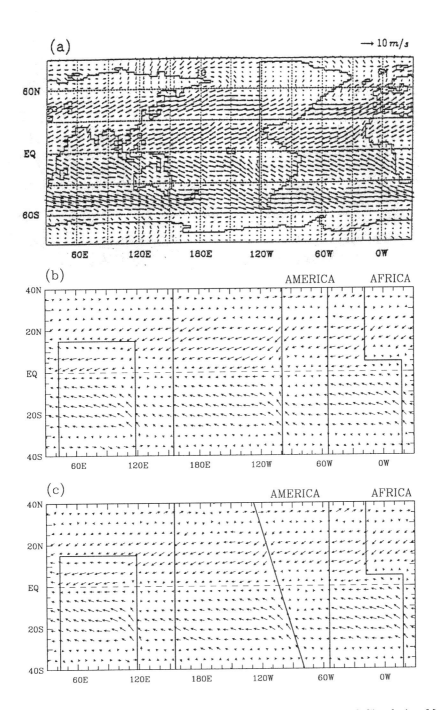

FIGURE 2. Surface wind vectors as calculated by an atmospheric General Circulation Model for different continental geometries. The specified sea surface temperatures vary only with latitude and correspond to those observed along the dateline. The maximum vector magnitude for panels (b) and (c) is 8.5 m/s.

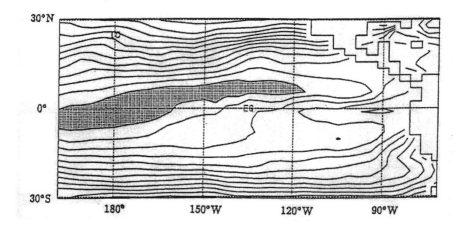

FIGURE 3. Results for the eastern tropical Pacific, winds (top) and sea surface temperatures in degrees Celsius (bottom), from a coupled ocean–atmosphere model forced with the time-mean solar radiation. The model includes a parameterization for stratus clouds.

other measurements – was adopted to enable the model to take the clouds into account. A coupled ocean–atmosphere model with such stratus clouds, forced with the annual mean solar radiation, which is symmetrical about the equator, reproduces a reasonably realistic asymmetrical climate as shown in Figure 3 [Philander *et al.*, 1996].

3. The Annual Cycle at the Equator

Various factors can cause an annual cycle at the equator. One is the ellipticity of the earth's orbit, which causes the earth to be closest to the sun in January, farthest away in July. Another is the asymmetry of the continents relative to the equator. (We just found that that asymmetry can give an asymmetric response to symmetric forcing.) A third possibility is the asymmetry of the time-averaged climate. To explore these possibilities, we rely on a simplified coupled ocean–atmosphere model, similar to that of Cane [1979] and Zebiak and Cane [1987], in which the time-averaged states of the ocean and atmosphere are specified. The forcing is seasonally varying solar radiation with a time average that is zero. Our results [Li and Philander, 1996] indicate that by far the most important reason for the annual cycle at the equator is the asymmetry of the time-averaged state. Consider, for example, how the response to the seasonal forcing changes when the time-averaged winds change from being symmetrical about the equator to asymmetrical. If the winds are symmetrical, then their meridional component, on the average, vanishes at the equator; it is southward during the southern summer, northward during the northern summer. (That is the case in the Indian Ocean.) However, if the time-averaged winds at the equator are northward, as they are in the eastern Pacific and Atlantic, then a superimposed seasonal cycle causes intense northward winds (in August), weak northward winds in (February). Hence the wind speed at the equator can have an annual cycle, and, because wind speed controls evaporation from the ocean, surface temperatures too can have an annual cycle. Figure 4 shows the seasonal variations of sea surface temperature in our model when a realistic, asymmetrical time-averaged state is specified, and when the forcing is strictly antisymmetrical about the equator. The dotted line is the result when, in the oceanic component of the model, only evaporation

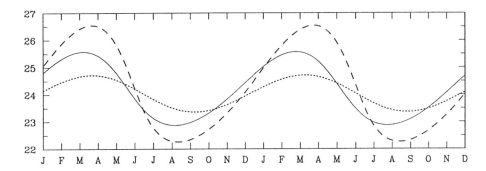

FIGURE 4. Seasonal variations in sea surface temperatures (in degrees Celsius) at 0°N 100°W as simulated with a coupled ocean–atmosphere model in which the specified time-mean state is realistically asymmetrical relative to the equator. The sea surface temperature variations are determined strictly by evaporation in the case of the dotted line, by evaporation and oceanic upwelling in the case of the solid line, and by evaporation, upwelling and presence of low-level stratus clouds in the case of the dashed line. The forcing is that component of the annually varying solar radiation that is strictly antisymmetrical about the equator. In this model the western coast of the Americas coincides with a meridian.

affects sea surface temperatures. The effects of oceanic upwelling are included in the solid line of Figure 4. The dashed line, which has a realistic amplitude for the annual cycle at the equator, is obtained when the model takes into account that the cold surface waters favor the formation of low-level stratus clouds which lower temperatures further. A prominent (and realistic) feature of the simulation is a signal (in sea surface temperature and the zonal component of the wind) that propagates westward at 60 cm/s along the equator. This signal involves ocean–atmosphere interactions that are symmetrical about the equator [Chang and Philander, 1994].

4. Conclusions

In summary, climatic asymmetries relative to the equator depend on several factors. First, the global distribution of continents determines where monsoons and where Trade winds prevail and hence determines variations in the depth of the warm surface layer of the ocean. In regions where the surface layer is shallow, ocean–atmosphere interactions plus a variety of feedbacks including those that involve stratus clouds, amplify modest perturbations that introduce asymmetries about the equator. The most important sources of asymmetric perturbations are the West African bulge to the north of the equator in the Atlantic and the slope, relative to a meridian, of the western coast of the Americas in the Pacific. Asymmetric time-averaged conditions permit a nonzero response at the equator to annual variations in solar radiation that vanish at the equator. Two factors that could affect climatic asymmetries of the time-averaged state were ignored in these calculations and remain to be explored: mountains, especially the Andes and Himalayas, and the effect of the seasonal cycle on time-averaged conditions. Both are likely to be very important in the Indian sector.

Acknowledgments. We thank Drs. S. Manabe, N.-G. Lau and D. Gu for valuable suggestions during the course of this work, which was supported by the National Oceanic and Atmospheric Administration (NA26G0102-01) and NASA (NASANAG 5-2224).

REFERENCES

Cane, M. A., The response of an equatorial ocean to simple wind stress patterns, I: model formulation and analytic results, *J. Mar. Res.*, **37**, 233–252, 1979.

Chang, P., and S. G. H. Philander, A coupled ocean-atmosphere instability of relevance to the seasonal cycle, *J. Climate*, **51**, 24, 3627–3648, 1994.

Halpern, D., Comparison of upper ocean VACM and VMCM observations in the equatorial Pacific, *J. Atmos. Oceanic Technol.*, **4**, 84–93. 1987.

Halpert, C. S, and C. F. Ropelewski, Atlas of tropical sea surface temperature and surface winds, NOAA Atlas No. 8, 1989 NOAA, Dept. of Commerce, Silver Springs, MD, 1989.

Knox, R. A., On a long series of measurements of Indian Ocean Equatorial current near Addu Atoll, *Deep Sea Res.*, **23**, 211–221, 1976.

Li, Tianming, and S.G.H. Philander, On the annual cycle of the eastern equatorial Pacific, *J. Climate* (in press), 1996.

Lindzen, R., and S. Nigam, On the role of sea surface temperature gradients in forcing low level winds and convergence in the tropics, *J. Atmos. Sci.*, **44**, 2240–2458, 1987.

Manabe, S., and D. G. Hahn, Simulation of atmospheric variability, *Mon. Wea. Rev.*, **109**, 2260–2286, 1981.

McPhaden, M. J., and M. E. McCarty, Mean seasonal cycles and interannual variations at 0°, 110°W and 0°, 140°W during 1980–1991. NOAA technical memorandum, Contribution No. 1377 from NOAA/Pacific Marine Environmental Laboratory, NOAA, Seattle, WA, 1992.

Neelin, D., F.-F. Jin, and M. Latif, Dynamics of coupled ocean–atmosphere models: the tropical problem, *Annu. Rev. Fluid Mech.*, **26**, 617–659, 1994.

Philander, S. G. H., El Niño, La Niña and the Southern Oscillation. Academic Press, 1990.

Philander, S. G. H., R. C. Pacanowski, N.-G. Lau, and M. J. Nath, Simulations of ENSO with a global atmospheric GCM coupled to a high-resolution Tropical Pacific Ocean GCM, *J. Climate*, **5**, 308–329, 1992.

Philander, S. G. H., D. Gu, D. Halpern, G. Lambert, N.-C. Lau, T. Li, and R.C. Pacanowski, Why the ITCZ is mostly north of the Equator, *J. Climate*, **9**, 2958–2972, 1996.

Weingartner, T. J., and R. H. Weisberg, A description of the annual cycle in sea surface temperature and upper ocean heat in the equatorial Atlantic, *J. Phys. Ocean.*, **21**, 83–96, 1991.

Zebiak, S. E., and M. A. Cane, A model ENSO, *Mon. Wea. Rev.*, **115**, 2262–2278, 1987.

On the Use of a General Circulation Model to Study Regional Climate

By Víctor O. Magaña-Rueda and Arturo I. Quintanar

Centro de Ciencias de la Atmósfera, Universidad Nacional Autónoma de México
Cd. Universitaria, 04510 México, D.F., México.

A statistical model is used to downscale General Circulation Model output to study regional precipitation anomalies over Mexico during winter and summer. Observations show that during winter, large-scale circulations modulate periods of active and inactive precipitation over Mexico, particularly over the northern and central parts. During El Niño (La Niña) events a quasi-stationary low (high), associated with the Pacific–North American (PNA) pattern, determines the intensity of precipitation anomalies over the northern and central parts of the country. During summer, above- and below-normal precipitation appears to be determined by the distribution of sea surface temperature (SST). Cold SST anomalies over the eastern Pacific enhance moisture flux convergence around the western coast of Mexico, leading to a stronger Mexican monsoon. The Community Climate Model 2 (CCM2), forced by observed SSTs, provides adequate large-scale circulation fields to diagnose regional precipitation over Mexico through a downscaling process with a statistical model. During summer, accurate SST data are necessary to diagnose precipitation. It is concluded that for regional climate predictions over Mexico, a coupled ocean–atmosphere model is necessary.

1. Introduction

Various methods have been developed to forecast seasonal precipitation. These methods are mostly based on statistical techniques using historical data [Barry and Perry, 1981; Jauregui, 1995]. The statistical models use lagged correlations between regional anomalies in surface meteorological variables and large-scale circulations. The observed relationships are translated into multiple regression equations to predict monthly or seasonal anomalies of surface temperature [Klein and Hammons, 1975] or precipitation [Klein and Bloom, 1987]. On the other hand, numerical models provide an alternative way of predicting climate. Among these, energy balance models such as Adem's thermodynamic model have been used to predict surface temperature and precipitation [Adem, 1982]. The development of General Circulation Models (GCMs) constitutes a fundamental step in modern Numerical Weather Prediction (NWP). GCMs have become an important tool in the study of large-scale circulations and global climate. Current GCMs are capable of reproducing the annual cycle or simulating anomalous events such as those associated with El Niño–Southern Oscillation (ENSO) [Philander, 1990]. However, GCMs by themselves are not capable of determining regional climate variability because of their lack of spatial resolution. For this reason, statistical methods have been developed to downsize or downscale GCM output in order to have regional prognoses of surface temperature or precipitation. Several of these methods are similar to the Perfect Prog or Model Output Statistics techniques used in NWP. More recent approaches to the problem of regional climate make use of GCMs and limited area models [Giorgi, 1990]. Although this approach appears to be physically more meaningful, it requires clear understanding of the processes that determine regional climate.

In the present study, we develop a statistical model to diagnose regional precipitation anomalies over Mexico during winter and summer. Winter precipitation is largely influenced by the midlatitude large-scale circulations, while summer precipitation is mostly associated with mesoscale tropical circulations. Our objective is to examine GCM output and consider its potential use as a regional climate prediction tool over Mexico. To this end, we examine El Niño and La Niña events and their impact during winter precipitation using observed data. Then, we analyze the results of a GCM experiment where observed SSTs are used as a forcing factor. During summer, we consider warm and cold episodes over the eastern Pacific since this seems to be a key element in Mexican precipitation during summer. In Section 2, we describe the data used in the present study as well as the numerical experiment we performed in support of this investigation. In Section 3, the results of an observational study on precipitation are presented. Here, we also describe the criteria followed to derive a statistical model for seasonal precipitation prediction. In Section 4, we present the results of downscaling GCM output and discuss the quality of our diagnoses. Finally, a summary and conclusions are presented in Section 5.

2. Data and Models

2.1. *Observational Data*

Various sources of data have been used in the present analysis. Mexican surface station data were compiled by Prof. Arthur Douglas from the University of Creighton, Nebraska, based on the observation network information of the Mexican National Weather System (available on the anonymous ftp server `ftp.cdc.noaa.gov`). These data include monthly total precipitation, surface temperature and Palmer indices. For climate classification purposes, Douglas divides the domain into 18 regions covering most of the Mexican domain. The regionalization considers station density, length of records, and geographic characteristics among stations. We will use these divisions for our regionalization of precipitation. The final value of regional precipitation is given by the average value of the data from the stations in the region. For this study, we focus on the period 1946–1987. The climatological precipitation data have been complemented with global precipitation data prepared by Legates and Wilmott [1990]. The large-scale data include geopotential height and temperature at 500 and 700 mb, as well as sea level pressure taken from the monthly data archives of the National Meteorological Center. These monthly hemispheric data span the period of interest with a resolution of 5° latitude × 5° longitude. For our analyses we will consider 500 mb relative vorticity as a large-scale variable instead of geopotential height. Finally, monthly global sea surface temperature (SST) data from the U.K. Meteorological Office have been used, not only to develop statistical relationships, but also to force the GCM used in our study.

2.2. *The Statistical Model*

A statistical diagnostic model for precipitation was developed using a linear multivariate regression technique between precipitation anomalies and four large-scale meteorological fields: 500 mb vorticity (z500), 700 mb temperature (T700), sea level pressure (SLP), and SST. To construct the model, two points were selected from within the domain of the large-scale fields whose time series yield significant correlation values with precipitation in each of the 18 regions described. In this manner, two sets of 18 equations were obtained for the summer (June through August) and winter (December through February) seasons. The statistical algorithm determines whether the large-scale input fields are redundant (i.e., fields are highly correlated with each other) and includes only the uncorrelated independent fields. The model was checked against observational data for the period 1986–1993.

2.3. *The Community Climate Model 2*

Numerical climate simulations were performed with the National Center for Atmospheric Research (NCAR) Community Climate Model version 2 (CCM2). This model includes parameterizations of cumulus convection, orographic effects, radiation, etc., and has been widely used for climate experiments. The version used has a spatial resolution of T42 (2.8 × 2.8 degrees approximately) and 18 levels in a vertical hybrid coordinate (see Hack *et al.* [1993] for model details). Considering that an important factor in summer and winter precipitation is the SST field, we force the NCAR CCM2 with observed SSTs. In this way, we concentrate on the atmospheric response of the model under ideal surface boundary conditions over the ocean. The experiment is run using data from 1948 to 1990. This period includes at least five ENSO (La Niña) events for our winter analyses, as well as five major warm and cold episodes over the eastern Pacific during summer.

3. Precipitation over Mexico

3.1. *Observed Winter Precipitation*

Winter precipitation over Mexico is generally confined to the northwestern part of the country and the southern part around the Gulf of Mexico (Figure 1). In the former region, midlatitude systems extending to subtropical regions contribute a large percentage of the observed precipitation. Around the southern Gulf of Mexico, lower tropospheric anticyclonic circulations (so-called Nortes) lead to intense precipitation. The largest amount of precipitation is observed throughout the year in this region. During the Northern Hemisphere winter, most of the interannual climate variability is determined by El Niño and La Niña events (e.g., Philander [1990], Diaz [1986]). The impact of these episodes on the world's climate has been extensively documented by many authors [Ropelewski and Halpert, 1987]. Winter precipitation over Mexico appears to be modulated by ENSO events [Cavazos and Hastenrath, 1990]. A correlation between the southern oscillation index (SOI) and precipitation anomalies (Figure 2) indicates that there are links between El Niño and precipitation over northern and central Mexico. By examining the quasi-stationary circulations associated with the El Niño events we can see how the Mexican climate is affected. For this purpose we have looked at the 500 mb difference in the streamfunction and wind fields during El Niño and La Niña events (Figure 3). The El Niño ensemble includes the events for 1983, 1987 and 1991. The cold events in the eastern Pacific and La Niña events include the 1981, 1985 and 1988 winters. During El Niño years, a low exists over the northern part of Mexico, leading to enhanced precipitation. This figure compares well with Figure 1, the correlation pattern between SOI and precipitation anomalies. On the other hand, the southernmost states of Mexico were not affected by ENSO events, which seem to be negatively correlated with warm SST anomalies over the eastern Pacific.

3.2. *Observed Summer Precipitation*

During summer, the precipitation regime over Mexico does not seem to be associated with major large-scale circulation patterns. Precipitation anomalies are basically related to SSTs. Figure 4 shows the correlation between SST anomalies over the eastern Pacific and precipitation anomalies. We can see that intense summer rainy seasons over the western states of Mexico are related to the presence of negative SST anomalies over the eastern Pacific. These negative anomalies combined with the warm summer SSTs over the western coast of Mexico lead to a direct circulation with a north–south low-level circulation that results in large moisture flux convergence over the western coast of

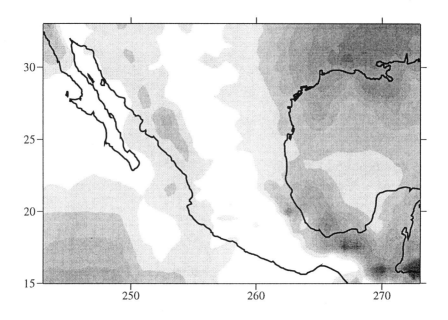

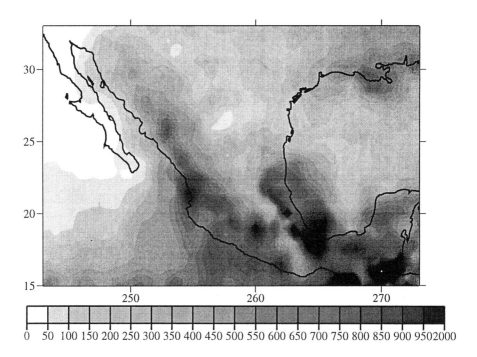

FIGURE 1. Winter (DJF, upper panel) and summer (JJA, lower panel) accumulated precipitation based on Legates and Wilmott climatology. Contour interval 50 mm.

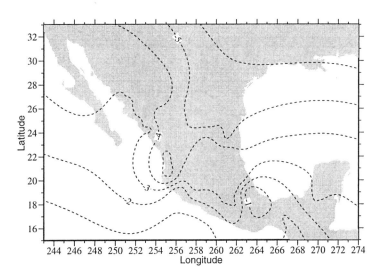

FIGURE 2. Correlation coefficient between the Southern Oscillation Index (SOI) and winter precipitation over Mexico.

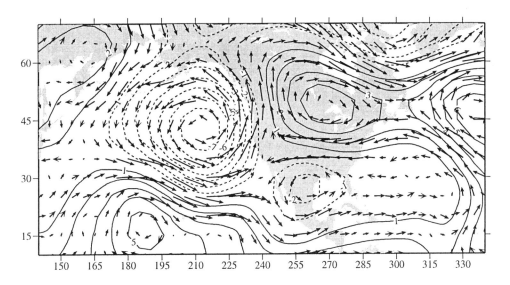

FIGURE 3. Difference in 500 mb streamfunction and wind field between El Niño and La Niña years.

Mexico (Figure 5). The availability of more moisture over this region is a key element for an intense Mexican monsoon and larger convective activity over the southern part of Mexico. Clearly, there are many more sources of moisture such as easterly waves, Atlantic tropical cyclones and other phenomena over the Caribbean and the Gulf of Mexico which have a profound impact on the summer rainy season over Mexico, particularly over the eastern states.

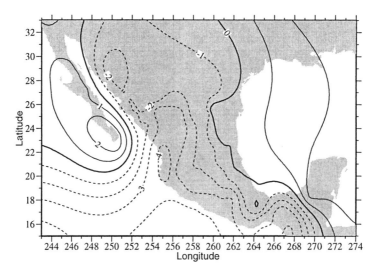

FIGURE 4. Correlation between SST in the eastern Pacific and precipitation over Mexico during summer.

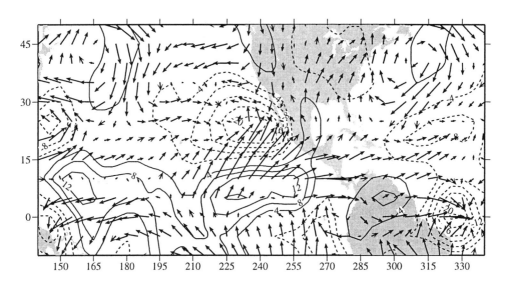

FIGURE 5. Difference in 1000 mb moisture flux convergence and winds between cold and warm events in the eastern Pacific during summer.

4. Diagnostic Results

As previously explained, on the basis of lag correlations between precipitation anomalies and large-scale patterns, we constructed a statistical model to diagnose precipitation anomalies for each of the 18 regions in Mexico. The model was evaluated using observed mean values corresponding to El Niño and La Niña (warm and cold) episodes during winter (summer). The diagnosed anomalies in precipitation were compared with the observed anomalies (not shown). In general, we found good agreement between diagnosed and observed precipitation; however, some significant errors (about 30%) were found over those regions with large seasonal precipitation. An example is the southern part of the Gulf of Mexico, where summer precipitation is an order of magnitude larger than that of the rest of the country (see Figure 1). Despite these limitations, the statistical model remains a useful tool for benchmarking and diagnostic purposes.

We now compare seasonal precipitation given by the CCM2 convective parameterization scheme and the diagnosed precipitation from the large-scale circulation patterns. Figure 6 corresponds to the difference in precipitation between El Niño and La Niña years as obtained directly from the CCM2. The CCM2 captures some of the overall structure in the precipitation anomaly fields partly because the general features of the large-scale circulation anomalies are well represented. However, the coarse resolution precludes the detection of regional features in the precipitation fields. The simulated difference in the 500 mb circulation between El Niño and La Niña events shows the existence of a PNA pattern and a major cyclonic circulation over the southeastern United States (Figure 7). This pattern compares well with that observed (Figure 3). On the basis of these large-scale circulation fields we diagnosed precipitation anomalies with our statistical model and obtained fields whose structure (Figure 8(a)) resembles that observed (Figure 8(b)).

During summer, precipitation fields are more closely related to SST and SLP fields. The difference in precipitation over Mexico predicted by the CCM2 during warm and cold summer events is not adequately represented. For instance, large increases in precipitation simulated by the CCM2 over the Yucatan Peninsula are unrealistic when compared with observational data. Such failure in the simulated precipitation anomalies is related to a poor simulation of the moisture convergence field over the Pacific west of Mexico. The differences in precipitation predicted by the statistical model (not shown) are much closer to the observed. The comparison between CCM2 and the statistical model simulations is a difficult one since the CCM2 simulation represents a one-way interaction between atmosphere and ocean. A better comparison would likely result if the SST fields were also part of the simulation, as in an atmosphere–ocean model simulation.

5. Summary and Conclusions

A statistical diagnostic model was developed to estimate precipitation anomalies over Mexico during El Niño and La Niña winters and warm and cold summer episodes over the eastern equatorial Pacific. This model is capable of reproducing major regional scale features of precipitation fields from large-scale fields. As a further test of the General Circulation Model performance, the statistical model is also used to "downscale" output from a simulation with the CCM2 forced at its lower boundary by observed SSTs for the 1950–1990 period.

During winter, precipitation over Mexico is highly influenced by ENSO events, when more precipitation is observed, particularly over the central and northern parts of the country. Although the CCM2 does not properly simulate winter regional precipitation, it successfully simulated the large-scale circulation fields. This allowed us to diagnose precipitation anomalies during El Niño and La Niña events properly.

During summer, precipitation over Mexico is largely determined by mesoscale phenomena and by the structure of the SST field over the eastern Pacific. A cold SST anomaly over this region leads to a stronger Mexican monsoon and to more active convective activity over the western states. Such enhancement in convective activity is related to increased moisture convergence over the western coast of Mexico. It is clear that good diagnostics in summer precipitation will depend on good simulations of ocean dynamics, and in particular SST simulations.

A stage has been reached where General Circulation Models are capable of simulating

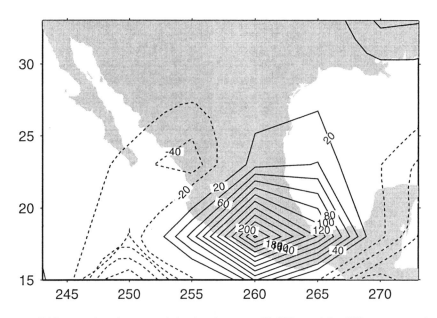

FIGURE 6. Difference in winter precipitation between El Niño and La Niña years as simulated by the NCAR-CCM2. Contour interval 20 mm.

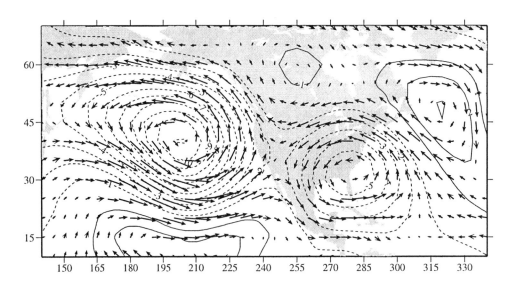

FIGURE 7. As in Figure 3 but for the NCAR-CCM2 simulation.

most of the large-scale circulations observed in the atmosphere. However, atmosphere–ocean coupled models constitute another avenue to improve predictions of ENSO events. It is necessary to develop tools that make use of such predictions if high-quality regional climate predictions are to be made. On the basis of the present study we believe that there is potential to predict regional climate over Mexico.

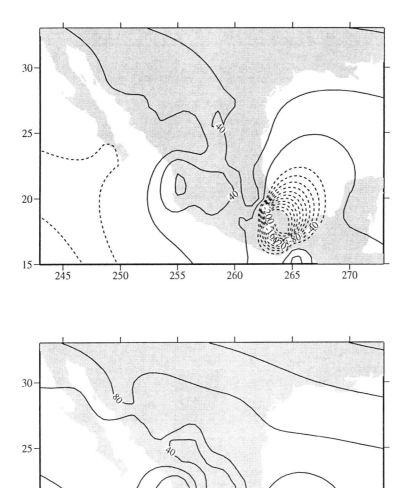

FIGURE 8. As in Figure 3 but for predictions with the statistical model using data from the NCAR-CCM2 simulations (upper panel) and with the statistical model using observed data (lower panel).

Acknowledgments. We are indebted to José Luis Pérez, Rodolfo Meza and David Langley for their collaboration in the development of this study. We also thank the Supercomputing Department at UNAM for invaluable support. This work has been supported in part by grant IN-105494 of the Program for the Support of Research and Technology Projects (PAPIIT) of the National Autonomous University of Mexico.

REFERENCES

Adem, J., Simulation of the annual cycle of climate with a thermodynamic numerical model, *Geofísica Internacional*, **21**, 229–247, 1982.

Barry, R. G., and A. H. Perry, *Synoptic Climatology, Methods and Applications*, Methuen, 1981.

Cavazos, T., and S. Hastenrath, Convection and rainfall over Mexico and their modulation by the Southern Oscillation, *Int. J. Climatology*, **10**, 377–386, 1990.

Diaz, H. F., Characteristics of the response of sea surface temperatures in the Central Pacific associated with warm episodes of the Southern Oscillation, *Mon. Wea. Rev.*, **114**, 1716–1738, 1986.

Giorgi, F., Simulation of regional climate using a limited area model nested in a general circulation model, *J. Climate*, **3**, 941–963, 1990.

Hack, J. J., B. A. Boville, B. P. Briegleb, J. T. Kiehl, P. J. Rasch, and D. L. Williamson, Description of the NCAR Community Climate Model (CCM2). NCAR Technical Note, NCAR/TN-382+STR, 1993.

Jauregui, E., Rainfall fluctuations and tropical storm activity in Mexico, *Erdkunde*, **49**, 39–48, 1995.

Klein, W. H, and J. B. Bloom, Specification of monthly precipitation over the United States from the surrounding 700 mb height field, *Mon. Wea. Rev.*, **115**, 2118–2132, 1987.

Klein, W. H, and G. A. Hammons, Maximum/minimum temperature forecasts based on model output statistics, *Mon. Wea. Rev.*, **103**, 796–806, 1975.

Legates, D. R., and C. G. Willmott, Mean seasonal and spatial variability in gauge corrected, global precipitation, *Int. J. Climatology*, **10**, 111–128, 1990.

Philander, S. G. H., El Niño La Niña and the Southern Oscillation, Academic Press, 1990.

Ropelewski, C., and M. Halpert, Global and regional scale precipitation pattern associated with El Niño/Southern Oscillation, *Mon. Wea. Rev.*, **114**, 2352–2362, 1987.

A Numerical Study of the Circulation and Sea Surface Temperature of the Gulf of Mexico

By Jorge Zavala, A. Parés-Sierra, J. Ochoa and J. Sheinbaum

División de Oceanología, CICESE, Ensenada, Baja California, México

The oceanic circulation in the Gulf of Mexico is studied with a $2\frac{1}{2}$ inhomogeneous layer model. Special attention is given to the seasonal component of the circulation of the western Gulf. The numerical calculations consider the influence of climatological winds, surface heat fluxes, vertical exchanges between the active layers or entrainment–detrainment, and open boundary conditions.

The results show that entrainment is crucial to the existence of cold surface temperatures during winter in the northern Gulf. When entrainment associated with buoyancy loss and turbulent kinetic energy is not included in the model, the temperatures in winter stay higher than the observations and the root mean square (RMS) temperature value, relative to observations, is 1.25°C. Including entrainment and detrainment the RMS of the temperature decreases to 0.66°C. The surface heat fluxes computed by using Comprehensive Ocean–Atmosphere Data Set (COADS) and model ocean surface temperatures result in an annual mean of 26 Wm^{-2} and 32 Wm^{-2} respectively.

1. Introduction

The Gulf of Mexico can be studied in different ways; we are interested in the mean circulation, the sea surface temperature (SST) and their variability on seasonal time scales. Our present interest is to identify the relative influence of the Loop Current eddies and the wind stress on the circulation of the northwestern Gulf and Campeche Bay; to study the seasonal circulation in these regions, and to review the heat fluxes in the Gulf and their relationships with the observed SST. In this paper we concentrate on the SST results. We discard topographic influences and restrict the domain under consideration to the nonshelf Gulf. Important Gulf features such as the Loop Current behavior and eddy shedding period will not be discussed here.

The seasonal heat budget has been studied by different authors [Hastenrath, 1968; Etter, 1983; Adem et al., 1991], who worked with climatologic atmosphere and ocean data and computed heat fluxes and heat storage using bulk formulae. Adem et al. [1991] also included climatic drift currents and their influence on the SST. Most authors [e.g., Etter, 1983] report an annual mean heat flux near zero, but slightly negative. Nevertheless, surface heat fluxes are intense and the SST has an important seasonal variability of more than 12°C. During summer the SST is very homogeneous in the entire Gulf, while in winter there are important temperature gradients. In other studies the advection terms in the heat equation have been estimated as a residual, and the subsurface fluxes have been neglected, assuming that they are not important below some prescribed depth. These assumptions close the heat equation but have not addressed the effects of advection and vertical exchanges.

The goal of this paper is to study the average and seasonal variation of the heat fluxes and SST in the Gulf of Mexico, with emphasis on the understanding of the relative importance of advection and entrainment components. We use a $2\frac{1}{2}$ inhomogeneous layer model similar to those used by McCreary and Kundu [1988] and Schopf and Cane

[1983]. This model allows for variations within each active layer of the horizontal velocity, thickness and buoyancy (temperature), all of them with dynamic consequences. In particular, the inclusion of laterally varying temperature allows for the modeling of nonuniform surface heat fluxes and realistic vertical exchanges between the layers. As forcing mechanisms we use the wind stress, via the data of Hellerman and Rosenstein [1983], and the transports for the Yucatan Channel and Florida Strait. To study the heat balance, we use 10 years of data from the Comprehensive Ocean–Atmosphere Data Set (COADS), covering the period 1980–1991. With these data we estimate monthly atmospheric climatological conditions to compute ocean surface heat fluxes. The COADS sea temperature was also used to verify model results. Several experiments were done in order to evaluate the importance of each mechanism in the resulting SST.

Section 2 presents a brief description of the model; Sections 3 and 4 describe results obtained for the SST and heat balance, respectively. Final remarks are included in Section 5.

2. The Model

We use a primitive-equation, inhomogeneous-layer model, similar to that of McCreary and Lu [1994]. The model consists of an active upper layer with variable thickness related to the mixed layer, an active intermediate layer and a deep motionless layer. The model allows the exchange of heat, momentum and mass between the upper and the intermediate layer. The most interesting characteristic of this model is that temperature may vary in space and time within each active layer. A description of inhomogeneous layer models can be found in McCreary and Kundu [1988] and Ripa [1993].

The model equations are not presented in this paper because of space limitations, but a similar set can be seen in McCreary and Lu [1994]. The main differences between our model and that of McCreary and Lu [1994] are that we use biharmonic damping in the momentum equations, we do not include dissipation terms in the continuity equation, and we use spherical coordinates.

The surface heat flux is computed using a Haney [1971] type of expression:

$$Q = Q^*(T_a^* - T_M) \tag{2.1}$$

where Q is a linear function of the difference between the model upper layer temperature T_M and an apparent air temperature T_a^*. In our model Q^* and T_a^* vary in space and time and are estimated by using bulk formulae and COADS data. We followed the bulk formulae from Castro *et al.* [1994], although with different constant parameters. The functions T_a^* and Q^* are defined such that Q is the linear term in a Taylor expansion of the bulk formulae [Haney, 1971], and T_M is the upper layer model temperature.

2.1. *Mixed Layer Physics*

In this model, the surface layer thickness varies in two time scales: a short time scale associated with eddies and wavelike perturbations and a long seasonal scale identified with the prevailing entrainment–detrainment mechanism.

To compute the entrainment $(w_e^{'})$ and detrainment $(w_d^{'})$ we use the following expressions:

$$w_e^{'} = \frac{(h_1 - H_e)^2}{H_e t_e}\theta(H_e - h_1) + D\frac{mu_*^3 - h_1 B_0}{g\alpha h_1 (T_1 - T_2 + \delta)} \tag{2.2}$$

$$w_d^{'} = -\frac{(h_1 - H_e)^2}{H_e t_e}\theta(h_1 - H_d) + D\frac{mu_*^3 - h_1 B_0}{g\alpha h_1 (T_1 - T_2 + \delta)} \tag{2.3}$$

The first term on the right side of (2.2) becomes important in regions of upwelling; eliminating numerical singularities prevents zero thickness of layers. It is different from zero only when h_1, the upper layer thickness, is thinner than H_e. This term has been used by McCreary (see, for example, McCreary and Kundu [1988]). The second term in the right side of (2.2) is a Niiler and Kraus [1977] type of term. It is associated with buoyancy flux, stratification and wind stress; it is weaker than the first one but remains active for long periods in wide areas, making its contribution more important in the seasonal time scale.

In the second term on the right side of (2.2), u_* is the drag velocity, m is the wind stirring coefficient, g is gravity acceleration, α is the coefficient of thermal expansion, δ is a constant to prevent overturning, and B_0 is the buoyancy flux, computed as

$$B_0 = \frac{\alpha g Q}{C_p}$$

Entrainment (detrainment) adjusts h_1 thickness to the Monin–Obukhov depth, given by

$$H_{\mathrm{MO}} = \frac{m u_*^3}{B_0}$$

The Monin–Obukhov depth corresponds to equilibrium between the terms in the numerator of the second term on the right of (2.2). The coefficient D controls the rate at which h_1 follows the Monin–Obukhov depth, and $\theta(x)$ is a Heaviside step function.

To restrict entrainment further to a determined mixed layer thickness, the following formulae for entrainment and detrainment are used:

$$w_e = w_e' \theta(w_e') \theta \left(H_{\max} - h_1 \right) \tag{2.4}$$

$$w_d = w_d' \theta(-w_d') \theta \left(h_1 - H_{\min} \right) \tag{2.5}$$

$H_{\max}$ and $H_{\min}$ prevent numerical instabilities that result from very thin or thick mixed layer depth due to extreme conditions in buoyancy or turbulent kinetic energy production. The numerical model was run on a C grid [Mesinger and Arakawa, 1976], with a zonal and meridional resolution of 1/6 of degree between the same variable points. A leapfrog scheme was used to integrate the equations in time, with a forward derivative every 99 timesteps to prevent time-splitting instability. The timestep was 20 min. For generality spherical coordinates are used. The parameters used in the model are listed in Table 1.

Open boundary conditions for the transports are prescribed at the Yucatan and Florida Straits, choosing equal amounts of inflow and outflow per layer (Table 1). Inflow temperature in the upper layer is prescribed with a seasonal (sinusoidal) variation. In each experiment we spin up the model for one year to reach stable conditions.

2.2. *Numerical Experiments*

To analyze the importance of the different terms of the heat equation we perform the following experiments: Experiment 1 includes all the terms in the heat equation and only the first term on the right side of (2.2); in Experiment 2, the advection term in the heat equation is excluded and only the first term on the right side of (2.2) is considered; Experiment 3 includes all the terms in both the heat equation and (2.2). In all the experiments the model is forced with the wind stress and the Yucatan Current. In the following sections we will refer to these experiments as E-1, E-2 and E-3, respectively.

TABLE 1. Model parameters.

Transports in open boundary upper layer		6 Sv
Transports in open boundary intermediate layer		6 Sv
Mean inflow temperature of upper layer		27.28°C
Amplitude of sinusoidal term of upper layer inflow temperature		1.94°C
Date of maxima upper layer inflow temperature		Aug 10
Inflow temperature of intermediate layer water	T_2	15.0°C
Temperature of deep ocean	T_3	4.0°C
Initial thickness of upper layer	H_1	75 m
Initial thickness of intermediate layer	H_2	200 m
Entrainment time scale	t_e	1 day
Detrainment time scale	t_d	1 day
Minimum detrainment upper layer depth	H_{min}	50 m
Maximum entrainment upper layer depth	H_{max}	160 m
Dynamic entrainment depth	H_e	50 m
Dynamic detrainment depth	H_d	150 m
Coefficient of wind stirring	m	2.5
Coefficient of second entrainment term	D	1.5
Coefficient of thermal expansion	α	0.00025°C

3. Sea Surface Temperature

To study the model seasonal behavior we smooth the interannual variability, due principally to propagating eddies, taking the mean spatial temperature over 4 years of monthly model data for each month.

3.1. *Response to Different Thermodynamical Forcing*

In E-1 the SST is determined basically by air–sea interaction and heat advection. Vertical fluxes between the upper and the intermediate layers are limited to small regions where there are strong boundary currents, and in this case there is no seasonal variability of the mixed layer thickness.

E-1 qualitatively reproduces the most important characteristics of the SST in the Gulf of Mexico well. SST model data and COADS data fits better in summer (between May and late August). In this period, model mean temperatures are lower than observations, with differences of less than 0.5°C and a root mean square error (RMS) of 1.0°C. In winter, model temperatures are higher than COADS–SST, and differences reach more than 2°C in the northern Gulf, with RMS between 1.6°C and 2.0°C (Figures 1, 2).

The maximum model temperature is reached in September, lagging one month that of the COADS data; the minimum corresponds to March, while in the observations this occurs in February. The lag in the SST cycle can be explained by considering that (1) we distribute the surface heat flux over the entire upper layer instantly instead of progressively diffusing it from the top, in which case the top SST responds faster, and (2) model temperatures remain very high in winter, so that in February, at the end of the cooling period, the difference of temperature in (2.1) may still be negative because of the high SST. This produces heat loss from February to March.

In E-2 the SST is determined basically by air–sea interaction and the mean spatial temperature is similar to that of E-1. The Loop Current and eddies temperature signal disappears, but north–south temperature gradients are more intense. In winter and spring there are lower temperatures at the northern Gulf than in E-1, and they remain

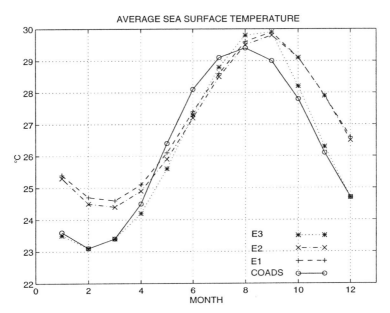

FIGURE 1. Mean monthly sea surface temperature from COADS data and three numerical experiments.

closer to observations, but in other regions there are higher temperatures and the final result is a similar RMS (Figures 1, 2).

The higher winter temperature of the model compared to that of the observations, from either E-1 or E-2, suggests that there are other important processes besides the interaction through the sea surface and advection.

In E-3, we include an entrainment–detrainment process associated with surface fluxes that simulate the seasonal deepening of the mixed layer. This process is due to the increase of turbulence due to strong winds and to a negative buoyancy flux in the upper layer, which results in an influx of cold water from the intermediate to the surface layer. The entrainment is stronger in the northern and western Gulf, and in October and November. The seasonal variation of the SST is much better reproduced when entrainment is included. The deepening of the surface layer begins in late September and ends in February, reaching its maximum rate in November. The average temperature difference of E-3 from observations is less than 0.5°C for most of the year and has an RMS of 0.66°C for the whole year while E-1 and E-2 have 1.35°C and 1.25°C, respectively (Figures 1, 2).

3.2. *Monthly Analysis*

In this section we describe the results from the most complete experiment (E-3). Figures 4 and 5 show the COADS–SST and the upper layer temperature in monthly maps. The SST cycle can be divided into rising and decreasing periods. The rising period starts in March, when the mixed layer thickness begins to decrease and the heat flux through the sea surface becomes positive (Figures 1, 3, 4, 5). In March, there are strong spatial gradients from north to south after the previous winter season. The anticyclonic eddies, filtered out in the monthly means, are clearly identifiable in SST because of their higher temperature than that of their surroundings. The Loop Current has the highest SST in

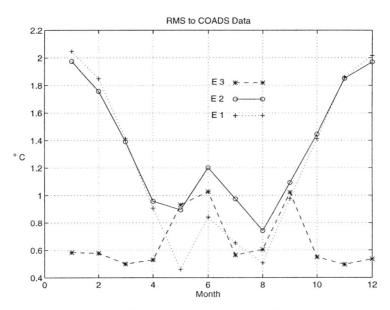

FIGURE 2. Root mean square of the sea surface temperature between numerical experiments and COADS data.

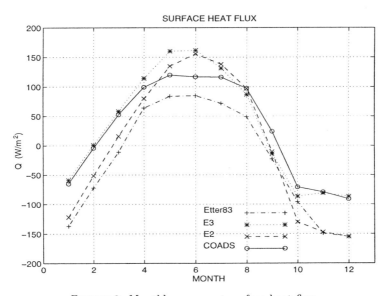

FIGURE 3. Monthly mean net surface heat flux.

the region and Campeche Bay maintains higher temperatures than those of the northern Gulf. The model data also has other features associated to the circulation, such as the signature of a semipermanent anticyclonic eddy in the northwestern Gulf. It also shows a spatial gradient in the Florida Strait due to the presence of different water masses. The southern area of the Strait has high-temperature water from the Loop Current, and the north Strait has a section with colder water from the northeastern Gulf (Figure 5).

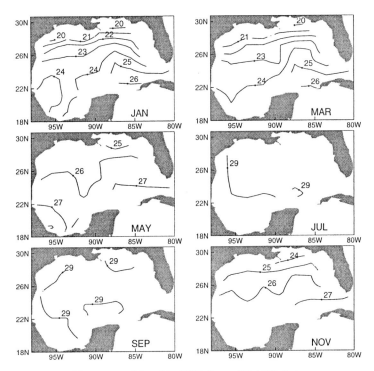

FIGURE 4. Monthly SST from COADS data.

In May the heat flux is positive in the Gulf and the SST still conserves some north–south gradients. In Campeche Bay the SST reaches 26°C, and the Loop Current, eddies, and the Florida Strait gradient are still evident.

In July almost the entire Gulf has the same SST, maintained between 28°C and 29°C. The anticyclonic eddies and the Loop Current are not distinguishable in individual maps of SST because of the lack of temperature contrasts.

The scenario described for July continues until September, raising the SST to 29°C or 30°C in the entire Gulf at the end of the temperature rise period. In October the heat flux through the surface changes and the Gulf starts to cool; also the winds are stronger and change direction, taking a south component. In November the heat loss through the atmosphere is strong and its combination with wind stirring gives higher levels of turbulent kinetic energy; this translates into entrainment, which deepens and cools the mixed layer. The two cooling mechanisms are not spatially homogeneous; entrainment is more intense in the northern and western Gulf, and atmospheric heat fluxes are stronger in the areas with higher temperatures or strong winds. As a result of cooling and the inflow of cold waters, spatial SST gradients increase and the anticyclonic eddy in the northwest Gulf, the Loop Current, and their shedded eddies have higher surface temperature than their surroundings. Also, a cold signature in western Campeche Bay, due to the cyclonic circulation in the region (Figure 5), can be seen. Entrainment in the upper layer continues until February, cooling and deepening the mixed layer and increasing the surface spatial gradients.

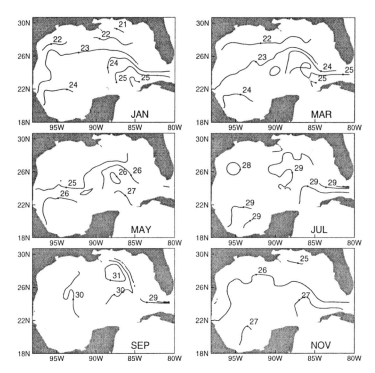

FIGURE 5. Monthly model upper layer temperature.

4. Heat Fluxes

In the Gulf of Mexico there are periods of positive and negative surface heat fluxes. These fluxes are not homogeneous in the Gulf area, with stronger variations in the north.

We estimated heat fluxes considering observed SST from COADS and model upper layer temperature. The results are compared with those of Etter [1983], obtained from data and bulk formulae. Model results show that from March to September the Gulf gains heat; it loses it from October to February (Figure 3). During autumn and winter the areas with stronger surface heat loss are in the northern Gulf, and those associated with the areas of higher temperatures: the Loop Current and the path of the anticyclonic eddies. However, the results of E-1 and E-2 indicate that the surface heat flux distribution cannot by itself explain the winter spatial variability of the SST.

From September to April, the E-2 spatially averaged heat flux values are lower than COADS values because of the higher model SST. Stronger discrepancies are found from January to March, reaching a difference greater than 50 Wm^{-2} in December (Figure 3). From August to April, the E-3 average heat fluxes are very similar to and slightly lower than those obtained using COADS–SST. Fluxes derived from E-3 are higher than those of E-1; this difference is associated with the SST because in the parameterization used for the heat flux in (2.1) the quantities associated with other variables do not vary. Heat fluxes from COADS and E-3 are higher than those estimated by Etter [1983] for the entire year (Figure 3). The annual net surface heat flux from COADS and E-3 is of 26 Wm^{-2} and 32.2 Wm^{-2} respectively, while Etter [1983] reports -24 Wm^{-2}. This difference may result from three causes: first, we do not consider the shelf, where there must be strong heat loss during winter; second, Etter [1983] has few data at Campeche

Bay, a region that maintains some upwelling of subsurface waters that produce heat gain through the sea surface; and third, we used different data source and parameters on the bulk formulae.

5. Summary

A $2\frac{1}{2}$ inhomogeneous layer model was used to study the seasonal surface temperature and heat fluxes of the Gulf of Mexico. It was forced by a constant inflow and outflow in the Yucatan and Florida Straits, respectively, and by monthly climatological winds. Different versions of the entrainment parameterization and the heat equation are used to determine the influence of different physical processes on the SST. Monthly heat fluxes through the atmosphere are parameterized as in [Haney, 1971].

In order to analyze the relative influences of the different processes on the SST, three experiments were done, including or removing terms in the model equations. When heat advection is removed from the model equation (E-2) the SST shows insignificant changes relative to the case in which they are included (E-1). On the other hand, when entrainment–detrainment processes associated to the wind stress and the buoyancy flux are included, the winter temperatures are better reproduced and the RMS of the data decreases to almost half of its value on E-1.

Acknowledgments. This work is supported by CONACYT grant 1002-T9111 and CICESE normal funding.

REFERENCES

Adem, J., V. M. Mendoza, E. E. Villanueva-Urrutia, and M. A. Monreal-Gomez, On the simulation of the sea surface temperature in the Gulf of Mexico using a thermodynamic model, *Atmósfera*, **4**, 87–99, 1991.

Castro, R., M. F. Lavín and P. Ripa, Seasonal heat balance in the Gulf of California, *J. Geophys. Res.*, **99**, C2, 3249–3261, 1994.

Etter, P. C., Heat and fresh water budgets in the Gulf of Mexico, *J. Phys. Ocean.*, **13**, 2058–2069, 1983.

Haney, R.L., Surface thermal boundary condition for ocean circulation models, *J. Phys. Ocean.*, **1**, 241–248, 1971.

Hastenrath, S. L., Estimates of the latent and sensible heat flux for the Caribbean and the Gulf of Mexico, *Limnol. Oceanogr.*, **13**, 322–331, 1968.

Hellerman, S., and M. Rosenstein, Normal monthly wind stress over the world ocean with error estimates, *J. Phys. Ocean.*, **13**, 1093–1104, 1983.

McCreary, J. P., Jr., and P. Kundu, A numerical investigation of the Somali Current during the Southwest Monsoon, *J. Mar. Res.*, **46**, 25–58, 1988.

McCreary, J. P., Jr., and P. Lu, Interaction between the subtropical and equatorial ocean circulations: The subtropical cell, *J. Phys. Ocean.*, **229**, 466–497, 1994.

Mesinger, F., and A. Arakawa, Numerical methods used in atmospheric models, *GARP Publications, No. 17.*, 1976.

Niiler, P. P., and E. B. Kraus, One dimensional model of the upper ocean. *Modelling and Prediction of the Upper Layer of the Ocean*, E. B. Kraus, Ed., Pergamon, 143–172, 1977.

Ripa, P., Conservation laws for primitive equations models with inhomogeneous layers, *Geophys. Astrophys. Fluid Dyn.*, **70**, 85–111, 1993.

Schopf, P. S., and M. A. Cane, On equatorial dynamics, mixed layer physics and sea surface temperature, *J. Phys. Ocean.*, **13**, 917–935, 1983.

Vukovich, F. M., An updated evaluation of the Loop Current's eddy-shedding frequency, *J. Geophys. Res.*, **100**, C5, 8655–8659, 1995.

Pacific Region CO$_2$ Climate Change in a Global Coupled Climate Model

By Gerald A. Meehl and Warren M. Washington

National Center for Atmospheric Research
Boulder, Colorado 80307-3000

Average climate change in the Pacific region from increased carbon dioxide (CO$_2$) in a global coupled ocean–atmosphere general circulation model resembled present-day El Niño conditions. This was a consequence of greater warming of the ocean surface in the eastern equatorial Pacific than in the west. Attendant increases in precipitation in the central equatorial Pacific were accompanied by precipitation decreases in the northern and southern tropical Pacific, Australasia, and eastern Indian Ocean regions. Associated effects in the midlatitudes were similar to El Niño conditions as well, with a deepened Aleutian low-pressure center in the north Pacific. These mean climate changes due to increased CO$_2$ in the model also resembled recent observed decadal-scale climate anomalies in the Pacific region. Rather than attribution of CO$_2$ climate change, the model results suggest that if there is a warming of SSTs across the Pacific for whatever reason, mechanisms partly due to cloud-albedo feedback effects would contribute to SST increases that are less over the western Pacific warm pool than over the tropical eastern Pacific, with attendant shifts in large-scale precipitation patterns and midlatitude circulation anomalies in the north Pacific that resemble ENSO events.

1. Introduction

It has been noted that decadal timescale climate fluctuations in the Pacific region have been characterized by a persistent warming of the surface waters of the central and eastern tropical Pacific Ocean during the 1980s and early 1990s [Nitta and Yamada, 1989; Trenberth and Hurrell, 1994; Kerr, 1994; Graham, 1994; Nitta and Kachi, 1994]. A climate model experiment has shown that these persistent warm ocean surface temperatures have probably contributed to observed global warming during this period [Kumar et al., 1994]. That study could not distinguish whether or not these warm conditions were themselves a product of global warming, or whether the warm tropical Pacific sea surface temperatures (SSTs) were a manifestation of increased frequency of warm events in that region or of some other low-frequency fluctuation of the climate system.

The balance between two processes, the super greenhouse effect and cloud-albedo feedback, has been shown to be important to SST response in the warm tropical oceans [Meehl and Washington, 1995]. Those results, and the suggestions of other studies [Ramanathan and Collins, 1992], allude to the importance of cloud-albedo feedback for Pacific region climate change. That is, if there is a warming of the tropical Pacific Ocean surface, the western Pacific warm-pool SSTs may increase at a slower rate than the SSTs in the tropical eastern Pacific, where this effect (i.e., cloud-albedo feedback) is not so strong. If this hypothesis is correct, a mean or decadal timescale warming in the Pacific region would be manifested by a relaxation of the SST gradient across the Pacific, not unlike what occurs in a present-day El Niño event. And, like what is observed during El Niño events [Ropelewski and Halpert, 1987; Kiladis and van Loon, 1988], attendant precipitation anomalies would feature enhanced precipitation in the central equatorial Pacific with precipitation deficits to the north and south in the tropical Pacific and over Australasia, as well as possible links to other regions [e.g., Pisciottano et al., 1994].

To test this hypothesis, we analyzed results from an experiment with a global coupled ocean–atmosphere general circulation model (GCM) with increased CO_2 in the atmosphere and attempted to gain insights into the recent observational results by studying model-simulated phenomena. The changes in mean climate due to increased CO_2 described here could also affect phenomena associated with El Niño–Southern Oscillation–(ENSO-)like variability in the model [Meehl *et al.*, 1993]. This will be the subject of a subsequent study with the present model.

2. The Global Coupled Climate Model

A second-generation global coupled general circulation climate model developed at the National Center for Atmospheric Research was integrated for 75 years with atmospheric CO_2 increasing at a rate of 1% per year compounded. The last 20 years of this experiment were analyzed (near the time of CO_2 doubling at year 70) and compared to a control integration with present-day amounts of CO_2. The atmospheric model had an approximate horizontal resolution of 4.5° latitude and 7.5° longitude with nine vertical levels. The ocean and sea-ice components had 1° latitude–longitude resolution with 20 levels in the ocean. Sea ice included a three-layer thermodynamic scheme along with dynamic sea ice. Of particular relevance to this experiment, the atmospheric model included a simple cloud-albedo feedback parameterization and a mass flux convective scheme. The former parameterization accounts for the observed relationship among very warm SSTs, deep convection, and bright clouds [Washington and Meehl, 1994]. The latter represents the super greenhouse effect in the model [Meehl and Washington, 1995]. A sensitivity experiment performed with this version of the coupled model where the cloud-albedo feedback was strengthened showed that (1) there was a large-scale response of the climate system combining radiative and dynamic feedbacks, and (2) the maximum values of tropical SSTs were a function of the strength of the cloud-albedo feedback as represented by the cloud-albedo feedback parameterization in the model [Meehl and Washington, 1995]. Thus, in spite of the limitations, systematic errors (e.g., see Mechoso *et al.* [1995] for typical errors in simulating the climate of the eastern tropical Pacific), and simplified parameterizations in the coupled model, important processes affecting the sensitivity of SSTs in the warm tropical oceans have been documented, compared to observations, and analyzed in a sensitivity experiment [Meehl and Washington, 1995].

3. Results

SST anomalies in the tropical Pacific region, increased CO_2 minus control, for the December–January–February (DJF) season (DJF is discussed here because of observational studies that focused on this season; other seasons in the model show similar results) showed least warming (less than 2°C) where mean SSTs were greatest in the model. These areas included the tropical Pacific near 20°N in the region of the Intertropical Convergence Zone (ITCZ), in the South Pacific Convergence Zone (SPCZ) region southeast of Papua New Guinea in the tropical southwestern Pacific, and over Australasia and the eastern Indian Ocean. These were the regions not only of warmest SSTs and greatest mean rainfall in the Pacific region, but also where the cloud-albedo feedback effects were greatest in the model [Meehl and Washington, 1995]. Meanwhile in the equatorial eastern Pacific, there was relatively greater warming of the ocean surface (2°C–4°C). Therefore, the CO_2-related surface warming was not uniform at the ocean surface across the tropical Pacific. There was a reduction of the meridional SST gradient not unlike what is seen during a present-day El Niño event [Meehl, 1987]. This was not due to a change of frequency of El Niño–like events in the model, since preliminary results from an analysis of El Niño frequency did not show significant changes between control and increased CO_2 experiments. Additionally, time averages over other periods showed a similar signal.

Observations from the Pacific region [Bottomly *et al.*, 1990] showed that such differential warming occurred during the decade of the 1980s. During this period when global temperatures also increased, the mean warming in the tropical western Pacific (about 0.15°C) was roughly less than half that in the tropical eastern Pacific (about 0.35°C), even though El Niño fluctuations continued and were superimposed on the warmer mean SST with no discernible change in frequency in relation to that warmer mean [Wang, 1995]. This relatively greater warming of mean SSTs in the eastern tropical Pacific compared to the western tropical Pacific during the 1980s has been noted in other studies as well [Nitta and Kachi, 1994; Houghton *et al.*, 1992].

Precipitation anomalies from the coupled model, increased CO_2 minus control, showed that precipitation increased in the central equatorial Pacific and decreased in the warm-water regime regions of the ITCZ, SPCZ, Australasia, and the eastern Indian Ocean. These mean climate-change patterns in the coupled model from increased CO_2 resembled those associated with present-day El Niño events in the tropical Pacific region, as well as similar decadal timescale changes observed during the 1980s.

Sea-level pressure (SLP) anomaly patterns from the coupled model, increased CO_2 minus control, also showed the El Niño–like feature seen in the observations of composite ENSO events of a deepened Aleutian low-pressure center in the North Pacific [van Loon and Madden, 1981] as well as the similar decadal timescale SLP anomalies in that region [Trenberth and Hurrell, 1994; Graham, 1994; Trenberth, 1990]. This similarity between the changes in SLP in the model in the North Pacific due to increased CO_2 and to those seen in the decade of the 1980s in the observations is associated with greater SST warming in the central and eastern Pacific compared to the western Pacific warm-pool region in both model and observations.

To examine in more detail the mechanisms producing these anomalies in the model, components of the surface energy balance and low-level moisture parameters were compiled for three tropical equatorial areas (eastern Pacific, western Pacific, and eastern Indian Ocean – Table 1). The changes in SST near the time of CO_2 doubling in the model for these three areas corresponded to similar changes noted earlier, with the eastern tropical Pacific area warming more (3.49°C) than the western Pacific (2.21°C) and eastern Indian Ocean (2.25°C) warm-pool regions.

First contrasting the eastern and western Pacific regions, absorbed solar radiation at the surface decreased in the western Pacific (-3.92 W m^{-2}) and increased in the eastern Pacific ($+2.23$ W m^{-2}), partly as a result of cloud-albedo feedback effects: increased convection over warmer water, increased cloud albedos, decreased sunlight reaching the surface [Meehl and Washington, 1995]. This would contribute to inhibited warming of the western Pacific compared to that of the eastern Pacific. There were small changes in net infrared radiation and sensible heat flux, but there was a large increase of latent heat flux (plus sign denotes heat removed from the ocean surface) in the eastern Pacific compared to the western Pacific ($+8.67$ W m^{-2} vs. $+0.81$ W m^{-2}). Changes in ocean advection (vertical and horizontal) were both large and positive for the eastern and western Pacific, indicating that the westerly anomaly surface winds and surface currents (not shown) resulted in less heat transported eastward (i.e., more heat stayed in the eastern Pacific, $+14.80$ W m^{-2}, compared to the western Pacific, $+11.82$ W m^{-2}). There was also a large increase in low-level moisture convergence in the eastern Pacific ($+62\%$) compared to the west ($+3\%$).

Thus, comparing the eastern and western Pacific, the changes in absorbed solar ra-

TABLE 1. Increased CO_2 experiment minus control, DJF, 20-year (positive sign acts to warm the surface).

	SST	ABS SOLAR	NET IR	SENS	LATENT	ADV	Q-CON	Q
E. Pacific	+3.49	+2.23	+0.92	+2.11	−8.67	+7.56	+62%	+22%
W. Pacific	+2.21	−3.92	+0.28	+0.88	−0.81	+5.28	+3%	+16%
E. Indian	+2.25	+2.80	−0.10	−1.59	+8.66	−13.02	+28%	+18%

E. Pacific = 5°S–5°N, 120°W–90°W; W. Pacific = 5°S–5°N, 140°E–170°E; E. Indian = 5°S–5°N, 75°E–95°E.

Units: SST is in °C; absorbed solar (ABS SOLAR), net infrared at the surface (NET IR), sensible (SENS), latent (LATENT), and upper-ocean layer heat advection (ADV) are all in W m^{-2}; surface moisture convergence (Q-CON) and surface moisture (Q) are percentage changes.

diation at the surface cooled the western Pacific compared to the east, and changes in latent heat flux cooled the eastern Pacific compared to the west, while ocean advection heated the eastern Pacific relative to the western Pacific. The larger SST rise in the east compared to the west contributes to the greater *relative* increase in latent heat flux in the east; yet the combination of increased latent heat flux and increased low-level moisture convergence in the east compared to the west only marginally increased the low-level moisture amounts (+22% vs. +16%). Thus, the initial effect of ocean warming due to increased CO_2 enhanced convection and decreased absorbed solar radiation at the surface and contributed to less SST warming in the west compared to the east. This redistribution of convective activity also altered the large-scale east–west atmospheric circulation such that surface easterly winds and currents were weaker (thus altering the SSTs) so that more heat remained in the eastern Pacific relative to the west. This change in the large-scale atmospheric circulation also altered the low-level moisture convergence such that there was enhanced moisture convergence into the eastern Pacific Ocean region compared to the west. However, in absolute terms, there were still greater latent heat flux and low-level moisture amounts in the western Pacific area compared to the eastern area as observed [Oberhuber, 1988; Zhang and McPhaden, 1995] because of the higher base-state SSTs in the west [Meehl and Washington, 1995]. Thus, the radiative changes of absorbed solar radiation (partly due to cloud-albedo feedback), plus the large-scale dynamical changes that reduced the amount of heat exported from the eastern Pacific, combined to make a major contribution to the greater relative warming of SSTs in the eastern Pacific than the western Pacific.

Even though the eastern Indian Ocean warm-pool SST changes were similar to those of the western Pacfic, there were quite different mechanisms inhibiting the ocean surface warming in the eastern Indian region. The precipitation changes can be used as analogs to changes in the large-scale vertical velocities. Thus, over the eastern Indian region there were suppressed convection and anomalous downward motion, fewer clouds, and an increase of absorbed solar radiation (+2.80 W m^{-2}) at the surface compared to the decrease over the western Pacific (−3.92 W m^{-2}). The negative value of sensible heat flux change (−1.59 W m^{-2}) cooled the surface (compared to the positive value of +0.88 W m^{-2} in the western Pacific). Of considerably greater interest were the latent heat flux anomalies, with negative latent heat flux differences for the Indian region

(-8.66 W m^{-2}) compared to positive differences in the western Pacific $(+0.81 \text{ W m}^{-2})$. This should have warmed the ocean surface in the Indian region. However, there was also a large difference in ocean advection, with negative anomalies in the Indian region $(-13.02 \text{ W m}^{-2})$ compared to positive values in the western Pacific $(+5.28 \text{ W m}^{-2})$. This would export more heat from the eastern Indian Ocean and inhibit surface warming. The large decrease of latent heat flux (producing the positive heating anomalies noted) can be explained by changes in the low-level moisture convergence. The eastern Indian region showed a sizable increase compared to the western Pacific $(+28\% \text{ vs. } +3\%)$. The net result for low-level moisture is an almost comparable increase in the eastern Indian and western Pacific $(+18\% \text{ vs. } +16\%)$.

Thus, the large-scale alterations of the atmospheric circulation associated with the increased convective activity over the equatorial Pacific produced enhanced low-level moisture convergence and increased low-level moisture in the eastern Indian region that inhibited evaporation and reduced latent heat flux. All the processes involving low-level moisture convergence, evaporation, and large-scale atmospheric circulation tended to balance, leaving the enhanced ocean advective flux out of the eastern Indian region (associated with increased and northward shifted southeast Trades in the Indian Ocean and strengthened westward surface currents, not shown), made a major contribution inhibiting further warming of the ocean surface. Thus, the eastern Indian and western Pacific Ocean regions were in the same category as far as SST warming goes (less compared to the eastern Pacific region), but significantly different mechanisms occurred in each of those regions to contribute to the SST anomalies.

4. Summary

The climate model results presented here and elsewhere [Knutson and Manabe, 1995] suggest that a warmer climate in the tropical Pacific is associated with greater warming of SSTs in the east than in the west, partly as a result of a combination of cloud-albedo feedback effects and attendant changes in surface energy balance and the large-scale circulation of atmosphere and ocean. Because of the greater warming of SSTs in the eastern tropical Pacific compared to the western tropical Pacific in the model, the mean changes in Pacific region climate resembled the climate anomalies associated with present-day El Niño events in many areas. Recent observations of warming in the Pacific showed that a mean warming of SSTs can occur with El Niño–like variability superimposed upon the warmer mean SSTs [Wang, 1995], as also occurs in the coupled climate model (i.e., periods of positive and negative SST anomalies in the tropical Pacific east of the dateline in relation to the warmer mean SSTs with increased CO_2; see Meehl *et al.* [1993] for a discussion of these types of definitions). This is not to say that we can definitively attribute the recent warming in the Pacific (and associated global warming of the 1980s) with increased CO_2 in the atmosphere, though the model results do lead to the possibility that CO_2 warming may have this signature. However, there may be decadal-timescale variability that could also have this type of signal. The model results suggest that if there is a warming of SSTs across the Pacific for whatever reason (increased CO_2, decadal variability, etc.), mechanisms partly due to cloud-albedo feedback effects could contribute to SST increases that are less over the western Pacific warm pool compared to the tropical eastern Pacific, with attendant shifts in large-scale precipitation patterns and midlatitude circulation anomalies in the North Pacific that resemble ENSO events. Further clarification of these effects awaits more definitive cloud-albedo feedback observational and modeling studies, improved cloud formulations, and better understanding of observed decadal-timescale climate fluctuations in the Pacific region.

Acknowledgments. A portion of this study was supported by the Office of Health and Environmental Research, U.S. Department of Energy, as part of its Carbon Dioxide Research Program. A portion of the computations was performed under the auspices of the Model Evaluation Consortium for Climate Assessment (MECCA) and Cray Research, Inc. The National Center for Atmospheric Research is sponsored by the National Science Foundation.

REFERENCES

Bottomly, M., C. K. Folland, J. Hsiung, R. E. Newell, and D. E. Parker, *Global Ocean Surface Temperature Atlas*, British Met. Office, Bracknell, U.K., 1990.

Graham, N. E., Decadal-scale climate variability in the tropical and North Pacific during the 1970s and 1980s: Observations and model results, *Clim. Dyn.*, 10, 135–162, 1994.

Houghton, J. T., B. A. Callander, and S. K. Varney, Eds., *Climate Change 1992: The IPCC Scientific Assembly Supplementary Report*, Cambridge University Press, 1992.

Kerr, R. A., Did the tropical Pacific drive the world's warming?, *Science*, 266, 544–545, 1994.

Kiladis, G. N., and H. van Loon, The Southern Oscillation. Part VII: Meteorological anomalies over the Indian and Pacific sectors associated with the extremes of the oscillation, *Mon. Wea. Rev.*, 116, 120–136, 1988.

Knutson, T. R., and S. Manabe, Time-mean response over the tropical Pacific to increased CO_2 in a coupled GCM, *J. Climate*, 8, 2181–2199, 1995.

Kumar, A., A. Leetma, and M. Ji, Simulations of atmospheric variability induced by sea surface temperatures and implications for global warming, *Science*, 266, 632–634, 1994.

Mechoso, C. R., and 19 coauthors, The seasonal cycle over the tropical Pacific in general circulation models, *Mon. Wea. Rev.*, 123, 2825–2838, 1995.

Meehl, G. A., The annual cycle and interannual variability in the tropical Pacific and Indian Ocean regions, *Mon. Wea. Rev.* 115, 27–50, 1987.

Meehl, G. A., G. W. Branstator, and W. M. Washington, Tropical Pacific interannual variability and CO_2 climate change, *J. Climate*, 6, 42–63, 1993.

Meehl, G. A., and W. M. Washington, Cloud albedo feedback and the super greenhouse effect in a global coupled GCM, *Clim. Dyn.*, 11, 399–411, 1995.

Nitta, T., and M. Kachi, Interdecadal variations of precipitation over the tropical Pacific and Indian Oceans, *J. Meteor. Soc. Japan*, 72, 823–831, 1994.

Nitta, T., and S. Yamada, Recent warming of tropical surface temperature and its relationship to the Northern Hemisphere circulation, *J. Meteor. Soc. Japan*, 67, 375–383, 1989.

Oberhuber, J. M., *An Atlas Based on the COADS Data Set: The Budgets of Heat, Buoyancy and Turbulent Kinetic Energy at the Surface of the Global Ocean*, Max-Planck-Institut für Meteorologie Report No. 15, Hamburg, Germany, 1988.

Pisciottano, G., A. Diaz, G. Cazes, and C. R. Mechoso, El Niño–Southern Oscillation impacts on rainfall in Uruguay, *J. Climate*, 7, 1286–1302, 1994.

Ramanathan, V., and W. Collins, Thermodynamic regulation of ocean warming by cirrus clouds deduced from observations of the 1987 El Niño, *Nature*, 351, 27–32, 1992.

Ropelewski, C. F., and M. S. Halpert, Global and regional scale precipitation patterns associated with the El Niño/Southern Oscillation, *Mon. Wea. Rev.*, 115, 1606–1626, 1987.

Trenberth, K. E., Recent observed interdecadal climate changes in the Northern Hemisphere, *Bull. Am. Meteor. Soc.*, 71, 988–993, 1990.

Trenberth, K. E., and J. W. Hurrell, Decadal atmosphere–ocean variations in the Pacific, *Clim. Dyn.*, 9, 303–319, 1994.

van Loon, H., and R. A. Madden, The Southern Oscillation. Part I: Global associations with pressure and temperature in northern winter, *Mon. Wea. Rev.*, 109, 1150–1162, 1981.

Wang, B., Interdecadal changes in El Nino onset in the last four decades, *J. Climate*, 8, 267–285, 1995.

Washington, W. M., and G. A. Meehl, Greenhouse sensitivity experiments with penetrative cumulus convection and tropical cirrus albedo effects, *Clim. Dyn.*, 8, 211–223, 1994.

Zhang, G. J., and M. J. McPhaden, The relationship between sea surface temperature and latent heat flux in the equatorial Pacific, *J. Climate*, 8, 589–605, 1995.

Prospects and Problems in Modeling the Impacts of Climate Change in Latin America

By Diana M. Liverman

Department of Geography, The Pennsylvania State University
University Park, PA 16802, USA

This paper presents selected results of a report on the possible impacts of global warming in Latin America, including a comparison of general circulation model (GCM) projections and evidence for climate change within the region. Several models project significant drying in Mexico, Northeast Brazil, and the Pampas – regions which are drought-prone even under existing conditions. A recent modeling study of global warming impacts on crop production in Latin America suggests that yields may decline in several key agricultural regions unless farmers can get access to irrigation, fertilizer and improved seed varieties. The problems with using the GCM results for estimating climate change are evident from the differing precipitation projections of models for the same region, and from the inability of several models to reproduce current climate. The main thesis of the paper, however, is that despite the uncertainty about how the physical environment may change, it is clear that the social environment of Latin America is creating conditions of severe vulnerability to climatic change and variation. Demographic and economic growth is placing stresses on water and land resources, and many farmers, hydroelectric facilities and poor urban dwellers are becoming increasingly vulnerable to droughts or warmer, drier climates. This increasing vulnerability will be illustrated with case studies from the Mexican states of Oaxaca and Sonora.

---

1. Introduction

Latin America is extremely vulnerable to interannual climate variability and to climatic change. Natural disasters, such as the droughts and floods associated with El Niño, can cause loss of life and millions of dollars in economic damages. Rain-fed agricultural yields and natural ecosystem productivity are very sensitive to year to year rainfall. Irrigated crops, drinking water supplies, and hydroelectric generation are also dependent on reliable precipitation. Climatic variations are also associated with epidemics and other stresses on human health. Rapid economic and demographic growth in Latin America is increasing demands for food and placing further pressure on water resources in the region. These growing vulnerabilities lend urgency to the need to understand whether and how global warming might affect Latin America. This paper describes some of the uncertainties in using the output of general circulation models to assess how global warming may affect Latin America. It describes some of trends which are making Latin America more vulnerable to climatic change and variation. The paper is based on a recent report to the Mexico City Office of Greenpeace, International [Liverman *et al.*, 1995].

2. Previous Studies of Global Warming Impacts in Latin America

Several previous studies of how global warming may change climate in Latin America suggest considerable disagreement among the results of different climate models. They rely on the results of general circulation model (GCM) experiments which simulate world climate under varying concentrations of greenhouse gases in the atmosphere. One set of experiments compares the climate of earth with lower levels of atmospheric CO_2 (abbreviated "$1 \times CO_2$") to a climate that might result with a doubling of those low-level CO_2 concentrations (abbreviated "$2 \times CO_2$"). That is to say, the experiments compare a climate with atmospheric CO_2 concentrations as they were before major human interference, estimated to be about 300 parts per million, to one with atmospheric CO_2 levels of about 600 parts per million. With equilibrium experiments, the greenhouse gas concentrations double instantaneously (rather than gradually over time) and the results are presented after the computer simulation has reached an equilibrium condition, that is, at an average or relatively stable climate for the increased CO_2 level.

Most studies focus on the differences (for temperature) and ratios (for precipitation) of results from the $1 \times CO_2$ and $2 \times CO_2$ experiments, rather than just looking at the results of the $2 \times CO_2$ simulations by themselves. This is the approach taken by the Intergovernmental Panel on Climate Change (IPCC), which, for example, projects an increase in average global temperature in the range of 1.5°C to 4.5°C (the $2 \times CO_2$ experiment results minus the $1 \times CO_2$ results) and a 3% to 15% increase in global average precipitation (the $2 \times CO_2$ results as a percentage of the $1 \times CO_2$ results).

For Latin America, the maps included in the 1990 IPCC report show temperature increases of more than 2°C throughout the region, with a 4°C increase projected by some models for the most southerly parts [Houghton *et al.*, 1990]. The IPCC maps of precipitation changes show large areas of Latin America where precipitation decreases, for example, in northeastern Brazil and northern Mexico. Soil moisture is also shown to decrease across large regions in some of the GCM experiments. The IPCC report also notes that several models show a weakening of the westerly winds in the Southern Hemisphere, and that there is some evidence that the frequency, intensity, and area of impact of tropical storms and hurricanes may increase. IPCC suggests that rainfall in Argentina, Chile and the north Andean region may decrease in the semiarid areas in the rain shadow of the Andes.

In 1991, Burgos *et al.* found very little agreement among climate models about how precipitation might change as a result of global warming, except for decreases in parts of Argentina and northern Mexico. They speculate about how the atmospheric circulation might differ in a warmer world and suggest that with higher temperatures, water would be evaporated more rapidly, fueling rising air. Atmospheric pressure would be lower over the hotter Amazon basin, and it would be easier for storms to move into the basin. More storms and convection would mean higher rainfall. The regions of corresponding sinking air – for example, over Argentina – could be displaced poleward, hence becoming drier. Another study, by Nuñez [1990], looked at the results of three GCMs (Princeton/GFDL, Goddard/GISS, and NCAR) for South America. Noting some disagreement among the models, especially for precipitation, he suggests that a doubling of carbon dioxide could produce temperature increases of 2°C to 4°C in South America, rising to as high as 10°C in the Austral winter in the most southerly regions. Because a warmer atmosphere will hold more moisture, global warming would mean higher rainfall in Amazonia. But the reduced temperature gradient from pole to equator could result in a weakening of the westerly winds and a decline in rainfall in south and central Argentina, Chile, Uruguay, Paraguay, and Brazil. The displacement southward of tropical anticyclones would increase this drying trend together with higher evaporation.

Liverman and O'Brien [1991] examined the simulations of five different GCMs for Mexico. They demonstrate the problems of the models in reproducing current climate as a result of factors like coarse grids and inadequate topographic data. Although the models disagree as to where and whether rainfall will change, the researchers suggest that increased evaporation associated with higher temperatures will result in less available surface water, even in regions where rainfall might increase. For example, the researchers estimate changes in potential evaporation in Mexico City, showing increases of up to 23%. A simple index of precipitation minus potential evaporation shows increases in water deficits of 11% to 23% despite an increase in rainfall in two model scenarios.

3. Recent Results from IPCC 1994 Scenarios

In the Greenpeace report we focus on the climate model results for the equilibrium and transient experiments produced in connection with the 1994 IPCC scientific assessment and other ongoing international studies. We analyze the results for the following general circulation models:

- Princeton: Geophysical Fluid Dynamics Laboratory, Princeton University, USA (referred to as GFDL in the figures)
- British: United Kingdom Meteorological Office (UKMO)
- Canadian: Canadian Climate Center (CCCM)
- Goddard: Goddard Institute for Space Studies, New York, USA (GISS)

4. Comparison of $1 \times CO_2$ to Observed Climate

The uncertainties associated with using model results are illustrated in Figures 1 and 2, which show the $1 \times CO_2$ average temperature and annual precipitation results of several equilibrium experiments for Latin America (designated as CCCM, UKMO, and GFDL in the figures) in comparison to the present-day observed climate (designated as CLIM). For the purposes of this discussion, only the Canadian (CCCM), British (UKMO), and Princeton (GFDL) model results are compared with present-day climate observations. The present-day climate observations have been interpolated to a grid for the purpose of comparison with the model results. The lines on the map join places of equal temperature (in degrees Celsius) or precipitation (in millimeters per day).

Most of the models are able to reproduce the broad-scale temperature patterns for Latin America, with cooler temperatures over the Andes, the Southern Cone, and the highlands of Mexico, and with warmer temperatures in the Amazon basin and Central America. The Canadian (CCCM) and Princeton (GFDL) models show temperatures too cool in Bolivia and northern Mexico. The British (UKMO) and Princeton (GFDL) models show temperatures a little too cool over the Southern Cone.

There are more dramatic errors in precipitation. For example, the Canadian (CCCM) model has precipitation reaching 14 mm per day for two grid points – one in northern Chile and the other in central Mexico – where the observed precipitation values are only about 5 mm per day. This probably results from an assumed elevation at those grid points that is much higher in the model than the observed average elevation over the region. The Canadian (CCCM) model also simulates rainfall that is too low in southern Chile and the Amazon.

The British (UKMO) model has precipitation too high in the Amazon and northeastern Brazil and too low in southern Chile. The Princeton (GFDL) model has higher than observed precipitation values in the Andes, Colombia, and central Mexico, and lower than observed in the Amazon, Southern Cone, and parts of Central America. These discrepancies indicate that global warming projections should be interpreted with some caution.

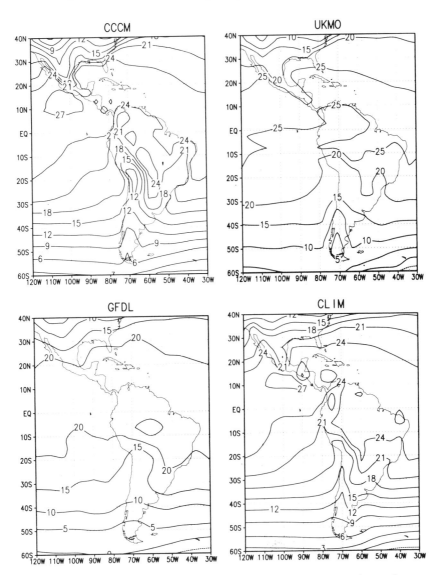

FIGURE 1. Comparison of simulated model temperatures to observed.

5. Projected Changes in Temperature and Precipitation

The results of the four equilibrium experiments are presented in Figure 3, expressing the projected changes in terms of degrees Celsius above the $1 \times CO_2$ simulations. All four of the equilibrium experiments suggest that if atmospheric CO_2 doubles, Latin America will warm significantly, with all areas experiencing annual average temperature increases of at least 2°C. The British (UKMO) model indicates the greatest local increases – 5°C or more in northern Mexico and in most of lowland South America south of latitude 10°S, and more than 4°C for most of the rest of the region. The Goddard (GISS) model indicates temperature increases of 4.5°C for northwestern Mexico and for a large area in southern South America. The Canadian (CCCM) and Princeton (GFDL) scenarios are more moderate, but show a pocket of higher temperatures in Paraguay, Uruguay, and northern Argentina. Thus, while all four model results show temperature increases

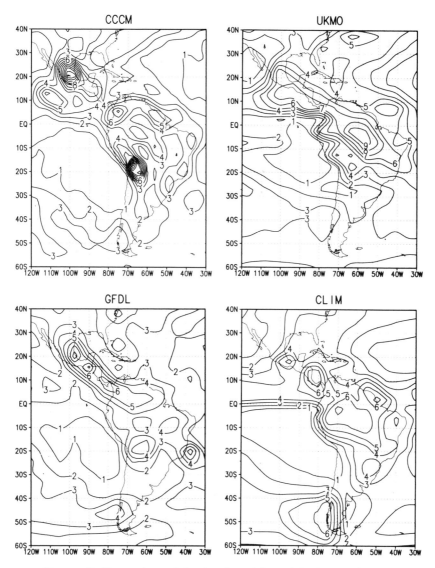

FIGURE 2. Comparison of simulated model precipitation to observed.

of at least 2°C for all of Latin America, there seems to be general agreement among the simulations of relatively higher increases (4°C to 6°C) in northern Mexico and in the south central part of South America (specifically northern Argentina, Paraguay, and southern Brazil).

The precipitation scenarios are less consistent. They are presented in Figure 4 and show the precipitation (in millimeters) simulated under $2 \times CO_2$ conditions as a percentage of that simulated for $1 \times CO_2$ conditions. Values higher than 1.0 indicate predicted precipitation increases, while values less than 1.0 indicate decreases. The Canadian (CCCM) model suggests that precipitation will decrease in most of Latin America, by up to 30% in parts of Colombia and Paraguay/northern Argentina. A dramatic increase is indicated at only one point in Bolivia. The British (UKMO) model indicates decreases in precipitation in eastern Mexico, Central America, and northern Brazil; increases of more

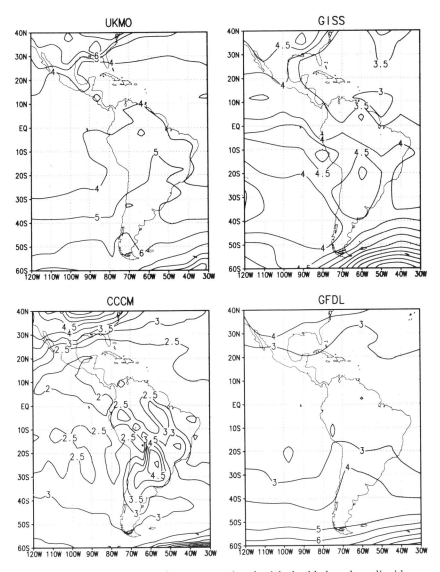

FIGURE 3. Temperature changes associated with doubled carbon dioxide.

than 20% in northwest Mexico, the Caribbean, and Argentina; and a 60% to 80% increase for parts of Peru and Ecuador. The Princeton (GFDL) model shows slight decreases in precipitation only in the Yucatan Peninsula of Mexico, eastern Central America and southeastern Brazil and suggests increases of more than 20% in central Mexico, Bolivia, and the southern tip of Chile and Argentina. The Goddard (GISS) model shows a slight decrease in precipitation for eastern Mexico, Venezuela, and the Guianas, and more than a 40% increase for western South America, including Bolivia, northern Argentina and Chile, and western Brazil.

Thus, by analyzing both the temperature and precipitation simulations together, we can see that one model, the Canadian (CCCM), indicates much drier and somewhat warmer conditions for most of Latin America. The high temperatures in the British (UKMO) model might result in much higher evapotranspiration, offsetting any increases

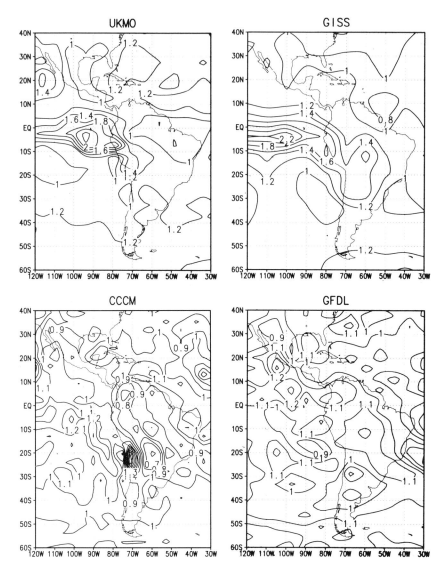

FIGURE 4. Precipitation changes associated with doubled carbon dioxide.

in rainfall. Even the Princeton (GFDL) model has 3°C temperature increases and therefore increased evapotranspiration in all regions where rainfall increases. And the GISS model shows temperature increases and precipitation decreases for eastern Mexico and the Guianas, while the area with the most precipitation increase is also the area of highest predicted temperature increase.

6. Impacts on Agriculture

Agriculture provides the basis for human subsistence and is one of the most climate-sensitive sectors of the economy. When droughts, frosts, or other climatic disasters affect Latin American agriculture, lives, health, jobs, economies, and political stability are at risk. It has been estimated that the climatic variations associated with a serious El

TABLE 1. Climate change and maize yields (in tons/hectare) in Mexico.

	Tlaltizapan rain-fed	Tlaltizapan irrigated	Poza Rica rain-fed	Poza Rica irrigated
BASE	4.02	3.72	3.18	3.07
GISS	3.07	2.84	2.97	2.84
GFDL	3.2	2.98	2.7	2.93
UKMO	1.56	2.9	2.35	2.79

BASE is the average yield with 15 years of observed climate data; GISS, GFDL, and UKMO are the average yields with 15 years of climate data generated by climate models for a $2 \times CO_2$ scenario and assuming 555 ppm of carbon dioxide available to plants.

Niño event cost billions of dollars in lost agricultural production and trade. Only a small proportion of Latin American agricultural producers have insurance; everyone else suffers, goes into debt, loses land, or migrates to the cities in years of harvest failure.

This extreme sensitivity to climate and climatic variations means that global warming may have dramatic and devastating effects on Latin American agriculture. Warmer and drier conditions and greater extremes or interannual variability could be especially disruptive. However, warmer temperatures, higher carbon dioxide levels, and higher rainfall could reduce frost and drought risks and increase agricultural productivity in some regions.

A recent study funded by the U.S. Environmental Protection Agency includes case studies of global warming and impacts on Latin American agriculture [Rosenzweig and Iglesias, 1994]. Scientists from Argentina, Uruguay, Brazil, and Mexico used several climate model scenarios to simulate changes in yields of maize, barley, soybeans, and wheat using the IBSNAT–CERES crop simulation models. These models include the effects of higher CO_2 and are sensitive to local soil conditions and farming practices.

In Mexico, maize yields were simulated at highland and lowland sites representative of major maize growing regions in Mexico, using typically low levels of fertilizer under both rain-fed and irrigated conditions. At the Mexican highland site, Tlaltizapan, the crop model simulated an average rain-fed yield of about 4000 kilograms per hectare with observed climate data from the 1973 to 1989 period. Irrigated yields were slightly lower because the added water leached nutrients from the soil. The GCM-based scenarios for climatic change produced significant declines in maize yields, ranging from a 20% drop with the Princeton (GFDL) model to a 60% drop with the British (UKMO) model. Irrigated yields also fell, despite allowing for extra water supplies, because warmer temperatures accelerated plant growth and reduced the time for the grain to fill and mature. Maize yields at the Mexican lowland site, Poza Rica, also declined by 6% to 26%. It is important to note that yields dropped with all GCM scenarios, even those which projected increases in rainfall, because of drought stresses produced by higher evaporation and accelerated plant development associated with higher temperatures. Under such conditions, reductions in yield could be offset by using fertilizer and irrigation and by switching varieties, but these adaptations are not always accessible or affordable to farmers (Table 1).

In Argentina, the EPA study looked at maize and wheat. In the Pampas, where the climate models estimate temperature increases of 4.5°C to 5.2°C, yields decrease by 19%

to 36% with global warming. The Uruguay study examined the possible impacts of climate change on barley production and found decreases in yields of up to 50% for this important export crop. Brazilian researchers simulated wheat, maize and soybeans at 13 sites from the equator to 30° south latitude. Wheat yields declined at all sites with global warming by up to 46%; maize yields fell by up to 26%. Only soybeans showed some positive responses, with yields staying about the same or increasing, depending on the climate model.

7. Vulnerability to Climate Change

Latin American societies are becoming increasingly vulnerable to climatic extremes because many people already live in regions where rainfall is low and variable, where soils are easily erodible or drought-prone, and where water resources are limited. Throughout Latin America and the Caribbean, deforestation, desertification and pollution are changing social and ecological vulnerability to climatic variation and changes. Local land use changes such as urbanization and cropland expansion are changing regional climate, hydrology and land quality, mostly toward warmer temperatures, more frequent floods, and erodible, less fertile soils. Human activities such as agriculture, mining, fishing, and hunting have encroached on ecosystems, resulting in serious losses of biodiversity and leaving remnant species and ecosystems which are highly vulnerable to climate change and hazards. Air and water pollution can make people and ecosystems more vulnerable to high temperatures, water scarcity, or sea level rise.

Inequality in access to resources means that many people cannot obtain enough land or water to ensure a harvest in dry years. People are often forced to live in hazardous environments (floodplains, slopes) where they are more vulnerable to floods and hurricanes. It is important to emphasize that the most vulnerable people may not be in the most vulnerable places – people can live in productive biophysical environments and be vulnerable because they are poor or have no land, and people can live in fragile physical environments and live relatively well if they have money or technology as buffers. Often the conditions are most critical where impoverished populations live in ecologically marginal environments.

In Latin America insurance against climatic hazards is rarely affordable or available. Modernization has eroded many of the traditional responses to drought and extreme events, such as terracing, intercropping, and community wealth sharing. New agricultural technologies such as irrigation, fertilization, and improved seeds can increase crop yields but may also increase vulnerability. Farmers need to borrow money for these inputs and may go into a cycle of debt when poor weather destroys the profit they would have used to repay the debt. As neoliberal economic policies remove credit, crop and input subsidies and reduce overall state spending, the safety nets which government has provided to help farmers, consumers, and industries cope with natural disasters and other crises have dissolved [Appendini and Liverman, 1994].

Population growth, migration, and economic development have created competition for water in many regions. Cities are becoming ever more dependent on hydroelectricity and on energy used for refrigeration, air conditioning, and other technologies which moderate the climate of human settlements. Many industries are increasingly sensitive to water or other resource shortages as they tune their production systems to a narrower range of environmental conditions.

Any shift to warmer, drier climates with more extreme events would make this situation much worse. Greenhouse gases are increasing as the rate of population is growing in many

Latin American countries, and over a period during which unequal economic growth may mean even greater vulnerability and a lack of resources to adapt to climatic change.

Technology can either increase or decrease vulnerability to climate change. For example, there is a debate over the success of the new agricultural technologies associated with the "Green Revolution" in improving environmental and social conditions. On the one hand, irrigation, high-yielding seeds, fertilizer, pesticides and machines were claimed to increase yields and decrease yield variability. Proponents show that the Green Revolution increased food self-sufficiency, reduced famine, decreased food imports, and increased average incomes in parts of South America.

However, the expansion of irrigation has also been associated with a loss of land and production from salinization and waterlogging in regions such as Mexico and Chile, and with an increased vulnerability to multiyear droughts if agriculture has become dependent on shallow wells, small reservoirs, or declining water tables. The new seeds have been shown to do well only in moist soils or irrigated environments, and to be vulnerable to diseases and weather extremes. A small range of improved varieties have replaced a diversity of traditional seeds adapted to a wide range of environments.

The twentieth century has integrated Latin America in a national and international market and financial system. Subsistence production of corn has been replaced by wheat, dairy, sugar, feedgrain and fiber production for domestic industrial and urban markets. In many countries such as Chile, Mexico, Costa Rica and Colombia, the agricultural system has increasingly become oriented to export production of fruit and vegetables and livestock. This has resulted in higher incomes for many farmers and has supported national trade balances. The increased acreage under fruit, vegetables and feed grains has increased the demand for water and hence vulnerability to drought. Alfalfa and lettuce, for example, have a much higher consumptive use than corn. Vulnerability to variable market prices and declining terms of trade has also increased, as well as vulnerability to natural disasters. Hurricane Gilbert in 1988 devastated Jamaica's economy when it destroyed 30% of the sugar, 54% of the coffee, and 90% of the banana and cocoa crops. The combination of a fall in world coffee prices and Hurricane Joan, also in 1988, intensified the economic crisis in Nicaragua and hastened the fall of the Sandinista government.

8. Conclusions

According to a study of natural hazards and disasters in Latin America between 1900 and 1988, 181 floods killed 12,000 people, affected 21 million, and cost a total of U.S. $7.6 billion. Drought affected the lives of 42 million and cost U.S. $4.1 billion. Hurricanes, 77 in all, killed over 33,000 people, affected 4.2 million, and cost U.S. $2.6 billion. The countries along the west coast of South America also showed drought and flood problems, likely resulting from disturbances associated with El Niño anomalies and human misuse of the land [Stillwell, 1992].

As the physical devastation and human suffering of some of these examples show, Latin Americans are quite vulnerable to unexpected and severe weather. Global climate change may increase the frequency of severe weather, alter rainfall patterns, and increase temperatures. Social trends such as environmental degradation, population growth and urban overcrowding, and repressive economic structures may further increase the vulnerability of Latin American populations. In a sense we can be more certain about demographic and socioeconomic trends than we can about climatic change. Unless there are major changes in birth rates, per capita consumption aspirations, technologies and economies, there will be more people, each demanding more water and food in a future warmer Latin America.

Acknowledgments. I would like to acknowledge the coauthors of the Greenpeace report upon which this paper is based: Mrill Ingram, Roberto Sánchez and Robert Merideth. Alejandro Calvillo commissioned the study, which was supported with funds from Greenpeace International.

REFERENCES

Appendini, K., and D. M. Liverman, Agricultural policy, climate change and food security in Mexico, *Food Policy*, **19**, No. 2, 149–163, 1994.

Burgos, J., H. Fuenzalida, and L. Molion, Climate change predictions for South America, *Climatic Change*, **18**, 223–239, 1991.

Greco, S., R. Moss, D. Viner, and R. Jenne, Climate scenarios and socioeconomic projections for IPCC WG II Assessment, CIESIN report for IPCC WG II Lead Authors, 1994.

Houghton, J. T., G. J. Jenkins, and J. J. Ephraums, Eds., *Climate Change: The IPCC Scientific Assessment*, Cambridge University Press, 1990.

Liverman, D. M., M. Ingram, R. Sánchez, and R. Merideth, Latin America and climate change. Report to Greenpeace, Mexico, 1995.

Liverman, D. M., and K. O'Brien, Global warming and climate change in Mexico, *Global Environmental Change*, **1**, No. 4, 351–364, 1991.

Nuñez, M., Cambio climático en Sudamérica: Uso de modelos de circulación general, *Revista Geofísica*, **32**, 47–64, 1990.

Rosenzweig, C., and A. Iglesias, Eds., *Implications of climate change for international agriculture: Crop modeling study*, U.S. Environmental Protection Agency, Doc. No. EPA 230-B-94-003, 1994.

Stillwell, H. D., Natural hazards and disasters in Latin America, *Natural Hazards*, **6**, No. 2, 149, 1992.

PIXSAT, a Digital Image Processing System in a CRAY–UNIX Environment

By Alfredo Cortés, Román Álvarez and Miguel A. Castillo

Instituto de Geografía, Universidad Nacional Autónoma de México, Apdo. Postal 20-850, 01000 México, D.F., México

The development of a digital image processing system in an MS-DOS Windows–CRAY multi-environment that is capable of processing images obtained by satellites or by manual scanning, mainly for remote sensing applications, is described. The design has as its core the hardware of the CRAY Y-MP/464 supercomputer. Previous experience with the design and implementation of the PIXSAT system on PCs, on minicomputers with UNIX (System V and BSD4.3) and on workstations (Silicon Graphics Onyx, Challenge and Iris; Sun SPARCstation) with UNIX (System V and BSD4.3) constitutes the basis of the system. The system was divided into seven principal modules: image display, utility functions, point operations, spatial domain filters, geometric transformations, and multispectral classifications. It is intended that the end-user interface will be Microsoft Windows on microcomputers and X-Windows on workstations. The model will be Client/Server, in such a way that PIXSAT will function in a local network; the protocols will be Internets TCP/IP and UDP/IP. The system will use the CRAY's parallelism in various ways such as multiprogramming (jobs), multiprocessing (tasks) and vectorization which is directly performed by the CRAY compiler. The code was compiled using maximum optimization and vectorization capabilities issued by the ANSI-C compiler (SCC). A couple of algorithms were tested on two systems, and the results are presented in this paper.

1. Introduction

Demand for large-scale information management related to environmental and natural resource problems is growing every day; hence it is necessary to have more powerful systems that can solve this problem. PIXSAT, the digital image processing system presented in this paper, represents a solution to this demand; its CRAY implementation warrants reduced execution times on large datasets. This system focuses on solving large-scale problems, such as forestry resources management and land cover and land use change analysis, among other environmental and natural resource topics.

PIXSAT is a digital image processing system that has been developed in the Instituto de Geografía, at the Universidad Nacional Autónoma de México, since 1992 in the X-Windows–UNICOS–CRAY environment. The system is divided into seven principal modules: (1) image display, (2) utility functions, (3) point operations, (4) spatial domain filters, (5) frequency domain filters, (6) geometric transformations, and (7) multispectral classifications. The code is written fully in ANSI-C to ensure portability.

In the first stage of the project, the command, or batch system, was developed. Approximately 40 different image processing functions, ranging from utility functions to multispectral classifications, were implemented. In the second part of the project we developed the system's graphical aspects, which consist mainly of the user-system's interface with X-Windows.

There have been other projects [for example, KHOROS, 1992] to implement image processing systems on supercomputers. The main difference between PIXSAT and these systems is the memory management subsystem. PIXSAT uses a buffer memory management subsystem that consists mainly of a temporary memory buffer that links between

FIGURE 1. Buffer memory management system used by PIXSAT.

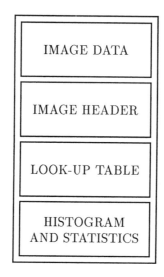

FIGURE 2. Structure of an image in PIXSAT.

load–process–write tasks on an image. Figure 1 shows this graphically. This feature makes PIXSAT a faster system than others.

The structure of an image in PIXSAT is shown in Figure 2. It consists of four files: an image data binary file, an image header ASCII file, a histogram image ASCII file, and a look-up table in ASCII file. Image data files store that data itself. Image header files store basic information about the image, including its width, height and data type; satellite or device which captured the image; and mean, median, minimum and maximum values and their standard deviation. Histogram files have one row for each pixel value in the image; each row contains a pixel value and its frequency of occurrence. Look-up table files store color tables for image display purposes. Although other files can be associated with an image (sample files, georeference files, etc.), those shown in Figure 2 exist for every image. These files are created automatically by the system when it imports an image.

2. Image Processes and Display

There are many image processes in PIXSAT that can be classified under the display concept. We consider the following operations as part of the image display module: image zooming, scrolling, histogram management, pseudocolor, look-up-table management, annotations, superimposition of grids, reading pixel values, and generating profiles. Each module is described following subsections.

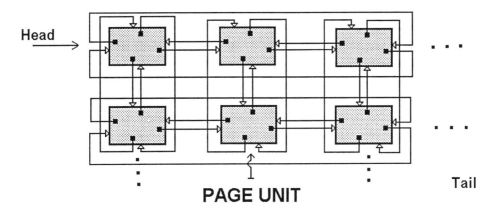

FIGURE 3. Memory management subsystem used by PIXSAT (paging technique) for image display purposes.

2.1. *Image Zooming*

Image zooming is a feature required in almost every image analysis system; consequently it is fully implemented in PIXSAT. Zooming and panning on an image, fitting an image to a window and one-to-one pixel mapping are image zooming activities that a user can do in a window. This function is very useful in jobs like feature identification, pixel sampling, image registration and making of geometric corrections.

2.2. *Scrolling and Memory Management*

Because of its importance, we separate discussion of scrolling and memory display management from the rest of the image display topics. Since its first version, PIXSAT was developed in environments where memory resource management was very important (MS-DOS and Windows), in such a way that the main approach was to use this resource as moderately as possible. In the CRAY version this is not so important because memory resources are not so restrictive. However, we consider that correct memory management will create a faster system for the end user; consequently we decided to preserve this memory management subsystem, as shown in Figure 3.

In Figure 3, memory management for image display purposes is based in a double-circular-xy-linked list, which consists of an array of pointers arranged in individual double-circular lists in both directions: horizontal and vertical. Each element of the list is called a node, which represents an image page. Each node has four pointers to its four neighbors and a pointer to a block of memory that contains the image data. Each image is divided into pages, in order to be loaded into memory.

This arrangement of memory is a highly efficient one, particularly on intensive scrolling activities, where the user is analyzing a full scene. One page is marked as ready to load when the user issues a scroll in the direction where the page is found, and ready to unload when a page fault occurs. A page fault occurs when a page from the disk is needed in memory and all the pages are occupied; in that moment a decision must be made to free a page in memory. The page that must be freed is in the opposite direction to that of the scroll.

TABLE 1. Utility functions of the PIXSAT system.

Operation	Description
Image format conversions	Header creation formats include BMP, BSQ, BIP, BIL, SunRasterfile
Data type conversions	BIT, BYTE, INTEGER, REAL images conversion to PIXSAT format (data type BYTE in length)
Users maintenance	Create user, change user, users report
Subimage extraction	Using pixel-image coordinates or ground reference coordinates (e.g., UTM)
Region extraction	Polygonal regions, where can be defined if it is internal or external
Printing	Images, histograms, and image data

2.3. *Histogram Management, Pseudocolor and Look-Up-Table Management*

Histogram management consists mainly of multilinear mapping of histograms to get the best contrast stretch over the displayed image. This mapping creates a look-up table, which in turn can be saved, loaded and modified. The file where this can be done is shown in Figure 2.

2.4. *Pixel Value Reading*

PIXSAT has a module that provides information about pixel position and value. It prints as many pixel values as there are bands present in the system. You can obtain elevation information if you have a digital terrain model [Burrough, 1986] instead of an image. The elevation value is round-up, however, because the image does not contain the elevation values per se; instead, it contains the coefficients obtained when the linear mapping function translated the digital terrain model to bytes.

2.5. *Profiles*

PIXSAT also provides reflectance profiles from any image. Users must define a sequence of points that define the profile, which can be any number of segments long. A profile from the sequence of points is displayed for the user.

3. Utility Functions

There are some image functions, called utility functions, that make it possible for the end user to interact with PIXSAT as easily as possible; however, they are not image processing functions proper. They are shown in Table 1.

4. Point Operations

Point operations act directly on each pixel; consequently, it is possible to vectorize and parallelize these operations. Given two images A and B, the general formula for this kind of operation is

$$C_{ij} = A_{ij} \langle op \rangle B_{ij} \qquad \text{for all } A_{ij}, B_{ij} \text{ in the image}$$

where C is the image obtained by applying the binary operation $\langle op \rangle$ to A and B. The preceding expression shows that each pixel operation is element-independent. PIXSAT

1/9	1/9	1/9
1/9	1/9	1/9
1/9	1/9	1/9

Low pass

0	-1/9	0
-1/9	5/9	-1/9
0	-1/9	0

High pass

0	1/9	0
0	-4/9	1/9
0	1/9	0

Gradient

-1/9	-2/9	-1/9
0	0	0
1/9	2/9	1/9

Horizontal

1/9	0	-1/9
2/9	0	-2/9
1/9	0	-1/9

Vertical

2/9	-1/9	0
-1/9	2/9	-1/9
0	-1/9	2/9

45° Diagonal

FIGURE 4. Different 3×3 kernels for spatial filters (upper three panels) and edge detectors (lower three panels).

can undertake the following point operations: histogram equalization; histogram linearization; adjust one histogram to a reference; add a constant factor to an image; add, substract, divide or multiply two images; AND, OR, NOT, XOR two images; and the bright index and the normalized and perpendicular vegetation indices.

Every histogram operation consists of a linear mapping applied to an image that has the general formula

$$A_{ij} = B_{ij}\mathbf{x} + \mathbf{c}$$

where B is the original image, A the linearly transformed image, $\mathbf{x}$ is a gain linear variable, and $\mathbf{c}$ is a constant. Both $\mathbf{x}$ and $\mathbf{c}$ are calculated for each image and their values depend on the histogram operation.

The vegetation index calculations require two spectral bands of an image [Richards, 1986], the red and infrared bands, and the basic formulae for the vegetation indices are

$$\frac{B_i - B_r}{B_i + B_r}, \quad \sqrt{\frac{B_i - B_r}{B_i + B_r} + 0.5}, \quad \text{and} \quad \sqrt{B_i^2 + B_r^2}$$

These formulae correspond to the normalized vegetation index, the perpendicular vegetation index, and the bright index, respectively, where B_i denotes the infrared band of a multispectral image and B_r denotes the red band of the same image.

5. Spatial Domain Filters

The image operations classified as spatial domain filters are more complex, computationally speaking, than point operations, since they are area operations or neighbor operations. They act over a neighborhood from the pixel of interest; therefore the vectorization and parallelization parts of this kind of operation will be harder to identify. The general operation for spatial filters is

$$C_{ij} = \sum_{k=-m}^{k=m} \sum_{l=-n}^{l=n} A_{i+k,j+l} \quad \text{for all } i, j \text{ in the image}$$

The system contains the following area operations: spatial filters (user defined): definition, change, delete and report; spatial filters (predefined): median, gradient, low pass, high pass and edge detectors. Figure 4 shows the kernels used in each one of these spatial filters [Mather, 1987].

6. Frequency Domain Filters

Following the sequence of increasing complexity, frequency domain filters constitute some of the most complex operations in PIXSAT because they require Fourier transforms. An efficient algorithm to execute the Fast Fourier Transform (FFT) [Gonzalez and Wintz, 1987] is implemented in the system. In order to apply a frequency domain filter on an image, the following steps must be done: (1) transform the image to the frequency domain using an FFT, (2) filter the image in the frequency domain, and (3) apply the inverse FFT to the image.

The Fourier transform of a sequence A_i (e.g., an image row) is defined as [Gonzalez and Wintz, 1987]

$$F_r = T \sum_{k=0}^{k=N-1} A_k e^{-j2\pi/Nrk}, \qquad r = 0, \ldots, N-1$$

and the inverse Fourier transform is defined as

$$A_k = \frac{1}{T_0} \sum_{k=0}^{k=N-1} F_r e^{j2\pi/Nrk}, \qquad k = 0, \ldots, N-1$$

where F_r is the Fourier transform of a row A_i from the original image; T is the period of the signal; and N is the number of columns of an image. The steps described generate a filtered image. We are attempting to vectorize and parallelize the FFT so that we will have a parallel algorithm for frequency domain filters.

PIXSAT has the following filters: (1) high pass filter, (2) low pass filter, (3) band pass filter, and (4) Butterworth filter. These filters act on an image in almost the same way as does a spatial domain filter; however, frequency domain filters are more precise in the sense they are made in the frequency rather than the spatial domain.

7. Geometric Transformations

The main characteristic of a geometric transformation is that it acts over the whole image, changing the original position of the pixels. There are places where geometric operations can be parallelized because they operate on single points. PIXSAT has the following geometric transformations: (1) color space transformations RGB $\leftrightarrow$ IHS, (2) principal components (Hotelling transformation), (3) mosaics, (4) geometric corrections, and (5) image registrations.

7.1. *Principal Components*

Given a multispectral image, A, principal components are a set of bands obtained from A by applying the transformation $b = Ga$, where a is a pixel vector of dimension n of A, n is the number of multispectral bands in the image, b is a pixel vector of the transformed image B, and G is an orthogonal matrix that transforms a into b [Richards, 1986]. G is the transposed matrix of eigenvectors of S, where S is the variance–covariance matrix of the original image and is defined as

$$S_{ij} = \sum_{k=1}^{k=N} \sum_{l=1}^{l=M} (A_{kli} - m_i)(A_{klj} - m_j)$$

where A_{kli} is the pixel in the kth row, lth column and ith band of A, and A has dimensions $N \times M$.

7.2. *Geometric Corrections*

There are at least two different methods of geometric correction: one defined by a model that characterizes the nature and magnitude of geometric distortions, and another that maps ground control points to points on the image. PIXSAT uses the second method in geometric corrections, defining different degrees of polynomials.

PIXSAT uses three different methods of interpolation on geometric corrections: linear, quadratic and cubic polynomial mappings. The expressions for these polynomials are as follows:

$$u = a_0 + a_1 x + a_2 y + a_3 xy + a_4 x^2 + a_5 y^2 + a_6 x^2 y + a_7 xy^2 + a_8 x^3 + a_9 y^3$$

$$v = b_0 + b_1 x + b_2 y + b_3 xy + b_4 x^2 + b_5 y^2 + b_6 x^2 y + b_7 xy^2 + b_8 x^3 + b_9 y^3$$

where the coefficients a_i, b_i ($i = 0, \ldots, 9$) are calculated using least squares interpolation. As can be noted, for the linear case it needs at least three points, six for the quadratic case and ten for the cubic case. Once coefficients are calculated, the process of mapping is carried out over each pixel on the image and one new (u, v) coordinate is obtained from it.

8. Multispectral Classifications

Multispectral classification operations are in high demand in image processing. A general Bayesian classification method consists of identification of characteristics present on the image (class), selection of a set of pixels (sample) representing each class to be classified, comparison with other pixels that were used as representative of the given class, and assignment of a label from the pixel to which it is nearest in a given metric.

To determine which pixel is the nearest, the system must make a distance calculation consisting of calculating a linear system of equations, multiplying the result by two vectors and adding a constant. These processes can be vectorized and parallelized in some places, in order to decrease the algorithm's execution time. Mathematically, it can be expressed as [Mather, 1987; Richards, 1986]

$$P(x) = (2\pi)^{-b/2} |C_i|^{-1/2} \exp\left(-\frac{1}{2}(x - m_i)^T C_i^{-1}(x - m_i)\right)$$

where C_i denotes the sample variance–covariance matrix for class i; $|C_i|$ denotes the determinant of C_i; and m_i is the mutivariate mean of class i. The probability $P(x)$ that a pixel x of b elements is a member of class i is given by the preceding multivariate expression. This probability function is evaluated for each sample class and the corresponding pixel is associated to the maximum probability class. With this method a pixel is assigned to a class which is the nearest in terms of probability. Once statistics are defined for each class, and each pixel classified, we get a category image, that is, a classified image.

The preceding expression can be reduced if multispectral classification by Mahalanobis's distance method is used [Mather, 1987; Richards, 1986]; the expression can be stated as

$$P(x) = (2\pi)^{-b/2} |C|^{-1/2} \exp\left(-\frac{1}{2}(x - m_i)^T C(x - m_i)\right)$$

where the sample variance–covariance matrix is the same for all the samples and can be calculated as a constant in the preceding expression. In the case of multispectral classification using the migrating means method [Mather, 1987; Richards, 1986], the

TABLE 2. CPU and system times (s) for two PIXSAT algorithms on two machines.

Host machine	Algorithm	CPU time	System time
SGI Onyx	med_migr	11011	74
(R4400, 150 MHz)	comp_pri	1492	99
CRAY YMP/464	med_migr	14829	155
	comp_pri	4387	46

preceding expression can be reduced to

$$P(x) = (x - m_i)^T (x - m_i)$$

that is, the euclidean distance, where the variance–covariance matrix is the same for all the samples and equal to the identity matrix, so it disappears from the expression. This equation is used for PIXSAT in the unsupervised multispectral classification method, whose performance in the CRAY supercomputer is evaluated in this paper.

PIXSAT can perform the following classification tasks: scattering diagram between two spectral bands; image sampling: sampling gathering, samples union, file samples union, samples classification; supervised classification: Bayesian distance, Mahalanobis distance, euclidean distance, and paralellepiped method; and unsupervised classification using a minimum-distance classifier.

9. Results and Conclusions

To test the performance of PIXSAT on CRAY, we prepared an image test set consisting of a full TM scene, which is 75 Mb per band, for each of seven bands (525 Mb total). We applied two PIXSAT algorithms, med_migr, migrating means (unsupervised classification), and comp_pri, principal components, analysis to the TM scene, first using an Onyx Challenge R4400 and then the CRAY Y-MP/464. The results obtained, in CPU time and system time, in both systems with both algorithms are shown in Table 2.

The med_migr algorithm is a nonsupervised classifier that used the seven bands of the TM scene; with a maximum of 100 initial classes; a radius of 10 pixels for each class (these pixels are counted from the center of a hypersphere which is the mean of each class); and a convergence percentage of 99.

Results obtained for both machines are very interesting. We should ask what is happening with the CRAY machine because its processing times are the greater of the two. For the med_migr algorithm, the CPU time is 34.67% greater for the CRAY than the Onyx and is 194.06% greater for the comp_pri algorithm. The CRAY required 107.62% of the Onyx's system time for the med_migr algorithm, but it was faster on the comp_pri algorithm, for which the Onyx used 114.64% of the CRAY's system time. The code used on both machines is exactly the same: full ANSI-C code. For these algorithms, the code was compiled using the maximum optimization and vectorization capabilities of the ANSI-C compiler (scc). There have been other reports of the same kind of problem [Davis, 1992]. It seems that the architecture of the CRAY is not well suited to image data management because such data are 8 bits long and words in a CRAY are 64 bits long, and the machine must pack and unpack words, a time-consuming task. The memory management subsystem needs adjusting to minimize data input–output activity when translating 8-bit data to 64-bit data. This is a good reason for us to begin the optimization phase of the project, and we will continue to work on these algorithms.

Acknowledgments. This project was sponsored by Cray Research, Inc., and the Universidad Nacional Autónoma de México.

REFERENCES

Burrough, P., *Principles of Geographical Information Systems for Land Resources Assessment*, Oxford Science Publications, 1986.

Davis, B., Using Cray computers to extract hydrology and topography from digital elevation data. *Proceedings of the Thirtieth Semi-Annual Cray Users Group Meeting: Grand Challenges*, 73–88, Washington, DC, September, 1992.

Gonzalez, R., and P. Wintz, *Digital Image Processing*, 2nd ed., Addison-Wesley, 1987.

Jensen, J., *Introductory Digital Image Processing, A Remote Sensing Perspective*, Prentice Hall, 1986.

KHOROS, *User/programmer manual*, University of New Mexico, 1992.

Lillesand, T., and R. Kiefer, *Remote Sensing and Image Interpretation*, John Wiley & Sons, 1979.

Lindley, C., *Practical Image Processing in C*, John Wiley & Sons, 1991.

Mather, P., *Computer Processing of Remotely-Sensed Images*, John Wiley & Sons, 1987.

Myler, H., and A. Weeks, *Computer Imaging Recipes in C*, Prentice Hall, 1993.

Richards, J., *Remote Sensing Digital Image Analysis: An Introduction*, Springer-Verlag, 1986.

Marine Productivity Seasonal Forecast along the Ecuadorian Coastal Zone Based on Geophysical Models of ENSO

By Gustavo Silva-Guerrero

Escuela Superior Politécnica, P.O. Box 09015863, Guayaquil, Ecuador

Monthly mean time series of coastal sea surface temperatures (CSSTs) of the Ecuadorian coastal zone were studied along with primary productivity anomalies (PPAs). PPA time series were derived from the monthly gridded chlorophyll data obtained by the coastal zone color scanner of the satellite NIMBUS-7. The objective of this work was to establish the applicability of geophysical model CSST calculations to estimating PPA tendencies. Applicability was evaluated in terms of the accuracy of PPA results when used to define productivity scenarios up to 1 year in advance.

This work was based on the hypothesis that for some Tropical Pacific marine environments (and especially for the Ecuadorian coastal zone) climate variability is the limiting factor of many productive activities. The significant skill that geophysical models of El Niño have developed lately for the zone studied gives us an alternative for assessing the probabilities of such productive activities. This approach could provide us with tools to address the related problems and socioeconomic impacts. The Cane and Zebiak geophysical model calculations for the CSST, as well as for the NIÑO3 region, were considered for this purpose.

Consistently with the upwelling-thermocline depth mechanism operating in warm events, the PPA time series derived from satellite data showed a negative correlation that might indicate a trend opposite to the observed CSST monthly anomalies time series. The large variability of the CSST, the applicability of geophysical models, and the uncertainties derived from the influence of these two parameters were specifically considered in this work. Multiple regression analysis showed a significant association of CSST with seasonal and planetary scale variability, in spite of the influence of local agents.

Occurrence frequency tables relating forecast CSST to PPA provide the probabilities of expected PPA as a function of forecast CSST. Tendencies derived from the statistical relationship between predictor and predictant can be very useful to production strategies. This is especially important for extreme events when the high risk of social impact can be reduced. Satellite data for 1982–1985 were used to evaluate the forecast scheme and the results are discussed.

1. Introduction

1.1. Major Oceanic Processes of the Ecuadorian Coastal Environment Related to Marine Productivity

The mean characteristics of Ecuadorian coastal sea surface temperature (CSST) are mainly determined by two major oceanic processes: (1) The northwest advection of the relatively cold water, related to the southeast trades and the Peru Current, and (2) the seasonal invasion of warm tropical surface water from the north, associated with low oxygen and salinity. These oceanic processes interact periodically and are modulated by a seasonal cycle which in normal conditions defines two different environments, each affecting marine species differently so that each scenario can be adequate for the development of some species and negative for others, and vice versa.

Fisheries related to the Peru Current marine environment are some of the world's largest in terms of catch landings volume. This is due to their high primary productivity levels induced by the effects of upwellings, both coastal and oceanic, which are the major inorganic nutrient reservoir of the ocean and are related to the Ekman drift of easterly wind stress. Seasonal changes of the coastal upwellings are highly correlated with sea surface temperature (SST), but upwellings due to processes other than Ekman drift are also involved and may be independent of the seasonal cycle, as are the oceanic upwellings of the Equatorial Undercurrent [Philander, 1990]. The depth variation of the thermocline plays a major role modulating the quality of the water entrained by upwellings and the amount of light needed for photosynthesis.

1.2. *Ecosystem Variations Induced by Climate Variability*

Observations at the latitude corresponding to the Chile–Peru area showed that the planetary-scale phenomenon referred to as El Niño introduced an element of disruption lasting longer than the event itself, altering the latitude distribution of the upwelling centers. The appearance of these upwelling centers was restricted to coastal pockets as a result of persistent invasion of subtropical waters. These waters are associated with high salinity and low oxygen and may be creating strong saline and thermal fronts [Caviedes and Fisk, 1992].

Rather than weakened, intensified upwellings are reported during these events, but water entrained is warmer and poorer in nutrients because of variations in the thermocline depth [Enfield, 1987]. Changes in the upwelling ecosystems include different water quality and low photosynthesis activity implying low primary production. This mechanism explains why primary productivity is clearly regulated by the physical changes of extreme events like El Niño 1982–1983. It is also suggested that larger species of fish, crustaceans, birds and mammals are severely affected. As a consequence of these environmental changes, severe impacts on social communities have been reported and largely documented.

2. Theoretical Justification

2.1. *Hypothesis*

This work is primarily based on the hypothesis that for some Tropical Pacific marine environments (and especially for the Ecuadorian coastal zone) climate variability is the limiting factor for many productive activities, justifying the use of CSST forecasts in defining future productivity scenarios.

The significant skill in SST forecasting that geophysical models of El Niño have developed lately for the zone studied gives us an alternative for assessing probabilities of productive activities that could provide tools to address the related problems and socio-economic impacts. Tendencies derived from the statistical relationship between predictor and predictant can be very useful in production strategies, especially for extreme events, when the high risk of social impact could be reduced.

This approach does not intend to model marine productivity by including all the biotic and physical processes involved; in fact, it ignores their direct relationship, assuming that these processes are largely governed by climate changes themselves [Cane *et al.*, 1994]. However, other climate parameters related to specific physical agents such as upwelling index and air temperature can be included in the statistical scheme to improve the skill of the linear model [Yanez, 1991; Caviedes and Fisk, 1992]. Thermocline depth should be considered in a shrimp larvae availability model because of its direct influence on the natural cycle of this animal. Shrimp culture is a very important activity to Ecuador and is highly dependent on availability of larvae and climate variability.

2.2. *Statistical Methods and Data Information*

A 15-year monthly mean time series of CSST for each of the four main shrimp larvae harvest centers (stations) along the Ecuadorian coastal zone was studied. The corresponding climatology was removed from the series. Results of one station (Salinas) are discussed in this paper.

A direct comparison of CSST with primary productivity was performed as the primary approach; the results are shown in Section 2.3. Primary productivity anomaly (PPA) time series were derived from the monthly gridded chlorophyll data provided by the NASA Physical Oceanography Distributed Active Archive Center at the Jet Propulsion Laboratory, California Institute of Technology. The data were obtained by the Coastal Zone Color Scanner of NASA's satellite NIMBUS-7. The data set runs from January 1982 to May 1986 and has a 1° resolution.

This paper includes a diagnosis of the oceanic influence on CSST based on linear regression analysis, using standard techniques [Ropelewski and Halpert, 1987; Weare *et al.*, 1976]. These results are discussed in Section 3. The possibility of using geophysical models to infer future conditions of CSST was tested. The Cane and Zebiak climate model [Cane *et al.*, 1986] SST forecast for the Ecuadorian coastal zone was evaluated in terms of the work of Barnett and Preisendorfer, which is a variation of the standard tercile approach [Barnett and Preisendorfer, 1987]. This technique compares the performance of the forecast SST with the results of random probability using a binomial distribution.

2.3. *Relationship between Sea Surface Temperature and Primary Productivity*

Primary productivity time series derived from the chlorophyll data described were used to obtain the corresponding statistical relationship with Salinas CSST. Chlorophyll is an index of phytoplankton abundance but not that of zooplankton, the major food of small fish. However, because phytoplankton are the major food of zooplankton, chlorophyll concentration does delimit the habitat of smaller and larger fish [Barber and Chavez, 1983]. The data were normalized by standard statistical methods, and the corresponding climatology was removed.

A negative relationship between the two series was observed when overlapping both, as shown in Figure 1. The chlorophyll data is represented with dotted lines. The negative correlation obtained ($r = -0.41$) is 95% significant according to Student's t distribution. Degrees of freedom were adjusted to remove the skill of the month-to-month dependency. This negative correlation might indicate an opposite trend compared to the solid line, which represents the CSST anomaly. The opposite relationship obtained is consistent with the upwelling-thermocline depth mechanism operating in warm events, as described previously. It is interesting to note that for some close to normal events the relationship is not so clear. This could indicate that the potential influence of other agents (i.e., productivity from river runoff) is stronger in normal events in this specific zone.

Considering terciles for the marine productivity conditions (above normal, normal and below normal), the hypothesis for declaring that the chlorophyll concentration anomaly at a given time would match the same category as the one of the negative value for the CSST anomaly with 1 month lag was tested for the 1982–1986 period. The result was a correct match for 52% of the cases. The significance of this result was evaluated by using the binomial distribution test described, comparing the observed distributions of chlorophyll concentrations with the expected random distribution. The probability that the result obtained was due to chance, after adjusting the degrees of freedom to remove the skill of the month-to-month dependency, was less than 15%. This last result was improved in a seasonal analysis when using 3 month average values. The categories were matched in 62.5% of the cases, giving a chance probability of less than 2%.

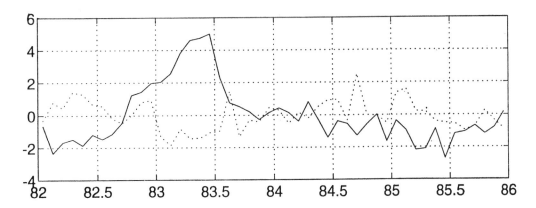

FIGURE 1. Time series of the sea surface temperature anomaly in the Ecuadorian coastal zone (solid line) versus the corresponding primary productivity time series (dotted line) derived from satellite observations of chlorophyll concentration. The negative correlation obtained ($r = -0.41$) is 95% significant according to Student's t distribution. This may indicate opposite trends in the two series.

3. Relationship between Oceanic and Coastal SST

The exposure to a large amount of different agents such as river inflows, response to geographic boundaries, and even anthropogenic factors like pollution is likely to give a high variability to the CSST at any specific station. If CSST forecasting is to be tested as a predictor of productivity, a better understanding of the observed CSST behavior is needed.

In previous work [Berri, 1995] spatial point to point correlation was performed, using the monthly anomalies of the CSST series for Salinas. The climatology reduced monthly mean gridded data set from the objectively analyzed COADS data [Pan and Oort, 1990] was used for the same periods as the CSST time series (from 1975 to 1988). Lag correlations for 3, 6 and 12 months were performed for each season. Results for Salinas showed an overall 99% significant simultaneous correlation of individual CSST series and the NIÑO1+2 region. Significant (99%) correlation spots apparently following the equatorial current track were found when a lag correlation analysis using the same technique was applied, showing an important share of variance from the NIÑO3 region for both seasons, especially when using a 3 to 6 month lead.

The NIÑO1+2 region also showed significant correlation in this analysis, but only NIÑO3 showed a greater signal ($r = 0.53$) than the simultaneous correlation ($r = 0.41$) when the lag correlation temporal distributions between the CSST anomaly series and the corresponding region indices were individually analyzed.

The current work shows a fit obtained by using multiple regression techniques representing the hindcast anomaly values for CSST. A relatively high correlation ($r = 0.73$) with the observed CSST was obtained when combining the NIÑO1+2 index and a 6 month lag NIÑO3 index. Figure 2 shows such a fit with a solid line and the observed values with circles. Figure 3 shows the total hindcast values for CSST, that is, after including climatology values with the hindcast anomalies. These results indicate that the CSST for Salinas constitutes an important signal associated with large time- and space-scale variability in spite of the influence of local agents.

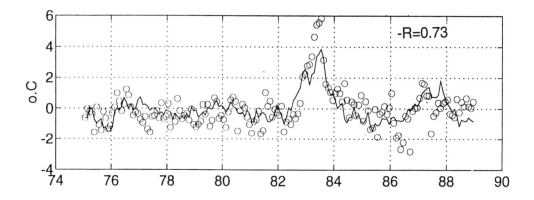

FIGURE 2. Fit of the hindcast anomaly values for the CSST (solid line) and observed CSST (circles), showing a relatively high correlation ($r = 0.73$).

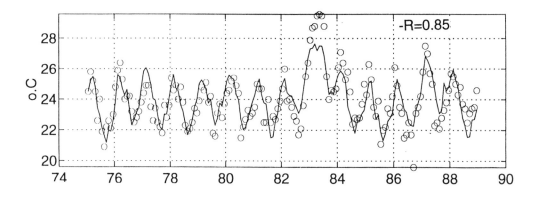

FIGURE 3. Fit of the total hindcast values (after including the climatology values with the hindcast anomalies) for the CSST (solid line) and observed CSST (circles), showing a high correlation ($r = .85$).

4. Applicability of the Forecast SST

Since fisheries related to the Peru Current environment are highly dependent on planetary-scale events like El Niño, so are those countries whose economy is highly dependent on these fisheries. It appears to be feasible to use forecast SST to estimate probabilities for marine productivity parameters influenced by El Niño. The significant skill that geophysical models of El Niño have developed lately gives us an alternative for addressing the climate related problems and socioeconomic impacts, especially for extreme events, when the high risk of social impact can be reduced.

Retrospective forecasting of mean monthly tropical Pacific anomalies has been done with the Cane and Zebiak model [Cane *et al.*, 1986; Zebiak and Cane, 1987] for the period from 1971 to 1985, and forecasts from 1986 onward. (The term "retrospective forecast" is used for the earlier 15 year period rather than "hindcast" because no information subsequent to initiation of each forecast influenced the forecast procedure [Simpson *et al.*, 1993].) The forecast CSST for Salinas was tested for two different regional outputs:

TABLE 1. Relationship between CSST and SST forecasts, using the Cane and Zebiak model, and observed CSST for Salinas, where r represents the linear correlation and s represents the significance level of such a correlation based on Student's t distribution. This table also shows the results of a binomial test where values of observed CSST anomalies and forecast SST anomalies have been divided into three categories, expressed in terms of percentage of occurrence, so that each group represents one-third of the total population. The occurrence frequency for each category (B = below normal; N = normal; AN = above normal) is displayed; sk represents the skill calculated as the ratio of the total success matches; rs corresponds to the probability of random occurrence.

Index	Lead (months)	r	s (%)	BN (%)	N (%)	AN (%)	sk (%)	rs (%)
Coastal region output	3	0.66	99	40	30	50	40	<17
Coastal region output	6	0.58	99	30	20	40	30	<60
Coastal region output	9	0.57	99	50	40	50	46.6	<5
Coastal region output	12	0.62	99	50	10	50	33.3	<50
Observed NIÑO3 values	-	0.48	99	40	30	60	43.3	<10
NIÑO3 region output	0	0.74	99	50	50	70	56.6	<1
NIÑO3 region output	3	0.61	99	40	40	60	46.5	<5
NIÑO3 region output	6	0.53	99	40	30	60	43.3	<10
NIÑO3 region output	9	0.48	99	40	30	60	43.3	<10
Observed 3 month average	-	0.57	99	40	20	50	36.6	<18
NIÑO3 region output	12	0.55	99	50	30	60	46.6	<5
NIÑO3 region output	15	0.35	<95	40	30	70	46.6	<5
NIÑO3 region output	18	0.13	<95	40	20	50	36.6	<30
NIÑO3 region output	21	0.09	<95	40	30	50	40	<17
NIÑO3 region output	24	0.15	<95	40	40	50	43.3	<10
Modified CRI	3	0.69	99	40	40	70	50	<2
Modified N3RI	0	0.71	99	50	60	70	60	<0.01

the NIÑO3 region and the SST average of the nearest to the station 2×2 degrees region (coastal). Correlation results had a 99% significance for the 3 month lead coastal output and an overall 95% significance for the two regions with different time leads.

Significant (99%) correlations were obtained in up to 12 month lead forecast CSSTs in a seasonal analysis, taking a 3 month average corresponding to the two seasons in Ecuador (wet and dry). The above normal and below normal categories showed an overall relatively high occurrence percentage of all the lead time forecasts, while the normal category was matched in a lower percentage, indicating better forecast scheme skill for the extreme events. Table 1 shows the relationship between the observed CSST and the geophysical model SST calculations for the two different areas in the Pacific, where r represents the linear correlation and s corresponds to the significance level of such a correlation based on Student's t distribution after adjusting the degrees of freedom to remove the skill of the month-to-month dependency.

5. A Forecast Scheme for Marine Productivity

A hypothetical case describing a forecast scheme using geophysical model outputs to predict future seasonal average marine productivity in this specific zone is discussed here. Considering terciles representing categories for the condition of the seasonal average marine productivity (above normal, normal and below normal), a hypothesis that the chlorophyll concentration seasonal anomaly, at a specific point in time, would match the same category as the one of the negative forecast value for the seasonal SST anomalies of the NIÑO3 region is tested for the 1982–1986 period. The Cane and Zebiak model 9 month lead calculations for the NIÑO3 region were used for this purpose. This case could represent a situation where the corresponding seasonal average marine productivity category was retrospectively forecast 9 months in advance and later compared to the observed conditions derived from the satellite data.

The result of this test was a correct match for 62.5% of cases. The significance is evaluated using the binomial distribution test described, comparing the observed distributions of chlorophyll concentrations with the expected random distribution. The probability that the result obtained is due to chance is less than 2%.

6. Conclusions and Discussion

In a manner consistent with the upwelling-thermocline depth mechanism operating in warm events, the primary productivity time series derived from the chlorophyll satellite data showed a negative correlation, which may indicate a trend opposite to that of the observed CSST monthly anomalies time series. Binomial tests confirmed the hypothesis, declaring that the chlorophyll concentration anomaly, at a specific point in time, would match the same tercile category as the one of the negative value for the CSST anomaly with 1 month lag, especially for the seasonal analysis when the random probability was less than 2% for the 1982–1986 period.

The results of this analysis can barely hold statistically for modeling purposes mainly because of the short length of the chlorophyll concentration time series, but nevertheless they give a clear signal justifying further efforts to improve the length and resilience of the data and consequently improve the inferred skill. Improvements should include analyzing extreme and normal events separately. A spatial analysis of the chlorophyll concentration distribution should also be considered, for it is possible to find a higher correlation for deeper water upwelling regions where the influence of oceanic processes is stronger and local agents such as river runoff are not present. Information derived from the spatial distribution analysis of chlorophyll could be of great interest for fishery related activities. Thermocline depth should be included in a shrimp larvae availability model because of its direct influence on the natural cycle of this animal. Shrimp culture is a very important activity for Ecuador and is highly dependent on larvae availability and climate variability.

The large variability of the CSST, the applicability of geophysical models, and the uncertainties derived from the influence of these two parameters have been specifically considered in this paper. Multiple regression analysis indicates an important signal for Salinas CSST associated with large temporal and spatial scale variability in spite of the influence of local agents. Occurrence frequency tables were constructed to establish the significance of different lead time Cane and Zebiak deterministic model calculations for the observed CSST. Significant (99%) correlations were obtained for up to a 12 month lead forecast CSST, especially in a seasonal analysis. The above normal and below normal categories showed an overall relatively high occurrence percentage for the lead time forecast, while the normal category was matched in a lower percentage, indicating better forecast scheme skill for extreme events.

The tercile hypothesis was tested as a forecast scheme using the model outputs to infer future seasonal average marine productivity in this specific zone, for the 1982–1986 period. The result of this test was a correct match in 62.5% of cases. The probability that the result obtained is due to chance is less than 2%.

This result can be used to predict future climate related problems. At the end of the 1980s, geophysical models performed poorly in predicting events, but the models' performance has improved since. Efforts of the scientific community to improve these models' performance continue, and better performance in the near future is expected. The combination of these factors may provide an alternative tool for addressing socioeconomic impacts in those countries whose economy is highly dependent on fisheries related to the Peru Current environment and planetary-scale events like El Niño.

Acknowledgments. Special thanks are given to the NOAA/OGP and IRICP pilot project for sponsoring this research, and to Dr. Guillermo Berri for his support and supervision. The work reported herein was supported by CATHALAC.

REFERENCES

Barber, R., and F. Chavez, Biological consequences of El Niño, *Science*, **222**, 1203–1210, 1983.

Barnett, T. P., and R. Preisendorfer, Origins and levels of monthly and seasonal forecast skill for United States air temperatures determined by canonical correlation analysis, *Mon. Wea. Rev.*, **115**, 1825–1250, 1987.

Berri, G. J., Applications and Training Activity Report N-1, International Research Institute for Climate Prediction – Pilot Project, June 1995.

Cane, M. A., G. Eshel, and R. W. Buckland, Forecasting Zimbabwean maize yield using eastern equatorial Pacific sea surface temperature, *Nature*, **370**, 204–205, July 1994.

Cane, M. A., S. E. Zebiak, and S. C. Dolan, Experimental forecasts of El Niño, *Nature*, **321**, 827–832, 1986.

Caviedes, C., and T. Fisk, The Peru–Chile eastern Pacific fisheries and climatic oscillation. *Climate Variability, Climate Change and Fisheries*, M. Glantz, Ed., NCAR, 1992.

Enfield, D., Progress in understanding El Niño, *Endeavour*, New Series, **11**, No. 4, 197–204, 1987.

Pan, Y. H., and A. H. Oort, Correlation analysis between sea surface temperature anomalies in the eastern equatorial Pacific and the world ocean, *Clim. Dyn.*, **4**, 191–205, 1990.

Philander, S. G. H., El Niño, La Niña and the Southern Oscillation, Academic Press, 1990.

Ropelewski, C. F., and M. S. Halpert, Global and regional scale precipitation patterns associated with El Niño/Southern Oscillation, *Mon. Wea. Rev.*, **115**, 1606–1626, 1987.

Simpson, H. J., and M. A. Cane, Annual river discharge in southeastern Australia related to El Niño–Southern Oscillation forecasts of sea surface temperatures, *Water Resources Research*, **29**, 3671–1680, 1993.

Weare, B. C., A. R. Navato, and R. E. Newell, Empirical orthogonal analysis of Pacific sea surface temperatures, *J. Phys. Ocean.*, **6**, 671–678, 1976.

Yanez, E., Relationships between environmental changes and fluctuating major pelagic resources exploited in Chile (1950–1988), International Symposium, Sendai Japan/Long-Term Variability of Pelagic Fish Populations and Their Environment, T. Kawasaki, 1991.

Zebiak, S. E. and M. A. Cane, A model El Niño–Southern Oscillation, *Mon. Wea. Rev.*, **115**, 2262–2278, 1987.

Land Cover Classification by Means of Satellite Imagery and Supercomputer Resources

By Roberto Bonifaz,[1] Román Álvarez[1] and Brian Davis[2]

[1] Instituto de Geografía, Universidad Nacional Autónoma de México, Apdo. Postal 20-850, 01000 México, D.F.

[2] USGS, EROS Data Center, Sioux Falls, SD, 57198-0001

Global observation of land and ocean surfaces, by means of satellite imagery, generates large data sets which vary from the gigabyte to the terabyte range. Observations are made in various frequency bands, from the visible to the thermal infrared, as well as the microwave region. Multitemporal observations, some of them made with high periodicity, increase severalfold the size of such data sets. Complete coverage of Mexico, for example, consists of 8.4 GB of multispectral scanner (MSS) imagery and 55.2 GB of thematic mapper (TM) imagery. Land cover classifications are required to extract regional information from satellite imagery. Although there are several algorithms for determining classifications, none of them has been specifically designed for supercomputer use. When codes are migrated from a reduced instruction set computer (RISC) workstation to a supercomputer, one often finds that they run slower. We tested five classification codes on five hosts: two high-end workstations and three CRAY supercomputers. Additionally, we tested generic and architecture-specific terrain-correction registration codes on three workstations and a CRAY C90; estimates were made for run times on a CRAY J90 and a CRAY T90. We conclude that, notwithstanding improvements when generic codes are migrated to and optimized with the tools of the supercomputers, such codes yield impractical processing times. With this type of data, the benefits of supercomputers are realized only when architecture-specific codes are used.

1. Introduction

Global change surveillance and evaluation require large data sets. Some are obtained by means of satellite acquisition systems [Townshend *et al.*, 1991]. Observations of land surface characteristics from space are usually made in various frequency bands of the electromagnetic spectrum; the number of bands and their corresponding bandwidths depend on the sensor used and the targets involved. The picture element (pixel) size is also of major importance in determining the amount of data required for the description of a given scene. As the spatial resolution increases, the pixel size decreases and the amount of data increases for each band. Spatial observation platforms often have high repeat cycles; thus, the same scene can be observed several times per unit time, increasing further the amount of data available for scene description. Land descriptions of global scale [Brown *et al.*, 1993] involve areas of tens of millions of square kilometers [Townshend *et al.*, 1987]. Even with space observations, the number of scenes required for full coverage is fairly large.

Satellite imagery must be registered to correspond with ground truth features. It must be cartographically correct under a given geographic projection to be used with precision. This process is often called the georegistration, or georeference, of the image. Another way of referencing multitemporal images of the same scene is image-to-image registration. Pixel (i, j) of an image taken at time t_1 corresponds to pixel (i, j) of the image taken at time t_2. If the former image was already georegistered, the latter will also

FIGURE 1. False-color composite of MSS bands 3, 2, 1. In this colorless representation, forests are shown in darker tones; image size is 4054 pixels by 3827 lines.

be. Evaluating change between two dates in a given area, when coregistration between images has been performed, is considerably simplified.

The pixel is the smallest element in a scene. Its reflectance can be described as the average of the reflectances of all the surface features contained in that unit area. If the pixel contains a water body, a forest patch, and a building, the reflectivity registered by the remote sensor will be a combination of the reflectivities of these features. Although different materials have different spectral signatures, when they are averaged in the unit area measured by the sensor, a certain degree of confusion is introduced. Sometimes it is possible to find areas of several pixels that contain essentially the same feature (for example, a forest, barren soil, or crop); it is possible to select them as training areas for a classification system. An enormous amount of processing is required to classify the contents of a scene. The process is called land cover classification [Rovinove, 1981] and requires computing resources and ground truthing in order to achieve acceptable degrees of accuracy. Although remote sensing, image registration, and image classification are not traditional supercomputing applications, as data sets increase in size and the multitemporal requirements for change detection grow, the need for powerful computer resources becomes unavoidable.

TABLE 1. Three sensors and their spectral and spatial resolutions.

Sensor		Spectral bandwidth (μm)	Resolution (m)
Landsat	TM	0.45–0.52	30
		0.53–0.61	30
		0.63–0.69	30
		0.78–0.90	30
		1.57–1.78	30
		2.08–2.35	30
		10.4–12.5	120
	MSS	0.50–0.60	79
		0.60–0.70	79
		0.70–0.80	79
		0.80–1.10	79
AVHRR		0.58–0.68	1100
		0.72–1.10	1100
		3.55–3.93	1100
		10.3–11.3	1100
		11.5–12.5	1100

2. Data Sets

Data have been collected from space platforms since approximately 1960, when the first meteorological satellites were deployed [Goward *et al.*, 1985]. The first systematic observations of the Earth's surface with a high-resolution sensor began in 1972 with the multispectral scanner (MSS), whose spatial resolution was 80 m. This sensor was operating in several satellites of the Landsat series until decommissioned in 1992. This 20-year period yielded an extremely valuable set of data that can now be analyzed with a historical approach in order to characterize land cover at specific times. This in turn serves as a baseline for change detection. In the early 1980s an improved sensor of the Landsat series, the thematic mapper (TM), was launched. Its spatial resolution was improved (30 m) with respect to the MSS, and the number of bands was also increased from 4 to 7, including a thermal infrared band. Table 1 shows the relevant characteristics of such sensors. Figures 1 and 2 show two scenes of the state of Chiapas in southeastern Mexico that correspond to the same area. Figure 1 was originally a false-color composite of MSS bands 3, 2, and 1, in which the forests are shown in red (the image size is 4054 pixels x 3827 lines, giving 15.5 MB); in this colorless version they appear as the darker areas. Figure 2 shows the same area, at a different date, with the larger resolution of the TM sensor. Bands 7, 4, and 1 are combined to produce this image. Herein they appear as lighter areas, as compared to Figure 1.

Another set of sensors are carried by NOAA satellites; they are known as advanced very high resolution radiometers (AVHRR), with a spatial resolution of 1.1 km; their coverage periodicity is severalfold that of the Landsat satellites. Since there is a series of these satellites in operation, it is possible to cover the same scene area several times per day. The AVHRR spectral coverage encompasses five bands, three in the visible and two in the thermal infrared. Images are typically 2400 by 5700 km. This spatial resolution and image size is most convenient for continental coverage [Loveland *et al.*, 1991]. Table 1 also summarizes the characteristics of these data. Figure 3 shows a cloud-

FIGURE 2. False-color composite of the TM image of a subarea of Figure 1; La Angostura dam, in Chiapas state, is clearly seen in both figures. The larger resolution (30 m) of the TM image is apparent. Forests also correspond to the darker tones.

TABLE 2. Sizes of three data sets for Mexico and North America.

(a) MSS coverage of Mexico
  - 1 scene: 4054 pixels × 3827 lines = 15.5 MB/band
  - 4 bands = 62 MB
  - Complete coverage of Mexico: 136 scenes = 8.4 GB
(b) TM coverage of Mexico
  - 1 scene: 8890 pixels × 8370 lines = 74.4 MB/band
  - 7 bands = 520 MB
  - Complete coverage of Mexico: 106 scenes = 55.2 GB
(c) AVHRR Coverage of North America
  - 1 scene: 11329 pixels × 7793 lines = 88.3 MB/band
  - 12 scenes (bands), one per month in 1 year = 1.05 GB

FIGURE 3. Normalized differential vegetation index (NDVI) of a cloud-free composite of a portion of North America, derived from an AVHRR image set from April 1 to April 10, 1992. Light tones correspond to areas with denser vegetation cover.

free composite of the normalized differential vegetation index [Turcote *et al.*, 1989] of Mexico, containing 20 MB of data. The cloud-free composite is built from an essentially cloud-free image, in which pixels with clouds are replaced by the corresponding pixels without clouds from other images. After cloud-free data sets are built, the classification process can be performed. Table 2 shows the sizes of the data sets that cover Mexico (MSS and TM) and those of the AVHRR sensor that cover North America.

3. Classification Process

There are a number of classification methods for multiband data [Kartikeyan *et al.*, 1994]. However, when one considers the variety of land-cover types involved, the extent of the area to be classified, and the size of the corresponding data set, the actual choices narrow down considerably. Performing supervised classifications [Townshend *et al.*, 1991; Chuvieco, 1990] over the whole of Mexico (2 million square kilometers) is not presently feasible because of the lack of sufficient training areas identified throughout the country for this purpose (i.e., areas of fairly homogeneous radiometric response involving tens to hundreds of pixels, where the land cover is univocally determined). For the same reason, the supervised classification of an area the size of the North American continent (approximately 23 million km^2) is not presently viable.

Unsupervised classification [Chuvieco, 1990] does not involve the use of training areas in order to carry out the classification process. Instead, an algorithm defines a group of

FIGURE 4. Unsupervised classification of the TM data shown in Figure 2. Water (black), clouds (white), and various vegetation types (varying tones of gray) are identified in the classification.

clusters of similar radiometric response in several bands, which are subsequently related to information classes (i.e., the classes identified by ground truthing). Often, the number of radiometric classes is higher than that of the information classes; thus, merging of the former, according to adequate criteria, becomes necessary. Figure 4 shows the unsupervised classification of a portion of the area shown in Figure 2; there are eight classes corresponding to water, clouds and various types of vegetation. Figure 5 corresponds to the portion of Mexico shown in Figure 3 and represents the corresponding unsupervised classification of such a data set.

4. Software

In order to perform the classification we used software that was developed for the determination of land use by means of satellite imagery, named Isoclass [Vanderzee and Ehrlich, 1995]. It has been thoroughly tested for several years in this type of application [Ehrlich *et al.*, 1994] at the EROS Data Center. It was originally programmed in Fortran in 1983, then converted to C in 1987, and subsequently modified for improved portability.

The Isoclass classification code uses the Isodata clustering algorithm [Duda and Hart, 1973]. The main parameters, in addition to the input data, are the maximum number of iterations, the minimum and maximum thresholds defined for classes to belong to the

FIGURE 5. Unsupervised classification of Mexico, corresponding to the type of data shown in Figure 3. A monthly vegetation index was used for 12 months between April 1992 and March 1993.

same cluster, the minimum and maximum numbers of classes to be found, the threshold for changing a class assignment, and the centroids of the initial cumuli. The output consists of a report on the data statistics, such as means, covariances, matrices, the number of samples for each cluster, and the cluster's assignments, in addition to the output image.

Figure 6 shows a partial report of a test run of the Isoclass code in the SGI2 (Table 3) for a window sample of the 1992 North American composite, consisting of 256 lines by 256 samples by 12 bands (months). Typical Isoclass processing is iterative; Figure 6 shows the results of iteration number 1 and, partially, those of subsequent iterations. After parameter definition and the first run of Isoclass, analysis of the statistical output data determines necessary parameter adjustments, and Isoclass is run again. These steps are repeated until the clusters are statistically balanced according to some predefined conditions, or compared to previous experiences. If problems are identified with one or more bands after processing, data must be reconstructed and the classification begun again. Consequently, any speedup of this process will help to make continental-scale image processing practical.

5. Hardware

A set of trials was run at UNAM's facilities using one 150 MHz R4400 processor of an SGI Onyx and one processor of UNAM's CRAY Y-MP 4/64 supercomputer. Another set of trials was run at the EROS Data Center facilities, using as host a SGI Challenge R4400, 150 MHz, 256 MB. A CRAY Y-MP 8/128 and a CRAY C90 8/512 at Cray Research Inc.'s Corporate Computer Network were used for software porting and optimization. One CPU was used in each case. Table 3 summarizes the results obtained at EROS Data

Classification Parameters

Threshold for margin clusters	4,00
Threshold for separating cluster	5,30
Split separation value for clusters	0,00
Minimum number of members allowed in a cluster	1000
Threshold for cluster chaining	3,2
Maximum number of clusters	64
Maximum number of iterations	20

Overall Statistics of Input Data

		1	2	3	4	5	6	7	8	9	10	11	12
Mean		117,21	132,58	133,36	113,21	125,25	121,04	118,46	109,33	106,75	100,30	102,69	111,30
Standard Deviation		7,29	9,19	10,61	9,79	9,58	8,54	7,80	8,27	7,59	5,61	6,79	6,78
Initial Cluster Means	Cluster 1	117,21	132,58	122,74	113,21	125,25	121,04	118,46	109,33	106,75	100,30	102,69	111,30
	Cluster 2	117,21	132,58	143,97	113,21	125,25	121,04	118,46	109,33	106,75	100,30	102,69	111,30

Number of clusters after iteration #1: 2
Number of pixels in cluster
Cluster 1   32137
Cluster 2   33399

		1	2	3	4	5	6	7	8	9	10	11	12
Mean	Cluster 1	115,81	128,87	126,65	108,91	122,23	118,41	116,29	109,13	106,53	99,69	100,62	110,08
	Cluster 2	118,56	136,15	139,81	117,35	128,16	123,57	120,56	109,52	106,96	100,88	104,68	112,47
Standard Deviation	Cluster 1	9,33	11,14	10,81	10,08	10,78	10,02	9,43	10,02	9,54	7,71	8,12	8,89
	Cluster 2	4,09	4,52	4,86	7,44	7,15	5,78	5,00	6,13	5,04	1,94	4,36	3,37

The process continues until the maximum number of iterations is reached generating the corresponding statistics for each cluster in each iteration.

Iteration # : number of clusters	1:2	2:4	3:7	4:11	5:13	6:15	7:14	8:15	9:14	10:15	11:15	12:16
	13:17	14:18	15:9	16:12	17:9	18:11	19:8	20:11				

FIGURE 6. Isoclass sample run on a $256 \times 256 \times 12$ AVHRR data set.

TABLE 3. Various hosts and software used in testing unsupervised classification of satellite images.

Host	Software	Data sets and run times		
		256×256×12	2048×2048×12	7793×11329×12
SGI 1	LAS	0:02:23	12:38:40	2̃ weeks
SGI 1	MEM	0:02:00	7:36:37	N/A
SGI 2	ISO	0:02:00	N/A	11d:23:37:25
CRAY A	ISO	0:02:20	N/A	N/A
CRAY A	ISOM	0:00:51	N/A	N/A
CRAY B	MEM	0:00:51	N/A	N/A
CRAY B	MEM/PACK	0:01:14	4:26:25	N/A
CRAY C	MEM	0:00:50	N/A	N/A
CRAY C	MEM/PACK	0:00:57	3:12:12	TBD

Host	SGI1	Challenge R4400, 150 MHz, 256 MB
	SGI2	SGI Onyx R4400, 150 MHz, 128 MB
	CRAY A	CRAY Y-MP 4/64
	CRAY B	CRAY Y-MP 8/128
	CRAY C	CRAY C90 8/512
Software	LAS	Original Land Analysis System (LAS) Isodata Algorithm
	ISO	Unoptimized Isoclass Non-Las Isodata Algorithm
	ISOM	Optimized Isoclass, modified with Cray's tools
	MEM	Optimized version of Isoclass, using memory arrays, not disk I/O
	MEM/PACK	Same as MEM, packing 8 values per word, to allow for larger data sets

N/A = Not Available
TBD = To Be Determined

Center and UNAM using the Land Analysis System (LAS) software, as well as optimized versions of the Isoclass algorithm. Initial runs of the baseline software revealed that the supercomputers took longer to execute the programs than the reduced instruction set computer (RISC) workstations. To reduce solution times on the supercomputers, we used some of the optimization tools available on the CRAY, such as vectorization, flow analysis, performance analysis, subroutine call tree, and procedure inlining. These modifications reduced execution times by a factor of 3.

Table 3 shows why processing continental-size data sets currently is not viable. SGI1 was used for the classification of the $7793 \times 11,329 \times 12$ pixel data set, with an unoptimized version of the Isoclass, using memory arrays instead of disk-based I/O. The total elapsed computer run time was 11 days, 23 hours, 37 minutes, and 25 seconds. Because of hardware and software limitations, however, the actual time from the beginning of the first attempts to final output was over 1 month. Only resources commensurate with the 512 MWord memory of the C90 are capable of retaining all values in memory, allowing more acceptable run times.

TABLE 4. Terrain-corrected registration of Landsat TM data (input is $5965 \times 6967 \times 7$ pixels).

Host	Software	Run times (s)
SGI A	G	6293
SGI B	G	4445
SGI C	G	4235
SGI C	A-S	655
CRAY A	G	5166
CRAY A	A-S	78
CRAY B	A-S	252*
CRAY C	A-S	39*

Host			
	SGI A	=	SGI R4400, 100 MHz
	SGI B	=	SGI R4400, 150 MHz
	SGI C	=	SGI R8000, 75 MHz, 8 CPUs
	CRAY A	=	CRAY C90
	CRAY B	=	CRAY J90
	CRAY C	=	CRAY T90
Software	G	=	Generic
	A-S	=	Architecture-specific, optimized software

* Times are estimates.

6. Terrain Correction

In addition to the classification processes tested with the Isoclass algorithms, we have used a second procedure, consisting of terrain-corrected image registration, to demonstrate the benefits of creation of architecture-specific optimized software. Satellite images require a series of corrections in order to achieve cartographic precision. Terrain correction takes into account the topographic characteristics of the area covered by the satellite image and adjusts the pixel coverage according to terrain slope. Control points of known geographic positions are required throughout the area to correct the image.

Terrain slopes can be derived from a digital elevation model (DEM) of the surveyed area [Jenson and Dominique, 1988]. A DEM is an ordered array of numbers that represents the spatial distribution of elevations above some arbitrary point in a landscape [Davis, 1992]. DEMs are a subset of digital terrain models (DTMs): ordered arrays of numbers that represent the spatial distribution of terrain attributes. Thus image registration can be performed using a DEM of the corresponding area and a set of control points, such as road intersections, water bodies, buildings, and the like, whose geographic positions are known. Typical registration accuracies are better than one pixel (i.e., better than 30 m in the case of Landsat TM data). Once the image is geographically registered, or georegistered, map features derived from the topography, such as depressions, flow paths, hydrologic features, and watersheds, can be readily and accurately generated.

Table 4 summarizes various terrain-correction runs on a $5965 \times 6967 \times 7$ Landsat TM data set. Note that the output terrain matrix is of a different size because of resampling of the original data after image rectification. Three SGI workstations and one CRAY supercomputer were actually used in this testing. Two estimated times are reported for a CRAY J90 and a CRAY T90.

7. Conclusions

The analysis of two procedures involving satellite image registration and image classification, using CRAY supercomputers and high-end RISC workstations, shows that architecture-specific optimized software is required for supercomputers to perform faster than RISC processors. When available code, running on traditional equipment, is directly migrated to the supercomputer, its performance is considerably below expectations. However, when code is specifically designed for them, supercomputers yield considerable savings in execution times. To improve significantly the adoption and use of supercomputers in this area of the geosciences, effort must be devoted to writing architecture-specific code that is capable of tackling the large processing demands of environmental and global problems.

Acknowledgments. We acknowledge material help from Alma L. Cabrera, Gabriela Gómez and Ian García, as well as supercomputer support from Dirección General de Servicios de Cómputo Académico, UNAM.

REFERENCES

Brown, J. F., T. R. Loveland, J. W. Merchant, B. C. Reed, and D. D. Ohlen, Using multisource data in global land-cover characterization, *Photogrammetric Engineering and Remote Sensing*, **59**(6), 977–987, 1993.

Chuvieco, E., *Fundamentos de Teledetección Espacial*, Ediciones Rialp, Madrid, 1990.

Davis, B., Using Cray computers to extract hydrology and topography from digital elevation data. *Proceedings of the Thirtieth Semi-Annual Cray Users Group Meeting: Grand Challenges*, 73–88, Washington, DC, September 1992.

Duda, R. D., and P. E. Hart, *Pattern Classification and Scene Analysis*, John Wiley & Sons, 1973.

Ehrlich, D., J. E. Estes, and A. Singh, Applications of NOAA-AVHRR 1 km data for environmental monitoring, *Int. J. Remote Sensing*, **15**, 145–161, 1994.

Goward, S. N., C. J. Tucker, and D. G. Dye, North American vegetation patterns observed with the NOAA-7 Advanced Very High Resolution Radiometer, *Vegetatio*, **64**, 3–14, 1985.

Jenson, S. K, and J. O. Domingue, Extracting topographic structure from digital elevation data for geographic information system analysis, *Photogrammetric Engineering and Remote Sensing*, **54**, 1593–1600, 1988.

Kartikeyan, B., B. Gopalakrishna, M. H. Kawborme, and K. L. Majumder, Contextual techniques for classification of high and low resolution remote sensing data, *Int. J. Remote Sensing*, **15**, 1037–1051, 1994.

Loveland, T. R., J. W. Merchant, D. O. Ohlen, and J. F. Brown, Development of a landcover characteristics data base of the conterminous U.S., *Photogrammetric Eng. and Remote Sensing*, **57**, 1453–1463, 1991.

Rovinove, C., The logic of multispectral classification and mapping the land, *Rem. Sen. Environment*, **II**, 23, 1–244, 1981.

Townshend, J. R. G., C. O. Justice and V. Kalb, Characterization and classification of South American land cover types using satellite data, *Int. J. Remote Sensing*, **8**, 1189–1207, 1987.

Townshend, J. R. G., C. O. Justice, Wi Li, C. Gurney, and J. McManus, Global land cover classification by remote sensing: Present capabilities and future possibilities, *Rem. Sens. Environment*, **35**, 243–255, 1991.

Turcotte K. M., W. J. Kramber, and K. Lula, Analysis of regional-scale vegetation dynamics of Mexico using stratified AVHRR NDVI data, *ASPRS*, **3**, 246–257, 1989.

Vanderzee D., and D. Ehrlich, Sensitivity of ISODATA to changes in sampling procedures and processing parameters when aplied to AVHRR time-series NDVI, *Int. J. Remote Sensing*, **16**, 673–686, 1995.

PART II
Dispersion and Mesoscale Modeling

Environmental Applications of Mesoscale Atmospheric Models

By Thomas T. Warner

The University of Colorado

and

The National Center for Atmospheric Research, Boulder, Colorado

Limited-area mesoscale atmospheric models have been applied widely for the simulation and prediction of environmental processes. These applications include air-pollution transport, regional climate change, and water-resources management. As the models become more accurate as a result of improved representation of physical processes and improved resolution, and as computers for running these models become faster and less expensive, the range of environmental applications will continue to grow. In this paper will be reviewed some of the present environmental applications of such models.

1. Introduction

Mesoscale atmospheric models have been used for a wide variety of applications in the environmental sciences during the last two decades. As greater computing power has become routinely available, and as the quality of the model simulations has improved as a result of our better understanding of physical processes, such use has recently become much more widespread.

The term mesoscale refers to atmospheric processes with horizontal length scales between 2 km and 2000 km. This broad range encompasses a plethora of phenomena, including small-scale baroclinic disturbances, mesoscale convective systems, individual thunderstorms, sea breezes, and terrain-induced effects.

These models are generally dynamically based and use relatively complete representations for the fluid dynamics and thermodynamics of the atmospheric processes. The exact configuration of a mesoscale model, in terms of its numerics and physical-process representations, depends on the specific purpose for which it is being used. However, in general, a model will include representations of boundary-layer turbulence, long- and shortwave radiation interaction with the atmosphere and surface, and moist processes such as convection and cloud microphysics. Applications of such models take two forms. One involves their use for producing model-generated data sets which can augment field-program data to enable us better to understand environmental processes that involve the atmosphere. That is, the very spatially and temporally dense model-generated "data" can supplement, or serve as a surrogate for, actual data. The other application is the use of these models to predict processes that have significant environmental consequences so that we can mitigate their effect on society.

In the next section, an overview will be provided of the variety of ways in which these models are used to study and predict our atmospheric environment. In the following section, a more detailed discussion of a few specific model applications can be found. Finally, the summary will review the current status of environmental applications of mesoscale atmospheric models and suggest trends for the future.

2. Review of Environmental Applications of Mesoscale Models

2.1. *Air-Pollution Transport and Dispersion*

Development of federal and state regulations to limit degradations in air-quality requires knowledge of the physical processes that control the transport and dispersion of air pollutants in a particular geographic area. Mesoscale models that can diagnose the intensity of the turbulence as well as simulate the complex mesoscale wind patterns that transport the pollution have played an important role in this regard. A typical application would involve the use of the atmospheric model to define the precipitation, transport winds turbulence intensity, and other variables. These meteorological data sets would then be employed by an "air quality" model that would use continuity equations for the various species of air pollutants [Pleim *et al.*, 1991]. Terms in the continuity equations include the transformations of the different pollutant species that result from dry chemical reactions, interactions with solar radiation, aqueous chemical reactions within clouds, and washout by precipitation.

A typical situation in which such model simulations would be required would be for preparation of an environmental impact study which must demonstrate the degree to which a new source of pollution, such as a coal-fired power plant or an industrial plant, will affect air quality in the region. Alternatively, if options are being assessed for improvement of the existing air quality, models can be used to simulate the effects on air quality of different mitigation strategies. For day-to-day control of regional air quality, the meteorological models can provide forecasts of a few days' duration of the conditions that cause elevated surface concentrations of pollution, such as low wind speeds and temperature inversions. These forecasts can be used as the basis for temporarily limiting sources of pollution such as automobile use, industrial output, and wood burning heating systems.

An especially complex air quality problem that has been studied with mesoscale models is acid precipitation. The models have been used to help us better understand the complex processes involved when specific chemical emissions react in an aqueous phase in clouds to produce the acidic precipitation downstream of the pollution source region. They have also been employed to help determine the sources of the pollution that are producing the environmental degradation associated with the acid precipitation in high-impact geographic areas.

There have been numerous specialized applications of mesoscale air-quality models. For example, such models have been used to help identify the environmental consequences of the Kuwait oil fires, and they have helped in understanding the nature of the dust dispersal associated with thermonuclear explosions, which is the basis of the nuclear-winter hypothesis.

Most mesoscale atmospheric models, and the diffusion and transport models coupled to them, are applied in a retrospective sense to historical events. However, some are run operationally as emergency response tools to predict the impact of accidental releases of gases and particles into the atmospheric environment. For example, if there were an accidental release of radioactive material from a nuclear-power reactor or a nuclear-waste facility, the models would be employed to predict the areal extent of the downwind hazard zone. Similarly, large facilities that store or manufacture toxic chemicals also sometimes have these transport and diffusion models ready for use in the event of an emergency.

In summary, these coupled modeling systems, composed of an atmospheric module that simulates the full dynamics and thermodynamics of the atmosphere and a transport and diffusion module that computes the transport and transformations of the pollutants, are used for a variety of air-quality applications. Industries use mesoscale air-quality models to determine their impact on the atmospheric environment, government regulatory agencies use the models to establish and maintain air-quality standards, and industries and other groups that work with hazardous chemicals employ the models as emergency response tools.

2.2. *Regional Climate Change*

Global models of the Earth's general circulation have been used for the last two decades to simulate potential climate-related changes in the large-scale atmospheric weather patterns. For example, a model whose computational domain spans the entire planet can be integrated for a multiyear period with increased concentrations of carbon dioxide in order to estimate the effects on the largest scales of atmospheric motion. Similar numerical experiments have been conducted to define physical processes and to estimate large-scale environmental impacts associated with increases in the concentration of stratospheric ozone. However, even though these simulations can yield a great deal of insight into the governing physical processes associated with climate change and serve as tools for assessing large-scale impacts of anthropogenically or naturally caused climate variations, the models do not have the horizontal resolution that is required to provide information about the regional variations in climate. And it is the regional response to the large-scale climate variations that directly impacts our environment and human activities. Thus, during the last 5 years, mesoscale models have been used in studies of how large-scale climate change can modify the mesoscale weather patterns. This is accomplished by performing multiyear simulations with a global model and then using the output from these simulations to drive the lateral boundaries of a regional model that is used for local-impact assessment [Giorgi, 1990].

The global and regional models used for these purposes must be developed and tested more carefully in some ways than are the models used for simulations of a few days' duration (such as for air-quality applications). For example, artificial energy imbalances that result from a model radiation budget that is not carefully formulated may result in only a slight drift in the simulated temperature during a period of a few days; however, the magnitude may be substantial if the model errors or biases are allowed to accumulate over a multiyear period. An additional complexity associated with multiyear model integrations is that there is sufficient time for a variety of natural systems to interact, and thus these systems should ideally be represented in some way in the simulations. For example, the atmosphere can interact with ocean circulations, the biosphere and the surface hydrologic conditions over long time scales, and thus these systems should be dynamically represented in the complete modeling systems that are used for long-range simulations.

Increases in computational capabilities have allowed this use of mesoscale models to increase significantly during the last few years. This research activity is driven by many practical environmental problems that humankind is facing. For example, there are underground facilities planned for long-term storage of low- and high-level nuclear waste, and it is essential that we understand the very long-term changes that may take place in the hydrology of these storage areas. Thus, mesoscale atmospheric models have been used to help estimate changes in regional precipitation distribution that are forced by large-scale climate change scenarios, and from these changes can be inferred the potential impact on water seepage and changes in the water table at the storage sites.

2.3. *Water-Resources Management*

Increases in population-related demand for water have necessitated that water-resources planners develop an improved understanding of the regional water cycle in their area.

Such an improved understanding of the detailed coupled atmospheric and surface water budget can be used in developing conservation strategies and in some cases can be used actually to increase the amount of available water. However, one historic problem has been that land-surface hydrologists have worked on the scale of small watersheds, and therefore they have had to parameterize the effects of the atmosphere, including precipitation, on the surface hydrologic processes. Atmospheric simulation models could not provide sufficiently high resolution information. However, our ability to perform reasonably accurate mesoscale simulations of the atmospheric hydrologic cycle now makes it reasonable to couple the atmospheric models and the surface hydrologic models [Leung et al., 1996]. This allows us to produce simulations of the complete hydrologic cycle, encompassing atmospheric, surface and subsurface processes. Biospheric models are also an important component of the system.

These coupled models can be used for a variety of purposes, including the following:

• Perform analyses of the different components of the water cycle to improve our knowledge of the water-transport pathways in different geographic regions

• Predict interseasonal or interannual changes in the water budget (e.g., river flow, aquifer level) to permit water-resource planners and managers to modify water conservation strategies

• Permit managers to address environmental riparian concerns associated with changes in the water levels in rivers, lakes and wetlands

• Predict flash floods so that evacuations and other damage mitigation measures can be performed

• Provide hydroelectric power company managers with advance information about future surface runoff and river discharge so they can discharge water from reservoirs in environmentally responsible ways

2.4. *Fire–Weather Prediction*

Forest and brush fires, both natural and initiated by man, have a significant impact on the environment. Recently, mesoscale atmospheric models and models of fire dynamics have been coupled, with the result that a simulation system may eventually be available that will allow more effective strategies to be formulated for fighting specific fires. Even though these numerical tools, and knowledge of how to use them, are far from mature, this application of mesoscale models has the potential of saving lives and allowing us to fight fires and minimize their environmental destructiveness more effectively.

3. Examples of Environmental Applications of Mesoscale Models

3.1. *Determining Atmospheric Conditions Associated with High-Ozone-Level Episodes in California*

During the summer, ozone levels measured in the San Joaquin Valley of California often exceed the federal limit of 120 parts per billion (ppb) mandated by the 1990 Clean Air Act. A combination of anthropogenic emissions and the unique meteorological conditions found in California are major factors contributing to photochemical production of ozone near the surface in the valley. Meteorologically and chemically based air-quality models can play an important role in defining optimal strategies for controlling the problem through regulation.

In order to understand the meteorological conditions in the San Joaquin Valley better, the Penn State/National Center for Atmospheric Research mesoscale model [MM5, Dudhia, 1993; Warner et al., 1992] was used to augment field-program data to provide a data set for detailed study [Seaman et al., 1995]. In this situation, where the model

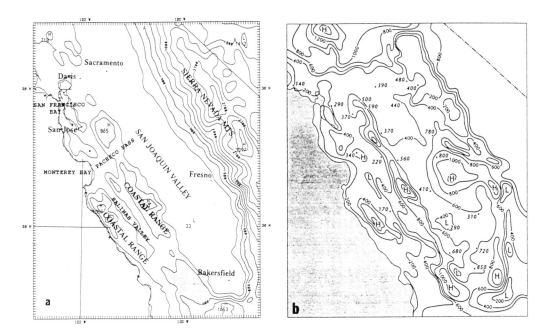

FIGURE 1. (a) Terrain elevation (m) for the high-resolution computational grid with a contour interval of 300 m, (b) mixed layer depth (m) at 0000 UTC 6 August 1990, as simulated by the model, with an isopleth interval of 200 m. Observed mixed layer depths are shown in italics [from Seaman *et al.*, 1995].

output and data are to be used to produce an optimal analysis of the four-dimensional (space and time) meteorological conditions, the model is run using special techniques that assimilate the data into the model solution during the integration. This technique effectively uses the model as a complex analysis system to produce gridded fields of all meteorological variables that are dynamically consistent with the data. Further discussion of mesoscale four-dimensional data assimilation (FDDA) can be found in Stauffer and Seaman [1990, 1994] and Stauffer *et al.* [1991].

The simulation shown here to illustrate the use of a mesoscale model with FDDA to simulate the meteorological processes affecting air quality utilized a version of MM5 with a grid increment of 4 km, and 30 computational layers in the vertical. The lowest computational layer is 35 m above ground level. The depth of the atmospheric mixed layer, which limits the vertical dispersion of pollution, is used here to illustrate the skill of the model. Figure 1(a) shows the complex topography of the region and Figure 1(b) illustrates the model-simulated mixed layer depth after approximately 4 days of simulation time. The observed mixed layer depths are shown in italics with a dot designating the location of the observation, where the values are typical of afternoon conditions during the episode. Although surface temperatures were as high as 39°C on this day, observed depths in the San Joaquin Valley range from only 310 to 850 m. The mean error in the simulation is only −14 m and the root mean square (RMS) error is 139 m. This model-generated mixed layer analysis and gridded fields of other atmospheric variables such as the wind are essential to improving our understanding of pollution transformation, transport, and diffusion in this meteorologically complex area. When this meteorological

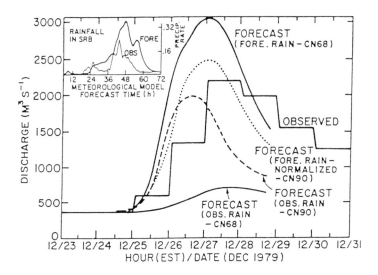

FIGURE 2. Plot of the observed and predicted Susquehanna River discharge at Sunbury, Pennsylvania. Inset at upper left depicts the atmospheric-model-based precipitation forecast for the watershed. See text for further details [from Warner *et al.*, 1991].

model is coupled with an air-quality model, the simulation system can be used to assess the impact of various strategies for managing the sources of pollution in an airshed.

3.2. *Predicting River Discharge with a Coupled Mesoscale-Atmospheric and Surface-Hydrologic Model*

MM5 has been coupled with a surface runoff modeling system as part of an effort to develop a comprehensive hydrologic modeling system that can be used for addressing water-resources management and related environmental problems. In this case, precipitation predictions from the atmospheric mesoscale model were provided to a surface runoff model in order to assess the skill of the coupled system in predicting discharge in the Susquehanna River in Pennsylvania [Warner *et al.*, 1991]. The atmospheric model produced 72-h precipitation predictions which were input into the single-event, distributed, HEC-1 Army Corps of Engineers runoff model for the large Susquehanna River Basin (SRB). Figure 2 shows the atmospheric model's forecast of area-average rainfall for the SRB (inset, upper left) in centimeters per hour for the first 3 days of the 8 day river-discharge forecast period represented on the abscissa. Even though the model overpredicted the rate of the precipitation in the SRB, the precipitation timing was not unreasonable. The other plots in the figure show the observed and predicted discharge of the river at Sunbury, Pennsylvania. When HEC-1 model parameters did not accurately reflect the existence of snow on the surface, and thus partitioned too much of the precipitation to infiltration rather than runoff, the discharge was greatly underpredicted when the observed rainfall was used (lower solid curve labeled "obs. rain – CN68"). The discharge forecast when the MM5-predicted rainfall was used with the same HEC-1 parameters is shown in the upper solid curve ("fore. rain – CN68") and illustrates how nonlinear the coupling can be between a runoff model and a precipitation-prediction model. That is, the 127% overprediction in the rainfall volume on the SRB caused a factor of 10 error in the river discharge because the precipitation amount influences the

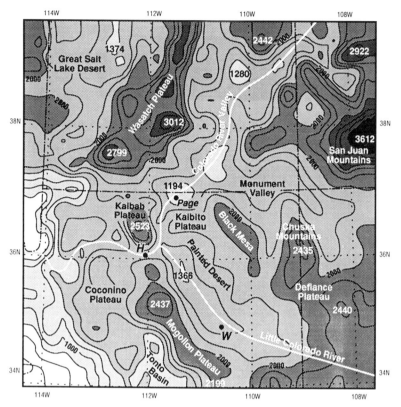

FIGURE 3. Terrain elevation (m) and geography for the computational grid mesh. The contour inteval is 200 m, with darker shading used for the higher elevations [from Stauffer *et al.*, 1993].

runoff/infiltration partitioning. When the HEC-1 parameters were adjusted to account for the presence of the snow on the surface, and a statistical adjustment was made to the rainfall prediction based on the known atmospheric-model biases, the discharge predictions based on the observed (obs. rain – CN90) and predicted rainfall (fore. rain – normalized – CN90) are fairly realistic. The coupled MM5/HEC-1 system predicted the timing and amplitude of the discharge peak reasonably well about 5 days in advance. If observed rainfall had been used with HEC-1, the forecast lead time would have been 3 days less.

3.3. *Study of the Causes of Visibility Degradation in the Southwestern United States*

Visibility in Grand Canyon National Park has shown degradation during the last decade, even though the air quality is federally protected in this and several other national parks on the Colorado Plateau. A first step in improving the air quality is to associate the pollution with specific sources, where possibilities are coal-fired power plants in the area, smelters, Los Angeles smog and other heavy industries in the United States and Mexico. To accomplish this, a high-temporal- and spatial-resolution four-dimensional analysis is required of the meteorological conditions. The MM5 model with the aforementioned FDDA capability was used for this purpose [Stauffer *et al.*, 1993]. The meteorological flow patterns are especially complex because of the extreme topographic variation, shown in Figure 3 for the model computational domain. A horizontal grid increment of 10 km and 20 computational layers were employed in 36 h model simulations. Figure 4 shows

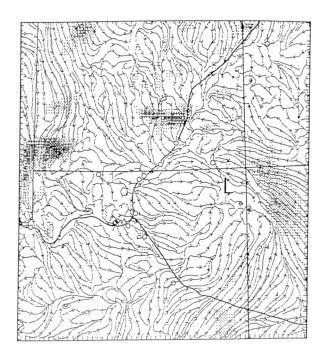

FIGURE 4. Streamlines near the Earth's surface (34 m AGL) defined by the atmospheric model after 36 h of simulation time for 0500 MST January 19, 1990. Streamlines are lines of flow that are parallel to the wind vectors throughout an instantaneous flow pattern. The "L" denotes the circulation center of a storm near the surface. Symbols plotted (mainly along the Colorado River) reflect special field-study wind observations near the surface. Shading indicates model-simulated wind speeds in excess of 5 m s^{-1} [from Stauffer *et al.*, 1993].

the model computed streamlines (lines drawn parallel to the wind) near the Earth's surface after 36 h of simulation time. Clearly, the flow pattern at low levels is complex and is strongly influenced by thermally driven circulations and topographic channeling of the wind. The available wind observations are shown and are generally in good agreement with the model-simulated wind directions. Given the relatively few meteorological observations in areas such as this, the model-generated fields are essential in defining the details of the air-pollution transport and diffusion.

3.4. *Regional Climate Change Associated with a Doubling of Carbon Dioxide Levels in the Atmosphere*

In this study by Giorgi *et al.* [1994], two continuous $3\frac{1}{2}$ year simulations were performed with a mesoscale model (MM4, a predecessor to MM5). The model's initial conditions and lateral boundary condition were provided by a global climate model that was integrated for the same period. The global model was the coarse-resolution Community Climate Model (CCM) of the National Center for Atmospheric Research, which employed a horizontal grid increment of over 500 km. One of the MM4/CCM simulations was for present-day conditions, and the other pertained to a doubling of the present carbon dioxide concentration. The horizontal grid increment of MM4 was 60 km, and 15 computational levels were used in the vertical. Note that less model resolution is used in the

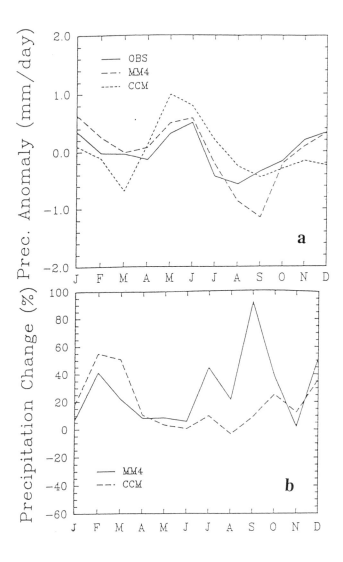

FIGURE 5. For the northwestern United States: (a) for present conditions, precipitation monthly anomalies from observations, MM4 and CCM, and (b) for a doubled atmospheric carbon dioxide, MM4 and CCM monthly precipitation expressed as a percentage change from present conditions [from Giorgi *et al.*, 1994].

vertical and horizontal dimensions in this kind of mesoscale model application because the model integration times are very long, and reducing the resolution allows the problem to be computationally tractable. The climatic response to the carbon dioxide doubling was determined for a number of different geographic areas in the United States, but results will be shown here only for the northwestern United States. Figure 5(a) shows the present seasonal variability in the precipitation for this region, based on observations, the CCM and MM4. The variability is displayed in terms of the monthly departures from the annual average for each source of data. In general, the MM4 performance is superior to that of the CCM in replicating the observed seasonal variability. Figure 5(b) shows the change in the monthly precipitation over this region as a result of a doubling in the carbon dioxide concentration, as predicted by MM4 and the CCM. Even though it is impossible to verify the accuracy of this result, the higher-resolution mesoscale model clearly shows a different seasonal variation in the precipitation amounts.

4. Conclusions

This paper has reviewed the various ways in which mesoscale atmospheric models have been and will continue to be used for environmental applications. One of the major factors that have limited the degree to which these models have been used in the past is the cost associated with their development and the computational cost of using them. Model development is no longer an issue because there are relatively mature mesoscale models that are in the public domain and are therefore available at little or no cost. In addition, relatively high-resolution versions of these models can be run on workstations that are becoming more inexpensive each year. Thus, it is anticipated that such models will find increasing use in many countries for addressing environmental problems and for use in operational weather prediction on small scales. For example, coarse-resolution interseason predictions are presently being produced by global models for El Niño–related variations in precipitation and temperature. However, mesoscale models are now sufficiently mature that, if they were forced by the coarse-resolution global forecasts, they could conceivably be capable of predicting the small-scale weather anomalies associated with El Niño that have the most direct impact on our environment and economy.

There are many factors that are causing increased pressures on our atmospheric environment, and its related physical and biological systems. In our efforts to understand better the dynamical interactions within and among the various systems that compose our environment, mesoscale atmospheric models have played and will continue to play a key role. This improved understanding of environmental sensitivity to natural and human factors will allow us to manage resources better and minimize our impact on natural systems.

REFERENCES

Dudhia, J., A nonhydrostatic version of the Penn State/NCAR Mesoscale Model: Validation tests and simulation of an Atlantic cyclone and cold front, *Mon. Wea. Rev.*, **121**, 1493–1513, 1993.

Giorgi, F., Simulation of regional climate using a limited area model nested in a general circulation model, *J. Climate*, **3**, 941–963, 1990.

Giorgi, F., C. S. Brodeur, and Gary T. Bates, Regional climate change scenarios over the United States produced with a nested regional climate model, *J. Climate*, **7**, 375–399, 1994.

Leung, L. R., M. S. Wigmosta, S. J. Ghan, D. J. Epstein, and L. W. Vail, Application of a subgrid orographic precipitation/surface hydrology scheme to a mountain watershed, *J. Geophys. Res.*, **101**, 12803–12817, 1996.

Pleim, J. E., J. S. Chang, and K. Zhang, A nested grid atmospheric chemistry model, *J. Geophys. Res*, **96**, 3065–3084, 1991.

Seaman, N. L. , D. R. Stauffer, and A. M. Lario-Gibbs, A multi-scale four-dimensional data assimilation system applied in the San Juaquin Valley during SARMAP. Part I: Modeling design and basic performance characteristics, *J. Appl. Meteor.*, **34**, 1739–1761, 1995.

Stauffer, D. R., and N. L. Seaman, Use of four-dimensional data assimilation in a limited-area mesoscale model. Part I. Experiments with synoptic scale data, *Mon. Wea. Rev.*, **118**, 1250–1277, 1990.

Stauffer, D. R., and N. L. Seaman, Multiscale four dimensional data assimilation, *J. Appl. Meteor.*, **33**, 416–434, 1994.

Stauffer, D. R., N. L. Seaman, and F. S. Binkowski, Use of four-dimensional data assimilation in a limited-area mesoscale model. Part II. Effects of data assimilation within the planetary boundary layer, *Mon. Wea. Rev.*, **119**, 734–754, 1991.

Stauffer, D. R., N. L. Seaman, T. T. Warner, and A. M. Lario, Application of an atmospheric simulation model to diagnose air pollution transport in the Grand Canyon region of Arizona, *Chem. Eng. Comm.*, **121**, 9–25, 1993.

Warner, T. T., D. F. Kibler, and R. L. Steinhart, Testing of a coupled meteorological-hydrological forecast model for the Susquehanna River Basin in Pennsylvania, *J. Appl. Meteor.*, **30**, 1521–1533, 1991.

Warner, T. T., Y-H. Kuo, J. D. Doyle, J. Dudhia, D. R. Stauffer, and N. L. Seaman, Nonhydrostatic, mesobeta-scale real-data simulations with the Penn State University/National Center for Atmospheric Research mesoscale model, *Meteor. Atmos. Phys.*, **49**, 209–227, 1992.

An Integrated Air Pollution Modeling System: Application to the Los Angeles Basin

By Rong Lu and Richard P. Turco

Department of Atmospheric Sciences, University of California, Los Angeles, Los Angeles, CA 90024-1565, USA

An air pollution modeling system, the Surface Meteorology and Ozone Generation (SMOG) model, has been developed for urban and regional air quality studies. The model, which couples mesoscale dynamics with air quality, has four major components: a meteorological dynamic model, a tracer transport code, a photochemistry and aerosol microphysics model, and a radiative transfer code. The meteorological model solves fluid dynamic and thermodynamic equations over complex terrain and incorporates physical processes such as turbulent diffusion, water vapor condensation and precipitation, solar and infrared radiative transfer, and ground surface processes. The tracer transport code computes the dispersion of gases and aerosols in the atmosphere, including emissions, and dry and wet deposition. The chemistry/aerosol module comprehensively treats coupled gas-phase photochemistry and aerosol microphysics and chemistry. Aerosol processes include nucleation, coagulation, condensational growth, evaporation, sedimentation, chemical equilibrium and aqueous chemistry.

The modeling system has been applied to Southern California. Comparisons of model simulation results with data collected during the Southern California Air Quality Study (SCAQS) have demonstrated that the modeling system is capable of reproducing three-dimensional wind fields, tracer transport features, and temporal and spatial distributions of pollutants in a basin with complex terrain. Complex three-dimensional distributions of ozone, especially ozone layers aloft, are commonly observed in the Los Angeles basin. Detailed three-dimensional distributions of ozone in the Los Angeles basin are simulated and analyzed. Photochemically aged pollutants trapped in the temperature inversion play an important role in pollution recirculation in the basin.

1. Introduction

On urban and regional scales, air pollution affects human health, damages vegetation, causes deterioration of materials, reduces solar radiation and visibility, and interferes with economic well-being. Air pollution is created by the emissions of pollutant precursors from anthropogenic and natural sources. Subsequently, a complex series of interactions occur during its evolution involving meteorological, physical, chemical, microphysical, and biological processes. In order to control air pollution effectively, it is necessary to understand in detail the physical and chemical processes that govern the formation, dispersion, transformation, and impacts of air pollution. Pollutant dispersion and transformation in the atmosphere are the key links between pollutant sources and the impacts they have at receptors.

Emissions, meteorological conditions, and chemical and physical transformations are the principal factors that determine concentrations of pollutants in ambient air and therefore the impacts on human beings and the environment. Severe air pollution problems on urban and regional scales are usually associated with heavy emissions and unfavorable meteorological conditions. Over the Southern California coastal regions, for example, urban traffic and industry are the primary sources of emissions. In the warm season, light winds, clear skies and a strong elevated temperature inversion are the normal synoptic scale weather in the region. The temperature inversion acts as a lid that restricts convective mixing, hence reducing the ventilation of air pollutants into the free troposphere. The high mountains that bound the Los Angeles basin act as physical barriers to the transport of pollutants farther inland. The elevated inversion layers and restricted advection are a prerequisite for the accumulation of pollutant precursors in the Los Angeles basin. Irradiated by sunlight, organic vapors and oxides of nitrogen undergo a complex series of reactions to produce secondary pollutants, including ozone, organic nitrates, oxidized organic compounds, and photochemical aerosols. The smoggiest episodes usually occur after several continuous stagnant clear-sky days.

A comprehensive air pollution modeling system should be capable of simulating the evolution of meteorological state and pollutant dispersion and transformations under a wide range of conditions. Meteorological dynamics and processes govern the motion and turbulence of the atmosphere. Driven by winds and turbulence, pollutants released into the atmosphere are dispersed and removed. Simultaneously, chemical and physical transformations take place as the airborne pollutants disperse. Detailed treatments of the coupled meteorology, tracer transport, chemistry and aerosol processes are ultimately necessary to simulate and predict concentrations and distributions of pollutants, especially secondary pollutants, in a region such as Southern California.

The Surface Meteorology and Ozone Generation (SMOG) modeling system, which couples meteorological dynamics, tracer transport and dispersion, and chemistry and aerosol processes, is developed for air quality studies on urban and regional scales [Lu, 1994; Lu et al., 1997a]. With treatments of these atmospheric processes, the coupled system has been successfully used to study photochemical smog in the Los Angeles basin.

2. The SMOG Modeling System

The SMOG integrated air pollution modeling system couples a meteorological dynamic model, a tracer dispersion model, treatments of atmospheric chemistry and aerosol microphysics, and a radiative transfer code [Lu, 1994; Lu et al., 1997a]. The dynamic and thermodynamic structures of atmosphere, e.g., wind and temperature fields, are predicted with the meteorological dynamic model. Driven by atmospheric winds and turbulence, the dispersion behavior of pollutants released into the atmosphere is modeled with the tracer dispersion model. The chemistry and aerosol microphysics model computes the transformations of pollutants as they simultaneously disperse in the atmosphere. The radiative transfer code calculates solar and infrared heating rates for the dynamic model and photodissociation rates for photochemistry computations.

The mesoscale meteorological model is a three-dimensional hydrostatic primitive equation code which solves fluid dynamic and thermodynamic equations over complex terrain [Lu, 1994]. The model incorporates atmospheric physical processes such as turbulent diffusion, water vapor condensation and precipitation, solar and infrared radiation transfer, and parameterization of ground surface processes. The meteorological model has been successfully used for regional mesoscale rainfall forecast and studies of atmospheric boundary layer dynamics and air pollutant transport [Lu and Turco, 1994, 1995].

The tracer transport model originated from NASA Ames Tracer Transport Code [Toon et al., 1988], which has been used in many regional and global simulations. Several numerical techniques have been applied in the transport code, including the time splitting algorithm, a finite element method for horizontal transport and a finite difference scheme for vertical advection and diffusion. These techniques provide efficient and accurate solutions for transport calculations with minimized computer memory demand.

The gas-phase chemistry uses a modified version of carbon bond extended mechanism (CBM-EX). The modified mechanism consists of 100 species and 223 kinetic and photo-chemical reactions. Concentrations of 95 active species are predicted at each grid cell. The "stiff" equations of chemical kinetics are solved by using the sparse-matrix vectorized Gear code (SMVGEAR) [Jacobson and Turco, 1994]. The Gear solver is a multistep, variable order numerical scheme, which provides robust and accurate solutions. SMVGEAR speeds up the original Gear code dramatically and becomes a fast and accurate solver for problems with stiff first-order ordinary differential equations. Aerosol processes include nucleation, coagulation, condensational growth and evaporation, sedimentation, chemical equilibrium and aqueous chemistry.

The efficient two-stream approximation radiative transfer model developed by Toon *et al.* [1989] for vertically inhomogeneous multiple scattering atmospheres is employed to calculate solar and infrared heating rates and photodissociation rates. The accuracy of the algorithm is usually better than 90% for photodissociation rate and heating rate calculations.

The modeling capabilities are greatly enhanced as these models are coupled together as an integrated air pollution modeling system. The internal coupling ensures consistency between submodels and eliminates some errors produced by data manipulation. The SMOG modeling system is a powerful tool for studying urban and regional air pollution problems. The model is robust and can be applied to most regions of the world if topography and emission data are available. In this study, the integrated system is applied to the Los Angeles basin. The ability of the modeling system to simulate pollutant dispersion over complex terrain can be seen in the simulation results.

3. Simulations for SCAQS, 1987

The Southern California Air Quality Study (SCAQS) was conducted during the summer and fall of 1987 in the California South Coast Air Basin (SoCAB); it is to date the largest air quality study carried on in Southern California. During the SCAQS summer sampling days, 36 surface sites reported hourly concentrations of O_3, NO, NO_x, CO, and SO_2, in addition to 24 hour averaged measurements of PM_{10}, particulate sulfates and nitrates. Surface meteorological data were collected at the SCAQS sampling sites and other locations throughout the basin. Upper air measurements at six rawinsonde and two airsonde sites were included in the summer study. Aircraft provided measurements of O_3, NO_x, SO_2 in the upper air, as well as images of aerosol back scattering recorded by lidar. A series of atmospheric tracer field experiments were conducted during the SCAQS period. A fixed surface sampling network was used to collect air samples and analyze the tracer concentration levels.

On August 27 and 28, 1987, the predominant pressure systems near the California coast region were the eastern Pacific high over ocean and a thermally induced low pressure stretching from northwestern Mexico to northern California. The intensifying thermal low increased onshore pressure gradient and strengthened onshore flow during the 2 days. Complex changes in surface winds occurred in the Los Angeles basin. On August 27, the flows were primarily westerly from the western coast to the eastern basin, except near Orange County, where southwesterly was found. On August 28, the southwesterly winds extended northward from Orange County to the San Gabriel Valley. Westerly flows were found near the western coast and in the eastern basin. Strong temperature inversion occurred during the period. The lifting of temperature inversion on August 29 improved air quality in the basin.

We simulated the high ozone episode during SCAQS between August 26 and 30, 1987

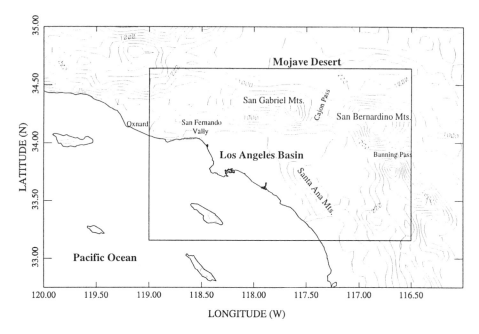

FIGURE 1. The model computational domain and topography. The mapped area is used for model calculation. The inner perimeter indicates the domain for air quality simulation. Heavy solid lines represent the coastline. Contour lines indicate terrain elevations. Contour intervals are 200 m.

[Lu *et al.*, 1997b]. The model simulation results are compared with observed data, i.e., surface air quality data, surface meteorological data, and upper air meteorological data, collected from a variety of sources including routine and special measurements during the period. In addition, on August 28, 1987, tracers were released at Southern California Edison's Alamitos Generating Station. The tracer experiment data collected from 40 sampling stations are compared with model results in this study.

The model computational domain for meteorology covers Southern California (Figure 1). It consists of 85×55 horizontal grid cells with grid spacing of 0.05° longitude (4.6 km) and 0.045° latitude (5 km). The domain for air quality simulation, which contains 51×34 horizontal grid cells covering the Los Angeles basin and adjacent areas, lies within the meteorology computational domain. Twenty nonuniform vertical layers with higher resolution in the lower troposphere are used in the simulation. The meteorological model makes a 24-hour prediction each day, starting from 0400 PST. The upper air soundings around 0400 PST are used to analyze the initial conditions for meteorology. The air quality simulation starts at 0400 PST August 26, 1987, and continues for 3 days until 0400 PST August 29. During the 3 day simulation the meteorological model is reinitialized at 0400 PST August 27 and 28. The current simulations include gaseous species and chemistry; aerosol microphysics is not considered.

Figure 2 shows the plot of observed and predicted surface wind vectors for 1600 PST August 28, 1987. The sea breezes and mountain flows were fully developed in the midafternoon. The predicted winds in the afternoon are in agreement with the observed winds in the basin. The westerlies in both the western coast and eastern basin are simulated. The southwesterlies extend northward from Orange County. The southeasterlies in the San Fernando Valley, which converge with the westerlies from Oxnard plain,

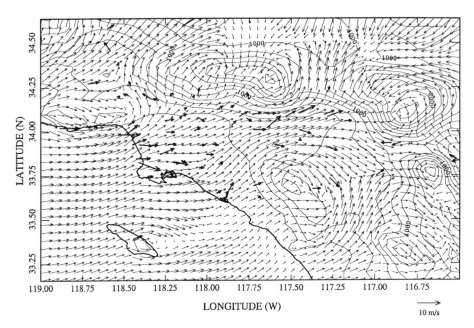

FIGURE 2. Comparisons of predicted and observed surface wind vectors at 1600 PST August 28, 1987. The thin wind vector fields are model prediction; the thick wind vectors are observations. Contour lines indicate terrain elevation.

are reproduced in the model simulation. Both observation and model prediction show that air in the basin moved up the southern slope of the San Gabriel Mountains. The easterlies across the Banning Pass are replicated in the simulation. The model simulation shows southerly winds across the Cajon Pass; however, a weak northerly was observed over mountains near the pass. The predicted vertical profiles of temperature and winds are generally consistent with upper air soundings (not shown), indicating that the model simulates well three-dimensional wind fields and boundary layer variations in the basin. Statistical comparisons demonstrate the model's skill in predicting the surface winds and temperature.

The integrated modeling system has been used to simulate the tracer experiment conducted on August 28, 1987. The tracers are released at the model grid box containing the Alamitos Generating Station (118.105°W, 33.781°E). The predicted tracer concentrations in the surface layer are compared with the observed surface concentrations. Figure 3 shows the predicted and observed PP-2 surface concentrations for 1600 PST August 28, 1987. The measurement data indicate that the tracer remained near the source and transported slowly toward the northeast before 1200 PST. The tracer was transported eastward into the eastern basin at 1400 PST. The northeastward and then eastward transport of the tracer is simulated well by the model. However, the head of the tracer plume is transported slightly more slowly than that shown in the measurements. The tracer transported much faster in the afternoon and reached the east end of the basin by 1600 PST. The model replicates the fast eastward transport in the midafternoon. The missing of the observed tracer peak in the eastern basin in the simulation may result from the time lag of transport at noontime. The predicted tracer peak moves to the San Bernardino region in the late afternoon. Both the observation and simulation show that the tracer remains in the eastern end of the basin during the night. During the evening,

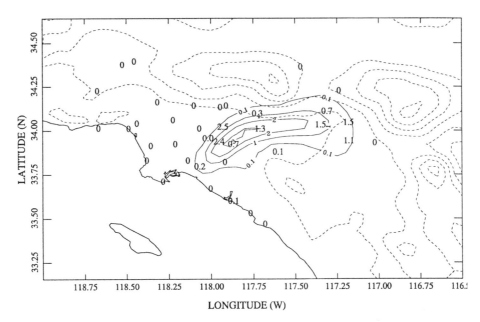

FIGURE 3. The predicted and observed tracer PP-2 surface concentrations at 1600 PST August 28, 1987. The solid contours are the prediction, and contour levels are 0.1, 1, 2, 3 pptv. The observed concentrations are indicated by numbers (pptv) at station locations. Dashed contour lines indicate terrain elevation.

the trace was transported through the Cajon Pass, and the transport event is replicated in the simulation. The consistency in the tracer concentrations and transport patterns between the prediction and observation indicates that the winds are predicted with good accuracy. Moreover, the agreement shows that the transport and dispersion processes are properly treated in the model. Elevated tracer layers are formed in the simulation. However, upper air measurements are not available for the tracer experiment.

The modeled concentrations of ozone and its key precursors are compared with observed data for August 27–28, 1987. The observed peak ozone concentrations in the basin occurred in the midafternoon. Figure 4 shows the spatial distributions of ozone concentrations for 1500 PST August 28. The observed 1-h averaged ozone concentrations at 1400–1500 PST are also plotted for comparison. The predicted surface ozone distributions were in close agreement with observations. An ozone concentration peak as high as 29 pphm was observed in the San Gabriel Valley at the Glendora station (GLEN). High ozone concentrations were measured along the southern slope of the San Gabriel Mountains and in the eastern basin. The model predicts the ozone peaks in both the San Gabriel Valley and the eastern basin. The prediction errors for peak values are within 15%. An ozone peak is predicted in the San Fernando Valley, but it is not shown in the observed data.

Diurnal variations of ozone concentrations can be depicted using time series plots. Figure 5 presents time series plots of predicted and observed ozone concentrations at the Los Angeles station (CELA, 118.23°W, 34.07°E), which represents the western Los Angeles basin, and at Glendora (GLEN, 117.85°W, 34.14°E) station in the San Gabriel Valley. At most stations, the concentrations and phases of ozone prediction are in close agreement with observations. The highest ozone maximum observed during the intensive

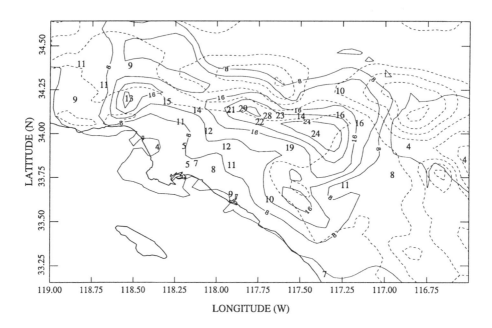

FIGURE 4. Comparisons of surface ozone concentrations for August 28, 1987. The solid contours are predicted ozone concentrations (pphmv) at 1500 PST. Contour intervals are 4 pphmv. The plotted numbers are observed 1-hour averaged ozone concentrations (pphmv) for 1400–1500 PST. The dashed contours are topographic elevation.

measurement days was 0.29 ppmv at GLEN. The model underpredicts the peak value by only about 0.04 ppmv.

Elevated ozone layers are created in these simulations [Lu and Turco, 1996]. For example, in the early afternoon August 27, 1987, the ozone distribution in the vertical cross section from Santa Monica Bay to the San Gabriel Mountains shows that an ozone layer aloft covers the western Los Angeles basin (not shown). The ventilation effects of the San Gabriel Mountains pump ozone into free troposphere. The coupled sea breeze and mountain flow circulation near the Santa Ana Mountains create ozone layers over the coast (not shown). In the evening, distinct ozone layers can be found in the vertical cross section across the basin (Figure 6). One is in the inversion layer, and another in the free troposphere. Vertical circulations associated with sea breezes and mountain flows are the principal mechanisms that lead to the formation of pollution layers in the basin. High mountains surrounding the basin act as barriers that confine pollutants in the basin. In the evening, as the boundary layer stabilizes over the coastal basin, a large portion of aged pollutants are injected into the temperature inversion. High levels of secondary pollutants, such as ozone, peroxiacyl nitrate (PAN), nitric acid and organic nitrates, appear in the elevated layers. Cut off from surface emissions, pollutants in the layers aloft can undergo fairly complete oxidation reactions. The aged pollutants can be mixed down the next day to enhance the surface concentration. Thus, the elevated pollution layers provide a mechanism for pollutant recirculation in the basin. The effect of downward mixing is responsible for the rapid increase of the surface ozone concentrations in the morning hours in the eastern basin, where emissions are relatively small. The

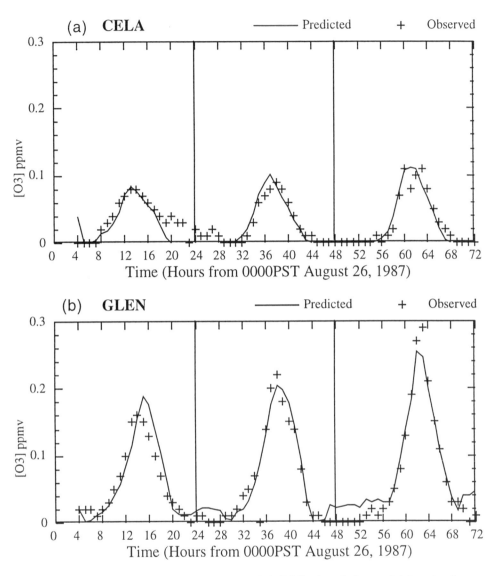

FIGURE 5. Time series plots of model predicted (solid line) and observed (plus signs) ozone concentrations: (a) Los Angeles (CELA), (b) Glendora (GLEN).

impact of photochemically aged pollution from the previous day on the surface ozone concentrations is estimated to be as high as 8 pphmv in the eastern basin.

4. Summary and Conclusions

An air pollution modeling system, which couples a meteorological model, a tracer transport model, detailed treatments of chemistry and aerosol microphysics, and a radiative transfer code, has been developed for air quality studies. The system is used to simulate meteorological conditions, dispersion of passive tracers, and pollutant distributions for August 27–28, 1987, SCAQS intensive days. The overall performance of the

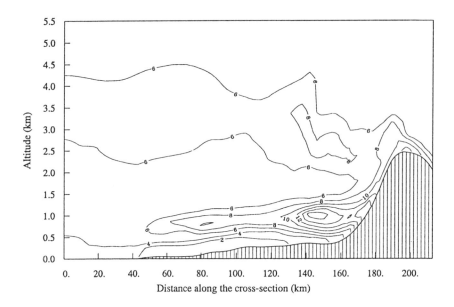

FIGURE 6. Simulated ozone concentrations (contours in pphmv) in the vertical cross section across the Los Angeles Basin from Santa Monica Bay to the San Bernardino Mountains for 2000 PST August 27, 1987. Contour intervals are 2 pphmv. The topography is indicated by striping.

modeling system is demonstrated by comparing the simulation results with observational data collected during the SCAQS period.

Three-dimensional winds and boundary layer variations over complex topography in Southern California are predicted well with the model. The predicted winds and temperature near the surface are consistent with observations. Vertical profiles of temperature and winds show the agreement between simulations and observations in the structure and evolution of the atmospheric boundary layer in the Los Angeles Basin. The concentrations and distribution patterns of passive tracers released in August 28, 1987, are replicated in the model simulations. The consistency between simulations and observations of tracer concentrations indicates that the meteorological conditions are predicted well and the tracer dispersion is treated reasonably in the modeling system. The predictions of ozone and other pollutants have been compared with SCAQS observations for the intensive measurement days. The spatial distributions and peak locations of ozone concentrations are consistent with observed data. Time series plots show that the ozone predictions are in good agreement with the observations in diurnal variations, peak concentrations, and timing of peak occurrence. The normalized gross errors for surface ozone concentrations during the daytime are 25% to 30% for the August 27–28 SCAQS episode.

Ozone layers aloft are commonly observed in the Los Angeles Basin and are created in our simulations. Photochemically aged pollutants trapped in the inversion can be mixed down the next day to enhance the surface concentrations, thereby contributing to the pollutant recirculation in the basin.

The overall agreements between predictions and observations show that the SMOG modeling system is able to reproduce the main features of mesoscale meteorology, pollutant dispersion and transformations in the atmosphere. The modeling system is capable of handling complicated situations such as the photochemical smog over complex topography in the Los Angeles Basin. The integrated modeling system is shown to be a powerful tool for studying coupled dynamical, chemical and microphysical processes on urban and regional scales.

REFERENCES

Jacobson, M. Z., and Turco R. P., SMVGEAR: A sparse-matrix, vectorized Gear code for atmospheric models, *Atmos. Environ.*, **28A**, 273–284, 1994.

Lu, R., Development of an integrated air pollution modeling system and simulations of ozone distributions over the Los Angeles basin, Ph.D. dissertation, Department of Atmospheric Sciences, University of California, Los Angeles, CA, 1994.

Lu, R., and R. P. Turco, Air pollution transport in a coastal environment. Part I. two-dimensional simulation of sea-breeze and mountain effects, *J. Atmos. Sci.*, **51**, 2285–2308, 1994.

Lu, R., and R. P. Turco, Air pollution transport in a coastal environment. Part II. Three-dimensional simulations over the Los Angeles basin, *Atmos. Environ.*, **29B**, 1499–1518, 1995.

Lu, R., and R. P. Turco, Ozone distributions over the Los Angeles basin: Three-dimensional simulations with the SMOG model, *Atmos. Environ.*, **30**, 4155–4176, 1996.

Lu, R., R. P. Turco, and M. Jacobson, An integrated air pollution modeling system for urban and regional scales. Part I. Model structure and formulation, *J. Geophys. Res.*, in press, 1997a.

Lu, R., R. P. Turco, and M. Jacobson, An integrated air pollution modeling system for urban and regional scales. Part II. Evaluation of model performance using SCAQS data, *J. Geophys. Res.*, in press, 1997b.

Toon, O. B., C. P. McKay, and T. P. Ackerman, Rapid calculation of radiative heating rates and photodissociation rates in inhomogeneous multiple scattering atmospheres, *J. Geophys. Res.*, **94**, 16287–16301, 1989.

Toon, O. B., R. P. Turco, D. Westphal, R. Malone, and M. S. Liu, A multidimensional model for aerosols: Description of computational analogs, *J. Atmos. Sci.*, **45**, 2123–2143, 1988.

An *h*-Adapting Finite Element Model for Atmospheric Transport of Pollutants

By Darrell W. Pepper and David B. Carrington

Department of Mechanical Engineering, University of Nevada, Las Vegas
Las Vegas, NV 89154-4027

An *h*-adaptive finite element model has been developed to simulate atmospheric flow and species transport. A diagnostic model is first used to create a mass consistent windfield based on meteorological data; this windfield is then used to initialize the prognostic model for calculating the time-dependent equations of atmospheric motion and species transport. The model employs a nonhydrostatic assumption for pressure, Petrov–Galerkin weighting for the advection terms, mass lumping, and reduced integration. The model runs on an SGI workstation and a CRAY Y-MP; a PC WINDOWS version is under development.

1. Introduction

Successful prediction of hazardous material trajectories following an atmospheric release resides predominantly with accuracy in forecasting and calculating advection terms [Pielke, 1984]. Efforts have been made to examine in closer detail the utilization of efficient numerical methods which automatically maintain dispersion error control [Mitchell and Herbst, 1986; Yu and Heinrich, 1986]. One particularly attractive method which can accurately resolve the highly irregular nature of atmospheric flow and pollutant transport over complex terrain is an adapting finite element method (FEM). The use of adaptive, unstructured grids is of recent origin and has been shown to be extremely powerful in precisely capturing sharp gradients and concentration fronts [Shapiro and Murman, 1988]. Application of this technique has not yet been fully exploited within the geophysical community.

In this study, a modified finite element method is used to solve the multidimensional equations of motion for atmospheric flow and pollutant transport. The numerical method is based on a hybrid finite element model [Pepper and Brueckner, 1992] used to simulate windfields over irregular terrain, and a mesh adaptation procedure [Pepper and Stephenson, 1995; Pepper and Emery, 1994] for contaminant dispersion. Because the method utilizes mesh refinement, steep gradients are accurately determined, even though one needs to begin with only coarse grid.

2. Governing Equations

The governing equations for multidimensional, time-dependent windfield and species transport can be written in vector form as

$$\frac{\partial V}{\partial t} + V \cdot \nabla V = -\frac{1}{\rho}\nabla p + \nabla \cdot (K \cdot \nabla V) + B$$

$$\frac{\partial \theta}{\partial t} + V \cdot \nabla \theta = \nabla \cdot (K_\theta \cdot \nabla \theta) + Q$$

$$\frac{\partial C}{\partial t} + V \cdot \nabla C = \nabla \cdot (D \cdot \nabla C) + S$$

where C is species concentration, ρ is density, K is the exchange coefficient tensor ($K \equiv K_h \cdot i + K_h \cdot j + K_z \cdot k$, where K_h is horizontal and K_z is vertical diffusion), K_θ is thermal diffusivity, V is the vector velocity field, θ is the potential temperature, D is the contaminant dispersion tensor, B represents velocity body force terms, Q is the source term for temperature, and S is the source/sink term for species concentration.

3. The Finite Element Method

The standard weak formulation of the Galerkin weighted residual technique is employed to cast the governing equations into integral form. Isoparametric quadrilateral (2-D) or hexahedral (3-D) elements are used to discretize problem domains. Applying Green's theorem to the integral equations, the matrix equivalent forms of the resultant weak statements can be expressed as

$$[M_V]\left\{\dot{V}\right\} + ([K_V] + [A(V)])\{V\} = \{F_V\}$$

$$[M_\theta]\left\{\dot{\theta}\right\} + ([K_\theta] + [A(V)])\{\theta\} = \{F_\theta\}$$

$$[M_C]\left\{\dot{C}\right\} + ([D] + [A(V)])\{C\} = \{F_C\}$$

where the overdot notation refers to time differentiation; M_V, M_θ, M_C represent the time-dependent banded sparse mass matrices; K_V, K_θ, D are the diffusion-stiffness matrices; and $A(V)$ is the advection matrix. F_V, F_θ, and F_C are the right-hand-side vector loads containing source terms. The variables have been replaced by the trial approximations

$$\hat{V}(x, t) = \sum N_i(x) V_i(t)$$

$$\hat{\theta}(x, t) = \sum N_i(x) \theta_i(t)$$

$$\hat{C}(x, t) = \sum N_i(x) C_i(t)$$

where x is the space vector and N_i is the shape (basis) function. Details on the formulation of the matrix coefficients are discussed in Pepper and Brueckner [1992].

The modeling of exchange coefficients is quite varied. Numerous examples are discussed in detail in the literature [Pielke, 1984]. In this study, a simple gradient diffusion approach based on relations developed by Smagorinsky *et al.* [1965] for horizontal diffusion is employed. Vertical diffusion is determined from K-theory and similarity assumptions [Blackadar, 1979; Businger *et al.*, 1971; O'Brien, 1970].

While finite element models for fluid flow and transport have been in existence for some time, most such models follow conventional approaches which can lead to large storage demands and long compute times. In this study, modifications are used to replace the conventional FEM global assembly operations with local formulations [Pepper and Brueckner, 1992; Pepper and Emery, 1994]. Performing the operations on a local level significantly reduces storage while enhancing overall solution speed. In cases where the element distortion is minimal, reduced integration is used (e.g., $2 \times 2 \times 2$ Gauss quadrature points are reduced to 1 point quadrature evaluated at the element centroid in the 8 node trilinear hexahedral element).

A Petrov–Galerkin scheme is used to weight the advection terms (only). The weight is set equal to the shape function plus a perturbation term based on velocity, i.e.,

$$W_i = N_i + \frac{\alpha h_e}{2V}(V \cdot \nabla N_i)$$

where h_e is the size of the element shown in Figure 1 and $\alpha = \coth \beta/2 - 2/\beta$ with

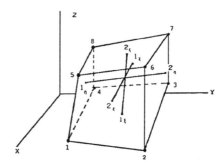

FIGURE 1. Hexahedral element.

$\beta = h_e |V|/2K_e$ [Yu and Heinrich, 1986]. K_e is an effective diffusion in the direction of the local velocity vector and is calculated using the components of K (also hold for D) as

$$K_e = \frac{V^T \cdot K \cdot V}{V^2}$$

where superscript T denotes the transpose of V. This weighting introduces selective artificial diffusion into the numerical scheme which acts along the local streamline [Yu and Heinrich, 1986].

For nonhydrostatic calculations, the pressure is obtained from the discrete momentum equations and a time-differenced version of the continuity equation. A potential function based on the velocity vector is subsequently solved from

$$[K]\{\Phi\} = C^T \{V\}^{n+1}$$

where Φ is the potential function. Once Φ is calculated, the velocities are updated according to

$$\{V\}^{n+1} = \{V\}^n + [M]^{-1} C^T \{\Phi\}$$

The remaining unknown variables are calculated, and the time step is advanced. The pressure is calculated from the discretized Poisson equation

$$[K]\{p\} = C^T [M]^{-1} [\{F_v\} - ([K] + [A(V)])\{V\}]$$

where $\{F_v\}$ is the body force term associated with the momentum equations. These pressure values are then used in the calculation for velocity.

Mass lumping is used to permit simple time integration and eliminate the need for total sparse-matrix inversion. An explicit second-order Runge–Kutta method is employed to advance the discretized equations in time. The Courant limits are calculated over each element, and the time step adjusted to the minimum value within the computational domain.

4. Objective Analysis

In order to obtain realistic windfield estimates quickly, a mass consistent, objective analysis approach is used [Smagorinsky *et al.*, 1965; Goodin *et al.*, 1980; Pepper and Brueckner, 1992]. Using interpolation of sparse measurements onto a mesh, the wind vectors at each node point within the computational domain are adjusted to minimize global divergence. Such techniques are simple to employ and have been shown to produce reliable diagnostic windfields suitable for atmospheric transport simulation. The surface windfield is constructed from measured data by interpolation to the initial mesh using inverse distance-squared weighting. A fixed radius of influence which indicates the distance beyond which the influence of a station's value is no longer felt is specified. A simple first-order inverse weighting scheme is used to produce a smooth upper level windfield and mixing depth. Once the surface level flow field has been established and the upper level wind data interpolated to the grid, the divergence in the total flow field is reduced. Horizontal divergence is first calculated, then the vertical velocity computed from the horizontal divergence. A final refinement reduces the remaining divergence globally over the entire domain. While *h*-adaptation can be applied to the diagnostic solution, adaptation is generally begun during the prognostic calculations (since the flow field changes with time).

5. Mesh Adaptation

Adaptation is the process by which the computational mesh changes in response to an evolving solution. Grid adaptation increases the number of grid points in regions of high gradients and reduces the number of grid points where the flow is smooth. The FEM adaptive procedure implemented in this study follows the technique discussed in Pepper and Emery [1994].

The starting point of the adaptation procedure is a mesh coarse enough to allow rapid convergence, yet fine enough to allow transport details to appear. An initial solution is computed on the crude mesh. Refinement indicators are computed in terms of the solution on the initial mesh, and elements that need to be refined or unrefined are identified. After all the mesh changes have been made, the grid geometry is recalculated, the solution is interpolated onto the new grid, and the calculation procedure is begun again. The procedure is repeated until a "converged" mesh is obtained and the concentration residuals are less than 10^{-4}.

Figure 2 shows an example of *h*-adaptation for the 4 node bilinear quadrilateral element. Suppose element A is marked for refinement, as shown in Figure 2(a). Element A is divided into subelements I, II, III, and IV as shown. The nodes marked with circles are virtual nodes (those with *x* are not virtual nodes). If element III is marked for further refinement, a problem arises since one of its neighbors, element B, is at a lower level. Thus, element B must also be refined, as shown in Figure 2(b). Node I is no longer a virtual node, but node j remains a virtual node. Element B becomes divided into elements V, VI, VII, and VIII, as shown in Figure 2(c). Finally, suppose that the group of elements V, VI, VII, and VIII become marked for unrefinement. This group is not eligible until the group of elements IX, X, XI, and XII has been unrefined. Element VIII has neighbors X and XI which are at a higher level. Hence, elements IX, X, XI, and XII become replaced by element III.

6. Preliminary Results

Transient two-dimensional (2-D) dispersion is modeled using mesh adaptation associated with the transport of contaminant from an elevated ($z = 50$ m) constant source ($S = 100$ g m^{-3}s^{-1}) upwind of two ridges. A constant wind ($u = 5$ m/s) is assumed to flow into the problem domain from the left boundary. The vertical and lateral dimensions are 600 m and 2000 m, respectively. The height of the small ridge is 100 m while the larger ridge is 300 m.

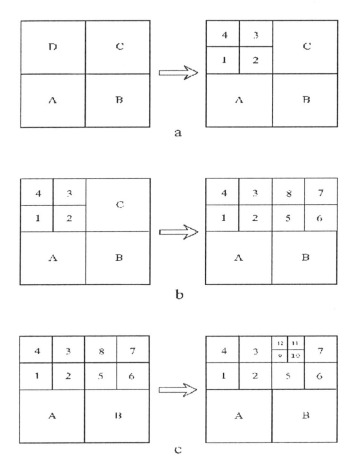

FIGURE 2. Element refinement.

Figure 3 shows the initial mesh, adapted mesh, velocity vectors, and isopleths with mesh adaptation. The original mesh consisted of 220 elements. The adapted mesh ultimately contains 884 elements. Several adaptations occurred during the transient solution of the contaminant transport. Observation of the adapted mesh shows those regions where the concentration gradient is high. A much finer mesh is normally required to obtain comparable accuracy using conventional finite element methodology; a revised mesh would then be regenerated to optimize element location and number – a rather time-consuming and expensive process.

Figure 4 shows three-dimensional (3-D) finite element model simulations of atmospheric flow over an irregular surface. A 3-D diagnostic windfield is first generated by utilizing available tower and upper air data; this procedure produces a mass consistent windfield over the entire 3-D domain. A prognostic windfield is then generated by using the diagnostic windfield to initialize the calculations. These simulations were performed on an SGI workstation and required about 15 minutes solution time; running on the CRAY Y-MP (vectorized) is nearly 150 times faster.

An example of 3-D mesh adaptation is shown in Figure 5 using several refinement passes for a region consisting of 8 noded trilinear hexahedral elements. Once adaptation begins, the ability to visualize and examine graphical displays of the 3-D results quickly

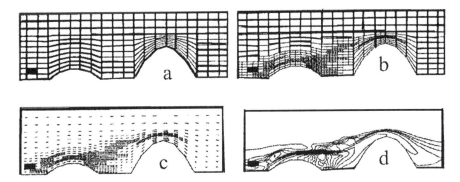

FIGURE 3. (a) Initial mesh, (b) adapted mesh, (c) velocities, (d) isopleths.

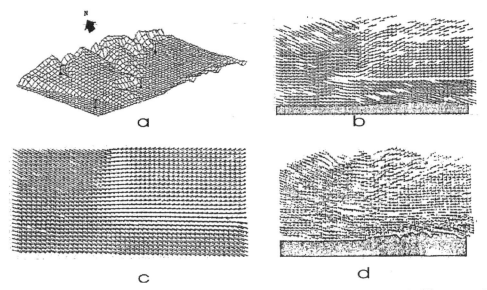

FIGURE 4. (a) 3-D terrain, (b) interpolated tower data, (c) diagnostic windfield, (d) prognostic windfield.

becomes complicated. FIELDVIEW, a commercial graphics package obtained from Intelligent Light, Inc., can be used to visualize the windfield and contaminant transport with mesh adaptation. FIELDVIEW allows one to display velocity vectors and contour lines on either structured or unstructured (FEM) meshes and generates streamlines and/or particle paths, isosurface plots, and movie animations of transient solutions utilizing an SGI workstation.

7. Conclusions

An adapting finite element algorithm has been developed for calculating windfields and contaminant transport within the atmosphere. The algorithm incorporates simple modifications to the basic Galerkin formulations to enhance speed and reduce storage; a Petrov–Galerkin weighting scheme is used for the advection terms. Mesh adaptation

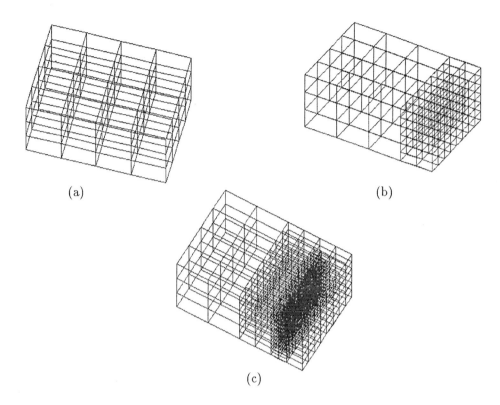

FIGURE 5. (a) Initial mesh, (b) level one adaptation, (c) level two adaptation.

is achieved by using interpolation-based commands and averaging to refine/unrefine the mesh. Objective analysis is used to create a mass consistent, initial windfield utilizing sparse tower data. This diagnostic solution is then used to the initialize the domain for the prognostic calculations. Preliminary simulations illustrate the ability of the finite element method with *h*-adaptation to model atmospheric transport accurately. The diagnostic and prognostic algorithms are quick, accurately resolve concentration fronts, and use less memory than more conventional approaches which must employ large numbers of nodes and globally fine meshes. These advantages are particularly significant when modeling three-dimensional transport.

Acknowledgments. We wish to thank Cray Research, Inc., and the Department of Energy, Nevada Operations Office, for their support during this project.

REFERENCES

Blackadar, A. K., High resolution models of the planetary boundary layer, *Adv. Environ. Sci. Eng.*, **I**, 50–85, 1979.

Businger, J. A., J. C. Wyngaard, Y. Izumi, and E. F. Bradley, Flux profile relationships in the atmospheric surface layer, *J. Atmos. Sci.*, **28**, 181–189, 1971.

Goodin, W. R., McGrae, and J. H. Seinfeld, An objective analysis technique for constructing three-dimensional urban scale wind fields, *J. Appl. Meteor.*, **19**, 98–108, 1980.

Long, P. L., and D. W. Pepper, An examination of some simple numerical schemes for calculating scalar advection, *J. Appl. Meteor.*, **20**, No. 2, 146–156, 1981.

Mitchell, A. R., and B. M. Herbst, *Adaptive Grids in Petrov-Galerkin Computations, Estimates and Adaptive Refinements in Finite Element Computations*, I. Babuska *et al.*, Eds., John Wiley & Sons, 315–324, 1986.

O'Brien, J. J., A note on the vertical structure of the eddy exchange coefficient in the planetary boundary layer, *J. Atmos. Sci.*, **27**, 1213–1215, 1970.

Pepper, D. W., and D. E. Stephenson, An adaptive finite element model for calculating subsurface transport of contaminant, *Ground Water*, **30**, 486–496, 1995.

Pepper, D. W., and F. P. Brueckner, A finite element model for calculating 3-D windfields over irregular terrain, *ASME FED*, **143**, 101–108, 1992.

Pepper, D. W., and A. F. Emery, Atmospheric transport prediction using an adaptive finite element method, Proceedings of the Fifth Annual International High Level Radioactive Waste Management Conference, 4, 1946–1952, 1994.

Pielke, R. A., *Mesoscale Meteorological Modeling*. Academic Press, 1984.

Shapiro, R. A., and E. Murman, Adaptive finite element methods for the Euler equations, AIAA Paper 88–0034, 1988.

Sherman, C. E., A mass-consistent model for wind fields over complex terrain, *J. Appl. Meteor.*, **17**, 312–319, 1978.

Smagorinsky, J., S. Manabe, and J. L. Holloway, Jr., Numerical results from a nine-level general circulation model of the atmosphere, *Mon. Wea. Rev.*, **93**, 727–798, 1965.

Yu, C. C., and J. C. Heinrich, Petrov–Galerkin methods for the time-dependent convective transport equation, *Int. J. Num. Meth. Eng.*, **23**, 883–901, 1986.

The Applicability of a Mesoscale Model in the Valley of Mexico during Extreme Air Pollution Episodes

By Ismael Pérez-García[1] and Everett C. Nickerson[2]

[1]Centro de Ciencias de la Atmósfera, UNAM, México

[2]Forecast Systems Laboratory, NOAA, Boulder, CO, USA

The long-term objective of this study is to develop a comprehensive meteorological model capable of making accurate predictions of local weather and pollution conditions in the Valley of Mexico. An important first step is to develop and demonstrate the capability of simulating the local wind circulations in the Valley and over the surrounding mountains. A mesoscale model which includes a detailed representation of the planetary boundary layer and which also includes parameterizations of soil, vegetation types, and the urban zone is presented. Preliminary results will focus primarily on airflow patterns in the Valley of Mexico. The model was used to study the atmospheric response to various types of surface (soil texture, vegetation cover, urban areas) over complex terrain.

1. Introduction

In winter, temperature inversions associated with air pollution over Mexico City have a major impact on the citizens. There is a need for both diagnostic and prognostic models of air circulation over the Valley of Mexico and the surrounding countryside in order to understand the relationship between local meteorological conditions and high-pollution events better and to provide improved forecasts of air quality over the city. Up to the present time, few studies of those winds have been carried out, because of the low density of observational stations and the short time interval of the existing data [Jáuregui, 1988].

The present study seeks to advance our understanding of the problem by presenting results of numerical simulations of the airflow over the Valley of Mexico using the meso-β scale (two- and three-dimensional) model SALSA [Nickerson et al., 1986; Mahfouf et al., 1987; Pinty et al., 1989], which treats the vegetative cover and the underlying soil. On the other hand, the meteorological conditions over an urban area are different from those over a rural area. In the urban area of Mexico City, which is the largest in the world, a region of high temperature known as an urban heat island is often found [Jáuregui, 1993], and a numerical model must include mechanisms for simulating such effects. The conditions at the lower boundary of the atmospheric surface layer over an urban area are very complex, and it is not possible to incorporate airflow and dispersion around individual buildings or vehicular traffic directly into a meso-β model. Such effects must be parameterized.

2. Temperature Inversions and Wind Circulation in the Valley of Mexico

Throughout the year, the Mexican high plateau is affected by distinct air masses which result in different meteorological conditions. By way of example, let us examine the temperature inversions over the Valley of Mexico during late winter and early spring: We take the daily soundings from the Mexico City airport at 12Z for the months of

TABLE 1. Average soundings from the Mexico City airport for February and March 1990–1994.

$\overline{P}$ (mb)	$\overline{T}$ (°C)	$\overline{Q}$ (g/kg)	$\overline{\theta}$ (K)	$\overline{U}$ (m/s)	$\overline{V}$ (m/s)
(a) February					
100	−77.32	0.00022	378.10	14.63	4.03
150	−65.67	0.00154	356.77	23.83	5.12
200	−53.82	0.00172	347.38	22.99	0.19
250	−43.95	0.00666	340.59	23.26	−0.42
300	−34.91	0.14512	336.05	19.18	1.82
400	−21.05	0.44159	327.54	16.24	2.36
500	−8.53	0.79081	322.58	11.02	1.87
700	9.08	4.52691	312.52	1.21	1.12
750	11.79	5.31548	309.37		
781	7.61	5.50108	301.32	−0.43	−0.08

$\overline{P}$ (mb)	$\overline{T}$ (°C)	$\overline{Q}$ (g/kg)	$\overline{\theta}$ (K)	$\overline{U}$ (m/s)	$\overline{V}$ (m/s)
(b) March					
100	−76.04	0.00052	380.57	12.92	1.27
150	−63.79	0.00252	359.99	16.01	1.39
200	−54.41	0.00395	346.45	20.70	0.80
250	−45.22	0.01217	338.71	20.60	6.23
300	−33.82	0.13465	337.59	17.81	2.33
400	−18.67	0.47390	330.64	14.24	1.21
500	−6.63	0.80667	324.89	9.10	1.55
700	7.60	4.74575	310.88	1.95	1.14
753	13.81	5.10837	311.20		
781	9.75	5.66259	303.61	−0.50	0.90

February and March for 4 years (1990–1993). We note that in February there existed 82 cases in which the difference in temperature ($\Delta T = T_g - T_i$) between the surface temperature, T_g, and the temperature at the top of the inversion fell into the interval $[-6, 0)°C = \{T \mid -6 \leq T < 0, \ T \text{ in degrees Celsius}\}$, of which 51 cases were in the range $\Delta T \in [-6, -3]°C$ and 31 in the range $\Delta T \in (-3, 0)°C$, where $\in$ is the usual set membership symbol.

The upper portion of Table 1 shows the average sounding for the case of $\Delta T \in [-6, -3]°C$, for which the average $\overline{\Delta T} \approx 4°C$, $\overline{T}_i = 11.7°C$, $\overline{T}_g = 7.6°C$; the average pressure at the top of the inversion was $\overline{P}_i \approx 750$ mb; and the average pressure at the ground $\overline{P}_g \approx 781$ mb, so that $\overline{\Delta P} = \overline{P}_g - \overline{P}_i = 31$ mb, which corresponds to an inversion height $\overline{H}_i \approx 300$ m.

In the four March months there were 83 cases in which $\Delta T \in [-6, 0)°C$, of which 40 were in the range $\Delta T \in (-3, 0)°C$ and 42 in the range $\Delta T \in [-6, -3]°C$. For the case in which $\overline{\Delta T} \in [-6, -3]°$ we see in the lower portion of Table 1 $\overline{T}_i = 13.8°C$ and $\overline{T}_g = 9.7°C$. Similar to that which was found for February here we have $\overline{\Delta T} \approx 4°C$ and $\overline{P}_g = 781$ mb, but $\overline{P}_i = 753$ mb, which corresponds to a $\overline{\Delta P} = 28$ mb, which is slightly shallower than for that which occurs in February.

In both soundings for the case in which $\Delta T \in [-6, -3]°C$ we observe a relatively dry layer (Q is the water vapor mixing ratio) around 500 mb and a very dry layer above 300 mb. Generally the wind direction at the ground is from the east; from 700 mb it is from the southwest and west, and the strongest winds begin at 500 mb and above (U, V

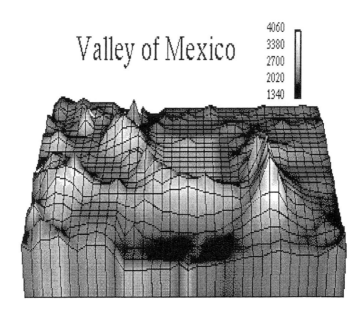

FIGURE 1. Perspective view of the 36 × 26 grid.

are the zonal and meridional components of the wind, θ the potential temperature; see Table 1).

Let us now examine climatologically the surface airflow over the Valley of Mexico, especially for the nested subregion shown in Figure 1, which generally covers Mexico City. In this preliminary work we use the hourly wind data from the Automatic Atmospheric Monitoring Network (RAMA; see SEDESOL [1992]) for the months of February and March (1990–1994), which consists of approximately nine stations that report hourly data. Figures 2 and 3 show the mean wind field. These wind fields were obtained using the objective analysis scheme of successive corrections of Tripoli and Krishnamurti [1975] or Pérez [1985], taking as a preliminary guess the analysis of the previous hour. The analysis mesh consisted of 8 × 9 ($i = 15, \ldots, 22, j = 11, \ldots, 19$) points with a separation between points of 5 min $\approx$ 11 km and is that which is shown nested in the mesh of 36 × 26 points (Figure 1).

In February the surface airflow from 0100 to 0700 LST is observed to consist of a cyclonic circulation (Figure 2(a) and Figure 2(b)) which is displaced from the east to the center of the analysis region during this period. A similar situation (Figure 2(e) and Figure 2(f)) for those morning hours is also observed for the month of March. The heat island in March during this period is located in the eastern part of the domain. These circulations are consistent with the nighttime drainage down the mountain slopes of cold air which flows toward the center of the Valley and later is reinforced by drainage from the less mountainous regions in the north and northeast. During the hours of 0800–1200 LST in February and March the predominant wind direction is toward the southwest, but at midday a zone of divergence is established slightly north of the city (Figure 2(c) and 2(g)), dramatically demonstrating the existence of a thermally driven mountain valley wind system, whose circulation cell includes ascent over the mountain slopes, and descending air over the center of the city, all of which are superimposed over the large-scale subsidence associated with an anticyclonic system. In February and

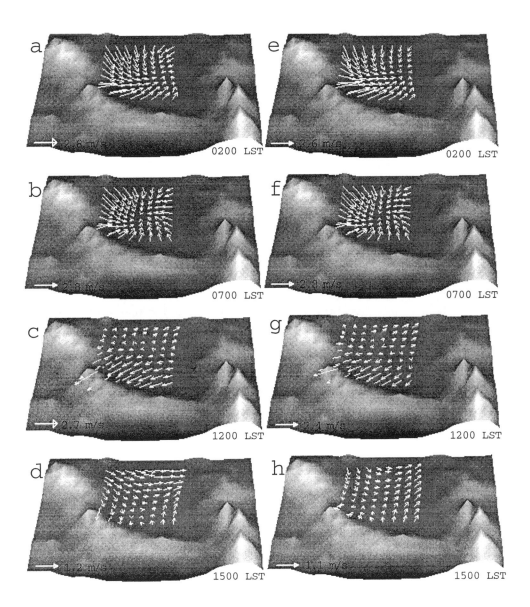

FIGURE 2. Surface wind in the Valley of Mexico for February (left) and March (right) 1990–1994.

March at 1500 LST the winds ascend the mountains of the southeast and east (Figure 2(d) and 2(h)). At 1800 LST they are directed toward the northeast, and in the last hours of the day (2100–2300 LST) a cyclonic circulation is established, at first close to the center of the eastern part of the analysis region and then, gathering strength, moving farther eastward. In March from 0900 to 1800 LST the warm zones of the city extend from the southeast to the northeast of the analysis mesh, with the warmest zone in the southeast. On the other hand, if we also analyze O_3 for the month of March (1990–1994), we observe that from 2000 to 0400 LST the most contaminated zone for this gas was the northwest; from 0600 to 0900 LST it was the north northeast; from 1000 to 1800 LST the southwest, being most intense at midday.

3. Description of the Meteorological Model

In this section we consider the problem of specifying initial and boundary values of the hydrostatic model developed by Nickerson *et al.* [1986], Mahfouf *et al.* [1987], Pinty *et al.* [1989], and Romero *et al.* [1995]. This model is a slightly simplified version of the Navier–Stokes system of equations for compressible fluids. It contains a parameterization of the microphysics of clouds, and considers the aerosol removal process, the processes of exchange which occur in the planetary boundary layer, vegetative cover, and soil.

Let $\Omega_s \subset \mathcal{R}^n$ ($n = 1$ or 2) be a domain of integration of the model and $\Omega_s = \Omega_{s_r} \cup \Omega_{s_u}$ be a portion of the soil at the surface, which may contain bodies of water, where Ω_{s_r} is a rural area and Ω_{s_u} an urban zone. Ω_s is limited by the boundary $\partial\Omega_s$; in this case the atmospheric part of the model is in $\overline{D}_s \times I_a$, where $\overline{D}_s = \Omega_s \cup \partial\Omega_s$, and $I_a = \{\nu \in \mathcal{R} \mid 0 \leq \nu \leq 1\}$, $\sigma = (p - p_t)/\pi = (4\nu - \nu^4)/3$, and ν is the vertical terrain following coordinate. The prognostic variables of the model are U, V, ϕ, π, θ, q, $\overline{e}$, where (U, V) are the horizontal components of the wind, ϕ the geopotential, $\pi = p_s - p_t$ the difference in pressure between the upper and lower boundaries, θ the potential temperature, and q the water vapor mixing ratio. In the planetary boundary layer we use the coefficients of turbulent diffusion, k_m and k_h, which are related to the mean turbulent kinetic energy $\overline{e} = (\overline{u'^2} + \overline{v'^2} + \overline{w'^2})/2$ by means of the relationship $k_m = c_\kappa l_\kappa \sqrt{\overline{e}}$ and $K_h = \alpha_n k_m$, where l_κ is the mixing length, α_n the reciprocal of the turbulent Prandtl number, and c_κ a constant. For more details about the equations of the model, see the references cited.

The heat balance equation at the soil–atmosphere interface or interface of the urban surface and the atmosphere is

$$R_n = H_\circ + L_v E_\circ + \Delta Q_\circ \tag{1}$$

where R_n is the net radiation, $H_\circ$ the sensible heat flux, $L_v E_\circ$ the latent heat flux, and

$$\Delta Q_\circ = \begin{cases} G_\circ & \text{in } \Omega_{s_r} \\ \Delta Q_s - Q_F & \text{in } \Omega_{s_u} \end{cases}$$

where $\Delta Q_s = \sum_{i=1}^{s} [a_{1i} R_n + a_{2i}(\partial R_n/\partial t) + a_{3i}]$ is the heat storage term which includes the conduction of heat from inside and outside the buildings, biomass, and soil of the urban zone; Q_F is the flux of anthropogenic heat; the a_s are constants [Grimmond *et al.*, 1991]; and $G_\circ = -\lambda(\partial T_s/\partial z)|_{z=0}$ is the heat flux into the ground surface. The soil part of the model is in $\overline{D}_s \times I_s$, where I_s is the depth of the soil layer. The primary prognostic variables in the soil are the temperature of the soil, T_s, and the volumetric moisture content of the soil, η. Those are related to the thermal conductivity, $\lambda = 0.167\eta + 0.1$, where $D_\eta = (-bK_{\eta_s} \psi_s^2/\eta\psi)(\eta/\eta_s)^3$ is the hydraulic diffusivity, $\psi = \psi_s(\eta_s/\eta)^b$ the moisture potential, $\rho c = (1 - \eta_s)\rho_i c_i + \eta\rho_\omega c_\omega$ the specific heat capacity of the soil, and $K_\eta = K_{\eta_s}(\eta/\eta_s)^{2b+3}$ the hydraulic conductivity of the soil. The calculation of the moisture flux at the ground surface, $(W_s - \rho\overline{w'q'})|_{z=0} = 0$, with $W_s = D_\eta\rho_w(\partial\eta/\partial z) + K_\eta\rho_w$; the energy and mass balance at the ground surface and beneath the vegetative cover; and the definitions of the other parameters are in McCumber and Pielke [1981] and Mahfouf *et al.* [1987]. The vegetative cover is parameterized according to the formulation proposed by Deardorff [1978], Mahfouf *et al.* [1987] and Pinty *et al.* [1989] where the Shielding factor $\sigma_f = 0.9$ and the leaf stomatal resistance are also included.

TABLE 2. Soil parameters; in all cases $c_w = 4.18 \times 10^6$ and $\eta/\eta_s = 0.2$.

Zone i	Soil type	Vegetation or area	σ_f	η_s (m/m^3)	K_{n_s} (m/s)	ψ_s (m)	c_i ($\times 10^6$)	b
1–11	Sandy loam	Rural		0.395	1.56×10^{-4}	−0.09	1.404	4.38
29–36	Sandy loam	Rural		0.395	1.56×10^{-4}	−0.09	1.404	4.38
16–24	Sand	Urban		0.413	1.22×10^{-4}	−0.143	1.402	4.4
12–15	Clay	Forests	0.9	0.466	14.9×10^{-7}	−0.349	1.207	10.7
25–28	Clay	Forests	0.9	0.466	14.9×10^{-7}	−0.349	1.207	10.7

4. Numerical Experiments

The latest version [Romero *et al.*, 1995] of the 3–D Mesoscale Model SALSA was run for a period of 12 hours in a 2–D mode over a west–east cross section AB through the Valley of Mexico ($i = 1, \ldots, 36$, $j = 15$, grid of Figure 1). The model was initialized with a single atmospheric sounding representative of special conditions for the month of March at 0600 LST (Table 1, lower part). The model atmosphere extends to an altitude of approximately 14 km. There are 15 computational levels in the vertical, with a nonuniform grid spacing that has the first grid point approximately 15 m above the lower boundary. The horizontal grid spacing is 11 km, and the horizontal domain is 385 km. The soil layer has 13 computational levels with a very high resolution near the ground which extend to a depth of 1 m below the surface. The initial soil temperatures are the same as the initial surface temperature, and the moisture content is initially to a depth of 1 m. The initial conditions for the soil and foliage variables are specified according to Clapp and Hornberger [1978] in Table 2. The astronomical parameters correspond to "15 March" at a latitude of 19.5° N and longitude of 99° W (Mexico City). The radiation model takes into account the slope of the terrain, as well as the latitude, longitude, and day of the year.

One important aspect of this work is the intent to incorporate urban effects into the meso-β model SALSA, which was originally developed to study the formation of clouds and airflow over complex terrain. The treatment of urban effects is in accordance with the results of the project of measurement of meteorological parameters to investigate the surface energy balance in Mexico City carried out by Oke *et al.* [1992]. The urban effect is achieved by solving the heat balance equation (1) in Ω_{s_u} (at the points $i = 16, \ldots, 24$) where the constants in the term ΔQ_s, the heat storage term, take the values $a_1 = 0.47$, $a_2 = 0.24$, $a_3 = -41$; and the anthropogenic heat flux attains the value $Q_F = 20$ W m^2 during the day and $Q_F = 5$ W m^2 at night. Also during the nighttime hours [Grimmond *et al.*, 1991] we used $\Delta Q_s = R_n + Q_F$ only when $R_n + Q_F < 0$.

In order to detect the urban influence, we carried out the following experiments with $\eta/\eta_s = 0.2, 0.3, 0.4, 0.5$ with an urban zone and without the urban zone (with type of soil sandy–loam), respectively. The most sensitive variable in the model is the temperature at the ground surface, T_s. We then calculated the mean temperature difference between the urban zone and a rural zone at the Valley of Toluca in the form $\Delta T_c = T_u - T_r$ where the mean temperature in the urban zone was $T_u = \frac{1}{9} \sum_{i=16}^{24} T_{s_{ij}}$, and in the rural zone $T_r = \frac{1}{9} \sum_{i=1}^{9} T_{s_{ij}}$. For the different cases of soil humidity, in the absence of the urban zone the average ΔT_c was $\approx 1°$C until 1100 LST, $\approx -2°$C from 1200 to 1300 LST,

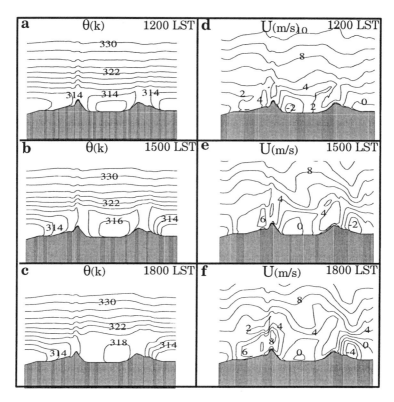

FIGURE 3. Two-dimensional numerical simulation of potential temperature (a)–(c) and horizontal wind (d)–(f) at 1200 LST, 1500 LST and 1800 LST, respectively.

and $\approx -5°C$ from 1400 to 1800 LST. On the other hand, the situation was just the opposite when the urban zone was included. The average ΔT_c was $\approx -2°C$ until 1100 LST, $\approx 2°C$ from 1200 to 1500 LST and $\approx -3°C$ from 1600 to 1800 LST. Consequently we see that the urban zone turned out to be slightly warmer than the rural zone. The thermal inversion aloft, which was close to the ground in the urban zone, disappeared between 1000 and 1100 LST with $(\eta/\eta_s) = 0.20$, between 1100 and 1200 LST with $(\eta/\eta_s) = 0.3$ and at 1300 LST with $(\eta/\eta_s) = 0.35$. On the other hand, in winter the "observed" thermal inversion in the urban zone usually breaks between 0900 and 1000 LST [SEDESOL, 1992].

Results of the 2-D numerical simulations for $(\eta/\eta_s) = 0.2$ and incorporation of the urban zone are shown every 3 hours up to a height of 7.5 km in Figure 3, beginning at 1200 LST. At noon, the winds over the urban zone are directed toward the mountains, consistently with Figure 2(g), and light upslope winds are established over the western hillsides (Figure 3(d)). The surface wind responds to the heating over the mountain slopes on either side of the Valley of Mexico. Over the western part of the valley there is an easterly flow, whereas over the eastern part there is a westerly upslope flow. A closed circulation cell of the thermally driven mountain and valley winds is manifested over the western slopes (Figures 3(d)–(f)). Above the crests there is subsidence, and above the slopes themselves the vertical velocities are positive. A weak subsidence (-0.05 m/s) is established over the central and eastern parts of the urban zone, covering a large area between 1200 LST and 1500 LST. The influence of the forests on the circulation patterns

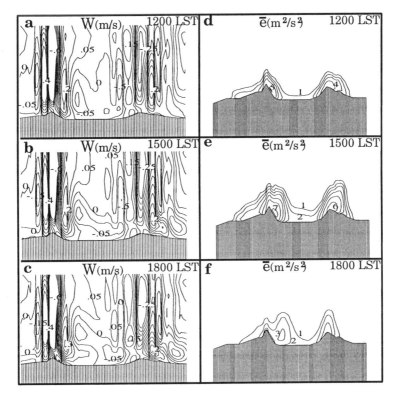

FIGURE 4. Two-dimensional numerical simulation of vertical velocity (a)–(c) and turbulence kinetic energy (d)–(f) at 1200 LST, 1500 LST and 1800 LST, respectively.

can be observed by means of the turbulent kinetic energy (TKE) predicted by the model and is marked by its near absence over the urban zone and its development upwind of the upwind mountain and downwind of the downwind mountain. TKE intensities reach their maximum at around 1500 LST and then begin to decline later in the afternoon. In Figures 4(d)–(f) it is shown that in the zone of weak winds (Figures 3(d)–(f)) that develop on the east side of the mountain crests the TKE has high values, and at 1800 LST, a zone of relatively high values of TKE appears above the western part of the urban zone (Figure 4(f)), which may be related to advection. However, we do not have observed data to be able to say for certain whether these quantities are real. The transverse sections of potential temperature are shown in Figures 3(a)–(c), and these fields are consistent with the distribution of turbulence.

5. Conclusions

Certain circulation patterns of local surface winds within the Valley of Mexico not previously described in the literature have been discussed in this paper. The distribution of observed O_3 is closely related to the analyzed wind patterns, which in turn are influenced very much by the mountains that surround the basin of the Valley of Mexico and which act as a barrier that confines the contaminants and the urban heat island. These wind analyses are preliminary, and we intend to improve them in the future with the incorporation of new data which will include the Valleys of Puebla and Toluca.

The mesoscale model SALSA has been used to simulate airflow over the Valley of Mexico, taking into account the urban effects of Mexico City, and it has helped in the theoretical understanding of the orographic processes that occur as a result of the mountains that are found on the outskirts of the Valley of Mexico. It has been found that the presence of a barrier induces a major turbulence zone and that the existence of a forest on the barrier causes even greater production of turbulent kinetic energy. The thermally driven fluxes of sensible and latent heat are very much dependent on the vegetation type, soil type, and soil moisture. The great quantity of sensible heat in an urban zone results in a boundary layer that is deeper and more turbulent. The model reveals that during the day, the most turbulent zone of the planetary layer is found in the west in association with a great deal of mixing of pollutants. The results presented show the first in a series of 2-D simulations designed to investigate the effects of different vegetation, soil properties and incorporation of an urban zone on local circulation patterns. These sensitivity studies will form the basis for 3-D simulations of airflow and its effects on the diurnal variability of pollutant distributions over the Valley of Mexico.

Acknowledgments. This study was financed by the DGSCA/UNAM–Cray Research, Inc., R&D Grant Program. We express our sincere appreciation for the collaboration of Luis M. de la Cruz for the graphics, of J. Zintzún, A. Aguilar, A. Solano and Ma. E. Palos for the gathering and assimilation of the data; of the Servicio Meteorológico Nacional for the radiosonde data; and of the DDF for the RAMA data. Special thanks go to Ma. Elena Castillo for composition and formatting of the manuscript.

REFERENCES

Clapp, R. B., and G. M. Hornberger, Empirical equations for some soil hydraulic properties, *Water Resources Res.*, **14**, 601–604, 1978.

Deardorff, J. W., Efficient prediction of ground surface temperature and moisture with inclusion of a layer of vegetation, *J. Geophys. Res.*, **20**, 1889–1903, 1978.

Grimmond, C. S. B., H. A. Cleugh, and T. R. Oke, An objective heat storage model and its comparison with other schemes, *Atmospheric Environment*, **25B**, 311–326, 1991.

Jáuregui, E., Local wind and air pollution interaction in the Mexico basin, *Atmósfera*, **1**, 131–140, 1988.

Jáuregui, E., Mexico City's urban heat island revisited, *Erdkunde*, **47**, 185–195, 1993.

Mahfouf, J. F., E. Richard, and P. Mascart, The influence of soil and vegetation on the development of mesoscale circulations, *J. Climate Appl. Meteor.*, **26**, 1483–1495, 1987.

McCumber, M. C., and R. A. Pielke, Simulation of the effects of surface fluxes of heat and moisture in a mesoscale numerical model. Part I. Soil Layer, *J. Geophys. Res.*, **86**, 9929–9938, 1981.

Nickerson, E. C., E. Richard, R. Rosset, and D. R. Smith, The numerical simulation of clouds, rain and airflow over the Vosges and the Black Forest mountains: A meso-B model with parameterized microphysics, *Mon. Wea. Rev.*, **114**, 398–414, 1986.

Oke, T. R., G. Zeuner, and Jáuregui, The surface energy balance in Mexico City, *Atmospheric Environment*, **26B**, 433–444, 1992.

Pérez, G. I., Un análisis semiobjetivo para el campo de viento en los trópicos, *Geofísica Internacional*, **24**, 3, 425–437, 1985.

Pinty, J.-P., P., Mascart, E. Richard and R. Rosset, An investigation of mesoscale flows induced by vegetation inhomogeneities using an evapotranspiration model calibrated against HAPEX-MOBILHY data, *J. Appl. Meteor.*, **28**, 976–992, 1989.

Romero, R., S. Alonso, E. C. Nickerson, and C. Ramis, The influence of vegetation on the development and structure of mountain waves, *J. Appl. Meteor.*, **34**, 2230–2242, 1995.

SEDESOL, Compendio, boletín informativo de la calidad del aire, México, D. F., Octubre 1986–Abril 1992, 1–34, 1992.

Tripoli, G. I., and T. R. Krishnamurty, Low-level flow over the GATE area during summer 1972, *Mon. Wea. Rev.*, **103**, 197–216, 1975.

Mexico City Air Quality Simulations under Different Fuel Consumption Scenarios

By Elba Ortiz, Jorge Gasca, Gustavo Sosa, Ma. Esther Ruiz and Luis
Díaz

Instituto Mexicano del Petróleo, Eje Central Lázaro Cárdenas 152, 07730 México, D.F., México

The effect in ozone diurnal evolution of changes in the hydrocarbon (RHC) and nitrogen oxide
(NO_x) emissions is simulated through the California/Carnegie Institute of Technology (CIT)
trajectory model, which calculates the evolution of smog pollutants by numerically solving the
atmospheric diffusion equation for a set of chemical species. Thus, it gives the description of
the physical and chemical processes responsible for the chemical transformation, transport and
fate of compounds in the atmosphere. The wind and other meteorological fields were obtained
by the use of the High Order Turbulence Model for Atmospheric Conditions (HOTMAC) model
adapted to the topography and conditions of the Mexico City Metropolitan Area. In this paper
Mexico City air quality under different fuel consumption scenarios is presented.

1. Methodology

The basis for the CIT model [McRae et al., 1982; Russell et al., 1988] is the atmospheric diffusion equation, which expresses the conservation of mass of each pollutant in
a turbulent fluid in which chemical reactions occur:

$$\frac{\partial C_i}{\partial t} + \nabla \cdot (\mathbf{u} C_i) = \nabla \cdot (K \nabla C_i) + R_i + Q_i$$

where C_i is the concentration of species i, $\mathbf{u}$ is the wind velocity vector, K is the eddy
diffusivity tensor (here assumed to be diagonal), R_i is the rate of generation of species i
by chemical reactions, and Q_i is a source term for elevated point sources of species i. If
the preceding equation is written repeatedly for each of the chemical species tracked by
the model, a system of coupled nonlinear equations is obtained.

To solve the equation system, a set of initial and lateral boundary conditions is established. The values of the different variables are set by using measured pollutant
concentration data. The ground-level boundary condition sets upward pollutant fluxes
equal to direct emissions minus dry deposition. Additionally the hourly diurnal variation
of the mixing height, given as input data to the model, incorporates the air entrainment
to dilute or concentrate species. A vertical concentration gradient for each species is
set so that it decreases to a zero value at the top boundary, ensuring that there is no
vertical transport of pollutants through the top of the modeling region. The effective
mixing heights were obtained from lidar measurements and the HOTMAC meteorological
model.

The version of the CIT model used in the simulation of O_3 formation in the Mexico City
Metropolitan Area (MCMA) contains a condensed version of the Lurmann, Carter, and
Coyner (LCC) mechanism [Lurmann et al., 1987]. An emission inventory categorized by
source, mobile and stationary, spatially and temporally distributed, indicates the hourly
emissions through the trajectory, with the RHC grouped according to the inputs required
by the lumped LCC mechanism that was also used.

The air quality and meteorology information from the MCMA Air Quality Monitoring
Network (RAMA) was validated before it was used as input data. Two scenarios, representing the emissions of 1991 and 1995, were considered. They included exhaust and

TABLE 1. 0600–0900 LST mean RHC concentration (ppm).

| Air quality | Year | |
network station	1992	1993
Xalostoc	4.81	6.76
La Merced	3.77	5.02
Pedregal	1.98	1.93

evaporative emissions from mobile and stationary sources, as well as projected leaded and unleaded gasoline consumption for 1995. Emission totals and emission characteristics including RHC speciation, for tailpipe and evaporative emissions, were obtained from a dynamometer using the FTP-45. The relative contributions of emission reductions by gasoline storage and commercialization were estimated.

The effects on ozone due to gasoline consumption are obtained by changing the input emission data in a base case and comparing the output results. The base case simulation represents the conditions of a typical winter day, associated with the highest ozone episodes in the MCMA. This selection is supported by the fact that the best information on the MCMA meteorology and air quality corresponds to February 1991, collected in the Mexico City Air Quality Research Initiative [MARI, 1993].

For the simulation the input information imposes a path for the air parcel governed by the typical wind pattern in Mexico City. It transports the pollutants released early in the morning from an industrial region in the Northeast; through downtown, characterized by commercial activities; to a typical receptor site in the Southwest, a residential area. Frequently the highest ozone concentrations are observed in this area. The trajectory is determined by working backwards from the site where the ozone maximum is observed. The procedure uses the data for wind speed and direction registered in that particular monitoring station. From such data it is possible to infer where the air parcel was 1 hour before. This process is repeated until the site corresponding to the air parcel in the morning hours is located. This selected trajectory coincides with the one predicted in MARI by the HOTMAC meteorological model for this time of the year [Williams *et al.*, 1993]. The RHC and NO$_x$ emission inventories used as input for the model correspond to those cells the air parcel occupies.

Along the trajectory of the air mass there are several monitoring stations which provide useful information to be used as input and to validate the model results. The starting point is the Xalostoc station at 0700 LST. Downtown, midmorning conditions are represented by the La Merced station, while Plateros and Pedregal stations provide the receptor characteristics in the afternoon and evening up to 2100 LST. The diurnal evolution of some parameters used as input for the trajectory model is shown in Figure 1. The parameters include mixing heights, monitored temperatures, relative humidity, and wind speeds. The radiation scale factors were calculated on the basis of the measured ultraviolet radiation at the National University of Mexico campus, in 1991 [Ruiz *et al.*, 1993]. A 3 ppm initial RHC concentration (0700 LST) was estimated from the results of two experimental ambient air sampling campaigns performed in February 1992 and February 1993 in a joint effort of the IMP and the USEPA. Air samples at various sites of the MCMA were collected from 0600 to 0900 LST, and the mean RHC concentrations determined in such campaigns are shown in Table 1.

The emission inventory is spatially distributed in a 5600 km^2 simulation grid (16 × 14

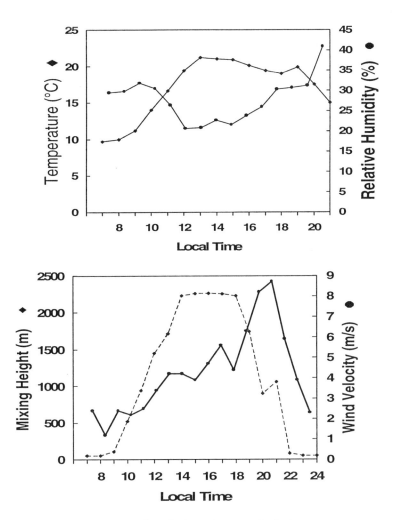

FIGURE 1. Diurnal evolution of meteorological parameters.

cells, 5 km × 5 km each). For each contaminant, the emission inventory splits into the 24 source categories, defined previously in the Air Quality Management Program for the MCMA [Mumme *et al.*, 1992]. In addition, each RHC source in the emission inventory is based on information specific to the MCMA, if available. Thus, the emission total in each grid cell is the sum of all the source categories within such cell. The emissions along the trajectory are simply the particular emissions of the respective grid cells.

To represent the emission variations with time of day, diurnal emission patterns for stationary and mobile sources were estimated from traffic flow information and industrial activity indicators. The resulting patterns are illustrated in Figure 2. Seven cases were simulated to evaluate the effects of gasoline-related actions: a base case for 1991 to serve as reference and six different projected scenarios for 1995. The main differences among the scenarios are, on the one hand, the characteristics of the gasoline in the market, and, on the other hand, whether additional measures were in place as well. One of the scenarios can be considered as a limit case, to test model sensitivity and to allow comparisons.

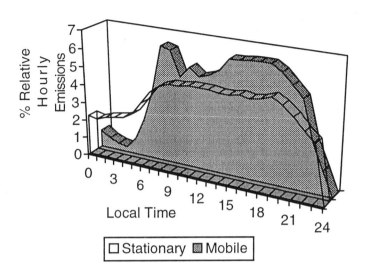

FIGURE 2. Temporal emission pattern.

TABLE 2. Basic characteristics of the gasoline used to simulate the scenarios.

Case no.	Year simulated	Fraction and gasoline type[b]	RVP[a] [psi]	Tail-pipe	Mobile evap.	Area evap.	% Changes in NO$_x$
				\% Changes in RHC			
Base	1991	100% LD	9.0	0	0	0	0
1[c]	1995	60% LD + 40% UL	8.5	−13	−27	−76	+25
2	1995	60% LD + 40% UL	7.5	−21	−42	−80	+27
3	1995	60% LD + 40% UL	9.0	−9	+8	+14	+18
4	1995	60% LD + 40% UL	8.5	−13	−5	+7	+25
5	1995	50% LD + 50% UL	7.5	−21	−23	−40	+26
6[d]	1995	100% LD	9.0	+20	+20	+20	+20

Notes:
[a] Reid vapor pressure.
[b] LD: leaded gasoline, UL: unleaded gasoline.
[c] Assumes that additional measures have been implemented.
[d] Assumes that no actions have taken place, and only reference gasoline consumption has increased 20%.

2. Results

The speciated RHC emissions obtained from tailpipe and hot soak test results are lumped according to the LCC mechanism. Table 2 outlines the characteristics of all the cases, the main assumption to build the scenario, and the departures from the base case emissions.

To assemble each case it is necessary to modify the emission inventory used in the base case simulation in several ways. It is not enough to account for the total emission changes alone but also for the way these alter the different categories, including the

TABLE 3. Predicted ozone concentration (ppm) for the simulated scenarios.

Hour	Monit.	Ozone (ppm)						
		Base	Case 1	Case 2	Case 3	Case 4	Case 5	Case 6
		RVP 9.0	RVP 8.5	RVP 7.5	RVP 9.0	RVP 8.5	RVP 7.5	RVP 9.0
8	0.015	0.003	0.003	0.003	0.003	0.003	0.003	0.003
9	0.019	0.012	0.011	0.011	0.012	0.011	0.010	0.011
10	0.057	0.047	0.047	0.040	0.058	0.050	0.039	0.048
11	0.123	0.107	0.111	0.085	0.128	0.121	0.086	0.113
12	0.219	0.184	0.192	0.145	0.218	0.206	0.147	0.197
13	0.262	0.265	0.273	0.217	0.286	0.282	0.220	0.283
14	0.305	0.297	0.299	0.252	0.300	0.303	0.256	0.314
15	0.331	0.314	0.311	0.279	0.304	0.311	0.284	0.328
16	0.280	0.262	0.254	0.242	0.245	0.251	0.247	0.271
17	0.178	0.190	0.188	0.176	0.179	0.183	0.176	0.200
18	0.094	0.149	0.135	0.125	0.127	0.143	0.126	0.133
19	0.050	0.101	0.095	0.076	0.074	0.101	0.077	0.086
20	0.028	0.063	0.070	0.054	0.054	0.067	0.045	0.057
21		0.036	0.041	0.037	0.037	0.049	0.025	0.028

particular changes for each species within the RHC emissions. However, the main effect of the gasoline formulation, leaded or unleaded, is produced in both the NO_x tail pipe emissions and the RHC emissions; the RHC emissions, in their turn, are due to tail pipe, mobile and area evaporative losses. The computer simulations produce the hourly concentrations of pollutants. In Table 3 the results for hourly ozone concentrations, from 0800 to 2100 LST, are listed, for the base case and the simulated scenarios. This table also contains the monitored value, registered by the RAMA station, representative of the grid cell along the trajectory for February 22, 1991: the date chosen to calibrate the base case.

Analysis of the results, particularly comparison with the base case, shows that the most noticeable effect is obtained when the RVP is reduced to 7.5 psi (cases 2 and 5). The results of cases 1 and 4 reveal that despite a formulation for obtaining an RVP of 8.5 psi to fuel a significant number of cars with catalytic converters (40%), the increased demand in gasoline and the associated NO_x emissions are not counterbalanced. However, it can be observed that if reformulation were not applied, and the 1991 reference gasolines were kept through 1995 (case 3), the ozone concentrations would increase even further, especially around noon. Obviously if one assumes that no actions have taken place in 1995, that is, the car fleet has no catalytic converters, and only reference leaded gasoline is offered (case 6), the 20% increase in consumption would lead to the least favorable scenario.

A plot of differences in ozone concentration, between selected cases and the base case, is shown in Figure 3. The resulting relatively small effects in ozone concentrations can be attributed to the very high RHC background concentrations used as initial conditions; reflected particularly in small sensitivity of the ozone peak concentration. In other words, the experimentally confirmed large amounts of RHC in the MCMA hinder the potential benefit of some specific control actions. It can be expected that if propitious initial conditions were present, the effects of emission control actions would be enhanced.

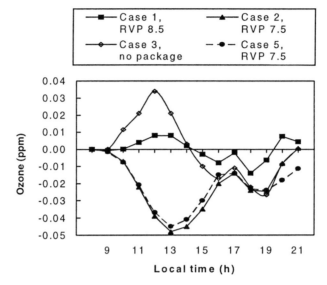

FIGURE 3. Ozone concentration differences.

REFERENCES

Lurman, F. W., William P. L. Carter, and Lori A. Coyner, *A Surrogate Species Chemical Reaction Mechanism for Urban-Scale Air Quality Simulation Models*. Volume I. *Adaptation of the Mechanism*. Volume II. *Guidelines for Using the Mechanism*. ERT and SAPRC EPA Contract 68-02-4104, USEPA, Research Triangle Park, NC, 1987.

MARI, Mexican Petroleum Institute and Los Alamos National Laboratory 1993 Mexico City Air Quality Research Initiative, IMP/LA-1269 México, DF, México and Los Alamos, NM, 1993.

McRae, G. J., W. R. Goodin, and J. H. Seinfeld, Development of a second-generation mathematical model for urban air pollution. I. Model formulation, *Atmos. Environ.*, **16**, 679–696, 1982.

Mumme, C. H., J. Cargajo, G. Eskeland, S. Margulis, K. Wijetilleke, P. Comwell, J. Cracknell, P. Glaessner, C. Weaver, and T. Yamada, Mexico: Transport air quality management in the Mexico City Metropolitan Area. Sector study, World Bank Report No. 10045-ME, Latin America and the Caribbean Regional Office, Washington, DC, February 1992.

Ruiz-Suárez, J. C., L. G. Ruiz-Suárez, C. Gay, T. Castro, M. Montero, S. Eidels, and A. Muhlia, Photolytic rates for NO_2, O_3 and HCHO in the atmosphere of Mexico City, *Atmos. Environ.*, **27 A**(3), 427–430, 1993.

Russell, A. G., K. F. McCue, and G. R. Cass, Mathematical modeling of the formation and transport of nitrogen-containing pollutants. I. Model Evaluation, *Environ. Sci. Technol.*, **22**, 263–271, 1988.

Williams, M. D., M. J. Brown, X. Cruz, G. Sosa, and G. Streit, Development and testing of meteorology and air dispersion models for Mexico City, Regional Photochemical Measurement and Modeling Studies International Conference, November 8–12, San Diego, CA, 1993.

Numerical Modeling of Pollutant Particle Diffusion in the Atmospheric Boundary Layer

By Gueorgui V. Mostovoi

Foundation for Meteorology and Water Resources – FUNCEME,
60.325-002 Fortaleza, CE, Brazil

A simple model simulating the instantaneous three-dimensional dispersion features of impurity particles emitted by a point source located in an arbitrarily stratified atmospheric boundary layer is presented. The algorithm is based on a random-walk model and also takes account of spatial correlation between particle velocity fluctuations. The correlation radius is derived from the corresponding Lagrangian integral time scale. Semiempirical similarity relations are used to approximate the vertical profiles of the velocity-fluctuation variance and the Lagrangian time scale within the atmospheric boundary layer. Two specific examples of the dispersion model under free convective conditions are presented.

1. Introduction

One method of atmospheric pollution dispersion numerical modeling is based on a calculation of trajectories of a great number of polluted airborne particles. Random-walk models are usually adopted in these calculations, which can easily reproduce observed atmospheric dispersion features, as demonstrated by Csanady [1973], Hall [1975] and Reid [1979]. This dispersion or mixing of the particles with the environment is a function of the wind velocity fluctuations, caused by the atmospheric turbulence of a specific spatial scale. Knowledge of the average and turbulent structures of the atmospheric boundary layer (ABL) is necessary in resolving various problems of the ABL pollution dispersion with the help of random-walk models. If a mesoscale atmospheric model is available, then it can be used to estimate diffusion parameters from the kinetic energy of turbulence, as suggested by Pielke [1984]. This approach is developed by Uliasz [1993] in the form of the computer dispersion modeling system. We attempt here another, more limited approach and use semiempirical similarity relations to describe the average and turbulent structure of the ABL. This data were accumulated during numerous field experiments.

2. Random-Walk Dispersion Model

Suppose that polluted particles (containing a passive scalar) are transported along the i-direction ($i = 1, 2, 3$) with velocity v_i. It is usually accepted that v_i can be represented as a sum of time average and turbulent components $v_i = \bar{v}_i + v'_i$. The mean velocity component $\bar{v}_i$ should be known from observations or calculated from a mesoscale atmospheric model. The turbulent component v'_i, in turn, is represented by the sum of time-correlated and purely random δ_i components

$$v'_i(t + \Delta t) = v'_i(t)\alpha_i + \delta_i \qquad (2.1)$$

where Δt is the time step. If $\alpha_i = R_L^i(\Delta t)$ is equal to the Lagrangian autocorrelation function value for the i velocity component and the time lag Δt, then relation (2.1) defines a stochastic time series, or the Markov chain. In this case, the time series v'_i will have

149

an exponentially decaying autocorrelation function. The definition of autocorrelation function gives $R_L^i(n\,\Delta t) = \alpha_i^n$, where n is an integer. Using $t = n\Delta t$, we have

$$R_L^i(t) = \alpha_i^{t/\Delta t} = \exp\left(\frac{-t}{T_L^i}\right) \tag{2.2}$$

The constant $T_L^i = \Delta t/\ln(1/\alpha_i)$ is the Lagrangian integral time scale for the series v_i'. It also follows from (1) that the purely random component δ_i has variance

$$\sigma_{\delta_i}^2 = \sigma_i^2\{1 - [R_L^i(\Delta t)]^2\} \tag{2.3}$$

where σ_i^2 is the variance of the i wind velocity component. It is evident from (2.1) that $\sigma_i^2(t)$ is constant, also assuming that δ_i has a zero mean. For the particle trajectory construction it is necessary to define appropriate spatial distributions of the σ_i and T_L^i parameters, which describe statistical properties of the atmospheric turbulence. That is the main difficulty in the application of such models. Within the ABL (assume horizontal homogeneity) these parameters vary mainly in the vertical direction. The vertical profiles of σ_i and T_L^i can be approximated by the well-known empirical similarity functions, which depend on the following scales: the Monin–Obukhov length of a surface layer (L), the height of a mixing layer (H), the friction velocity (u_*), the roughness (z_0) of a surface and the velocity scale (w_*) for the convective ABL turbulence. These parameters can be estimated from the standard gradient observations of mean wind velocity and air temperature. Here we choose a very simple way to estimate them, which relates the Pasquill classes of atmospheric stability with the L values. The values of σ_i and T_L^i are calculated according to the formulas recommended by Hanna [1982].

The necessary spatial resolution Δ_i of the particle concentration field in any i-direction is prescribed by the user. Then the time step Δt may be estimated from the following constraint

$$\Delta t \leq \sqrt{\frac{\sum_i \Delta_i^2}{\sum_i v_{i,\max}^2}} \tag{2.4}$$

where $v_{i,\max}$ is the total typical particle velocity value along the i-direction. The last condition provides a relatively slow exchange of airborne particles between the adjacent resolution elements.

Scheme (2.1) reproduces the random movement component of each particle, which has no spatial correlation. Turbulent eddies in the ABL will cause spatial correlation between consequent values of v_i. It will provide the structure of a plume, consisting of propagating isolated air volumes called puffs. This kind of a plume structure can be observed frequently in the lower part of the ABL. For that reason random time series δ_i are generated to have additional spatial correlation intervals, proportional to $\bar{v}_1 T_L^i$. They are equal to the average dimension of the turbulent eddies responsible for the most efficient plume concentration dilution far from the source position. A special procedure, proposed by Shimanuki and Nomura [1991], is used to generate the random numbers δ_i possessing the spatial correlation interval. That is the main difference between the algorithm described and that of the pure random-walk model. More information about the model is presented in a paper by Mostovoi [1993].

3. Results

The model has the following input parameters chosen by the user: the grid resolution, which can vary from $1 \times 1 \times 1$ m^3 to $100 \times 100 \times 100$ m^3; the height of the point source above the surface (0–1000 m); heights of the gradient measurements (two) of mean air

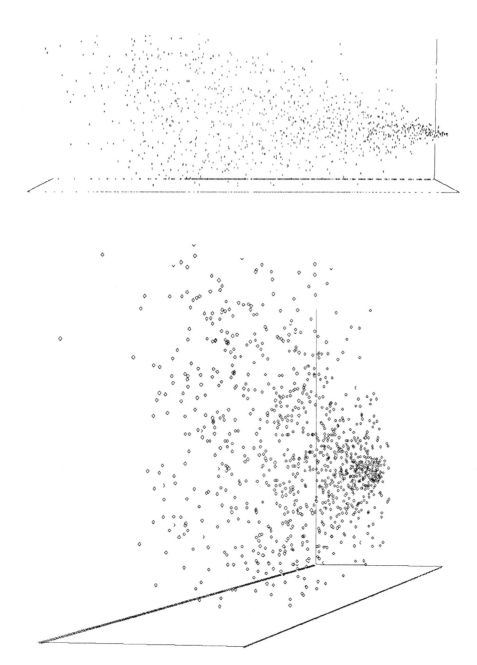

FIGURE 1. The perspective view (for two different eye positions) of instantaneous polluted particle distribution emitted by a point elevated source of 250 m height. It corresponds to a high convective instability ($L = -9$) within the ABL (2 and 5 m/s, 18.4° and 18.0°C). The first two numbers are mean wind speed; the next two, air temperature. All are related to the measurement levels of 2 and 10 m above the surface. The box size is equal to $2807 \times 427 \times 707$ m^3 and emission time is equal to 4 min.

temperature and wind velocity; the measured values of the air temperature and the wind velocity at these levels; and the period of time for averaging plume concentration.

Output information is available for each time step in the form of instantaneous images of particle cloud emitted by the point source. To demonstrate the program's capability we will describe here two examples of plume–particle development in different conditions of convective instability within the ABL. One case of plume dispersion corresponds to a high convective instability in a lower part of the ABL ($L = -9$ m). Figure 1 shows the perspective view (for two different eye positions) of the instantaneous contours of the plume originating from an elevated source. The other case of plume dispersion under a moderate convective instability ($L = -201$ m) is shown in Figure 2.

The grid resolution is equal to $7 \times 7 \times 7$ m^3, and the height of the source is 250 m above a surface. All emitted particles have initial vertical velocity of 0.3 m/s.

Figures 1 and 2 illustrate the wavelike pattern of a propagating plume observed frequently in the natural environment. In a situation of moderate convection (Figure 2) polluted particles cannot overcome the resistance of the shear wind turbulence layer. They are nearly stopped about the level of $|L|$ height. Below this level, wind shear turbulence dominates and above it convection dominates. On the other hand, the situation with the high convective instability is more favorable for rapid polluted particle propagation to the surface.

4. Conclusions

We consider the current version of the dispersion model a preliminary one, mainly for educational purposes. For example, it demonstrates clearly for students how relatively small differences in the mean temperature and wind velocity stratification can significantly alter the plume pattern within the ABL and illustrates the plume pattern transitions under a daily variation of ABL stability conditions.

REFERENCES

Csanady, G. T., *Turbulent Diffusion in the Environment*, D. Reidel, 1973.

Hall, C. D., The simulation of particle motion in the atmosphere by a numerical random-walk model, *Q. J. R. Met. Soc.*, **101**, 235–244, 1975.

Hanna, S. R., Applications in air pollution modeling, *Atmospheric Turbulence and Air Pollution Modelling*, F. T. M. Nieuwstadt and H. Van Dop, Eds., D. Reidel, 275–310, 1982.

Mostovoi, G. V., A model of impurity diffusion from a point source in transitional layer of the atmosphere, *Vestnik of Moscow State University*, Ser. 5, 54–62, 1993 (in Russian).

Pielke, R. A., *Mesoscale Meteorological Modeling*, Academic Press, 1984.

Reid, J. D., Markov chain simulations of vertical dispersion in the neutral surface layer for surface and elevated releases, *Boundary Layer Met.*, **16**, 3–22, 1979.

Shimanuki, A., and Y. Nomura, Numerical simulation of instantaneous images of the smoke released from a chimney, *J. Met. Soc. Japan*, **69**, 187–196, 1991.

Uliasz, M., The atmospheric mesoscale dispersion modeling system, *J. Atmos. Sci.*, **32**, 139–149, 1993.

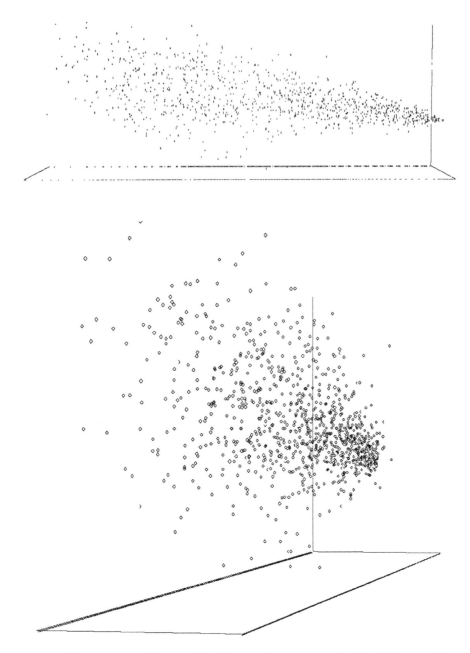

FIGURE 2. The same as in Figure 1, but for a moderate convective instability condition ($L = -201$ m) within the ABL (2 and 5 m/s, 18.1° and 18.0°C).

Investigating the Dispersion Inside Idealized Urban Street Canyons Using a k-epsilon Turbulence Model

By YanChing Q. Zhang

ManTech Environmental Technology, Inc., 2 Triangle Dr., P.O. Box 12313
Research Triangle Park, NC 27709, USA.

In this paper, we apply a numerical model to investigate dispersion inside idealized (two-dimensional) urban street canyons with upstream flow normal to the canyon. Five two-dimensional cases were studied and compared to investigate the effects of the building geometric configurations on concentration levels in the vicinity of the street canyon. The results provide some insight into the generalized flow structure as well as the dispersion patterns. Only emissions from a ground source are examined herein. The effects of the canyon geometric configurations and number of the canyons in the wind direction are discussed as a function of the concentration field.

Pollutants emitted within a canyon will generally not be transported to the front of the upstream buildings. The recirculating flow above the upstream building depends more strongly on the upstream shear level than on the relative canyon configurations. The flow and concentration pattern inside the canyon depends on both the upstream flow conditions (shear level and turbulence level) and the relative heights of the buildings on both sides of the canyon. The dispersion fields behind the downstream building depend on both the flow characteristics within the canyon and the downstream building height.

The high concentration stays on the upstream side of the building, except in the step-down canyon cases.

When there is more than one canyon in the downwind direction, the upstream flow conditions can be very critical to the dispersion fields inside the first canyon, but they are not as important to the concentration fields inside the consecutive canyons. The dispersion fields in the vicinity of the first canyon are very similar to the corresponding cases involving only one canyon and the same upstream boundary conditions. The fields in the vicinity of the following canyons depend only on the relative geometric configurations on two sides of the canyon.

1. Introduction

The term street canyon refers to a relatively narrow street between a row of buildings. The street canyon is the basic geometric unit constituting the urban canopy. Air flow within the urban boundary layer is dominated by microscale influences. Most studies on pollutant dispersion in an urban area using numerical methods have been primarily site-specific. In this paper, we apply a numerical model to investigate dispersion inside idealized (two-dimensional) urban street canyons with upstream flow normal to the canyon. Five two-dimensional cases were studied and compared to investigate the effects of the building geometric configurations on concentration levels in the vicinity of the street canyon. The uniform approaching flow is perpendicular to the street canyon. The results provide some insight into the generalized flow structure as well as the dispersion patterns. Only emissions from a ground source are examined. The effects of the canyon geometric configurations and number of the canyons in the wind direction on the concentration field are discussed.

2. Numerical Simulation

2.1. *Introduction to the TEMPEST Model*

The Transient Energy Momentum and Pressure Equations Solution in Three Dimensions (TEMPEST) model is a three-dimensional, time-dependent, nonhydrostatic numerical model that was developed at Battelle Pacific Northwest Laboratory; it has been applied to a broad range of engineering and geophysical problems [Trent and Eyler, 1989]. The model includes the ability to account for small density variations through the Boussinesq approximation; cylindrical, Cartesian, or polar coordinates may be used. It has the ability to use variable grid spacing along any coordinate, and the inflow/outflow boundaries can be either specified or computed. Turbulence is treated by using a turbulent kinetic energy/dissipation (k-ϵ) model. The solution technique in TEMPEST is similar to the Simplified Marker-and-Cell (SMAC) technique [Amsden and Harlow, 1970], whereby at each time step, the momentum equations are solved explicitly and pressure equations implicitly; temperature, turbulent kinetic energy (TKE), dissipation of kinetic energy (DKE), and other scaler transport equations are solved by using an implicit continuation procedure. The standard formulation for the k-ϵ turbulence model [Trent and Eyler, 1989] is used in our simulation, and a staggered grid system is adopted.

The governing equations for the k-ϵ model (as used in TEMPEST) in a Cartesian coordinate system are presented. We neglect molecular diffusion in comparison with turbulent diffusion in the momentum equations and confine our simulations to the atmospheric surface layer over a small domain (say, 5 km × 5 km), so that Coriolis effects can also be neglected. The governing equations, subject to Boussinesq approximations and Reynolds averaging, are

Continuity:

$$\frac{\partial U_i}{\partial x_i} = 0 \tag{2.1}$$

Momentum:

$$\frac{\partial U_i}{\partial t} + U_j \frac{\partial U_i}{\partial x_j} = -\frac{\partial}{\partial x_i}\left[\frac{\delta P}{\rho_0} + \frac{2}{3}k\right] + \frac{\partial}{\partial x_j}\left\{\nu_t\left[\frac{\partial U_i}{\partial x_j} + \frac{\partial U_j}{\partial x_i}\right]\right\} - \frac{\partial p}{\rho_0}g\delta_{3i} \tag{2.2}$$

State:

$$\frac{\delta\rho}{\rho_0} = -\frac{\delta T}{T_0} \tag{2.3}$$

Thermal energy:

$$c_p\frac{\partial T}{\partial t} + c_p U_j \frac{\partial T}{\partial x_j} = \frac{\partial}{\partial x_j}\left[\frac{\nu_t}{Pr_T}\frac{\partial T}{\partial x_j}\right] + Q \tag{2.4}$$

Concentration equations:

$$\frac{\partial C^i}{\partial t} + U_j \frac{\partial C^i}{\partial x_j} = \frac{\partial}{\partial x_j}\left[D_j \frac{\partial C^i}{\partial x_j}\right] + S^i \tag{2.5}$$

where repeated subscripts denote summation, U_i is the ith mean velocity component, t is time, $\delta\rho$ is the deviation of density from its reference value ρ_0, δP is the deviation of pressure from its reference value, ν_t is an effective viscosity which is the sum of molecular viscosity (which is neglected in our simulations) and turbulent eddy viscosity, g_i is the ith component of acceleration due to gravity, c_p is the specific heat of air at constant pressure, Pr_T is the turbulent Prandtl number, Q is the volumetric heat generation rate, C_i is the mean mass fraction of the ith constituent, D_j is the effective mass diffusivity in the jth direction (which is the sum of turbulent and molecular mass diffusivities), S_i is the mass generation rate, and k is the turbulent kinetic energy (TKE), which is defined as

$$k = \frac{1}{2}[\overline{(u)^2} + \overline{(v)^2} + \overline{(w)^2}] \tag{2.6}$$

where u, v and w are the velocity fluctuations in the x, y and z directions, respectively. The preceding equations treat density as a constant (incompressible flow) except in the body force term of the momentum equation, allowing us to simulate the stratification.

The Reynolds stresses and fluxes have sometimes been modeled or parameterized using the gradient transport relations to close the preceding system of equations. TEMPEST uses a k-ϵ turbulence model to close the system of equations by providing estimates of effective turbulent viscosity and mass diffusivity. This is accomplished by using transport equations for the turbulent kinetic energy k

$$\frac{\partial k}{\partial t} + \frac{\partial k U_j}{\partial x_j} = \frac{\partial}{\partial x_j}\left[\frac{\nu_t}{\sigma_k}\frac{\partial k}{\partial x_j}\right] + (S + G) - \epsilon \tag{2.7}$$

and the dissipation of turbulent kinetic energy

$$\frac{\partial \epsilon}{\partial t} + \frac{\partial \epsilon U_j}{\partial x_j} = \frac{\partial}{\partial x_j}\left[\frac{\nu_t}{\sigma_t}\frac{\partial \epsilon}{\partial x_j}\right] + \frac{\epsilon}{k}(C_1 S + C_3 G) - C_2 \frac{\epsilon^2}{k} \tag{2.8}$$

where ν_t is the kinematic (molecular) viscosity

$$\epsilon = \nu_t{}^2 \overline{\left(\frac{\partial u_i}{\partial x_j}\right)^2} \tag{2.9}$$

Here S is the shear production term, and G is the buoyancy term in the TKE equation (2.7), defined as

$$S = \nu_t \left[\frac{\partial U_i}{\partial x_j} + \frac{\partial U_j}{\partial x_i}\right]\frac{\partial U_i}{\partial x_j} \tag{2.10}$$

$$G = \frac{\nu_t}{\sigma_t}\frac{g}{\rho_0}\frac{\partial \rho}{\partial z}\delta_{3i} \tag{2.11}$$

where $\sigma_t, \sigma_\epsilon, \sigma_k$ are three of the empirical constants in k-ϵ models. The effective diffusivity of momentum is estimated as $\nu_t = C_\nu k^2/\epsilon$. The effective diffusivity of mass is $D = \nu_t/Sc_T$, where Sc_T is the turbulent Schmidt number. The standard values of constants which have been used for most engineering applications [Gibson and Launder, 1978; Zhang et al., 1993] used in the TEMPEST are $(\sigma_t, \sigma_k, \sigma_\epsilon, \sigma_\nu, C_1, C_2, C_3, Sc_T, Pr_T) = (0.9, 1.0, 1.3, 1.44, 1.92, 1.44, 0.09, 0.77, 0.9)$.

2.2. Simulation Configuration and Model Setup

The uniform (no shear) and laminar (no turbulence) approaching flow with the uniform velocity of $U = 10$ m/s is used in our simulations.

The buildings were arranged to provide cases having one, two, or four street canyons. The influences of the two-dimensional building heights are examined by using combinations of a 60 m building height (B1) and a 90 m building height (B2). In all cases, the building width is constant at 120 m and the urban street canyon (SC) width is 60 m. The pollutant source is located at the center of the street with a width of half of the street canyon (i.e., $L_{\text{source}} = 30$ m) (Figure 1).

A variable-spaced grid of 37 cells in the vertical direction is used in all simulations; the variable-spaced grid in the longitudinal direction depends on the length of the domain, varying from 86 cells (for one canyon) to 170 (for four canyons). The smallest cell dimension (3 m) in the domain is one-twentieth of the building height.

The computed fields of U_i, k, and those from TEMPEST, as discussed, are used to

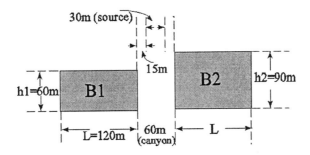

FIGURE 1. The side view of the generalized 2-D buildings and the street canyons for numerical simulations.

TABLE 1. Geometric configuration for numerical simulations.

Case	Geometric configuration
U1	B1–SC–B1
U2	B1–SC–B2
U3	B2–SC–B1
U4	B2–SC–B1–SC–B2
U5	B2–SC–B1–SC–B2–SC–B1–SC–B2

calculate the concentration field for the street level sources under five different street canyon types in each case group (see Table 1). The concentration field was calculated by solving the concentration transport equation [Zhang *et al.*, 1993] with the first-order closure scheme.

Considering the computing time and machine memory requirements, these idealized two-dimensional street canyon simulations are possible on either a personal computer or a workstation.

3. Results and Discussion

3.1. *Dispersion Structure Inside a Single Street Canyon*

Five simulations with two different building type (B1 and B2) combinations were conducted to investigate the influence of geometric configurations on the dispersion inside the idealized urban street canyons.

Only cases consisting of one canyon (U1, U2, and U3) are presented here. Figures 2, 3, and 4 display the concentration contour fields around a two-dimensional street canyon. All of the one-canyon cases (U1, U2, and U3) have a very similar flow pattern before the upstream edge of the upwind building. The flow structure before the upstream building had negligible effects on the overall dispersion fields. A recirculation zone exists on the top of the upstream building in all three cases. In both cases U1 and U2 (Figure 2, Figure 3), the flow just above the first building recirculation zone passes smoothly downstream, tending to remain at the same level above the ground. In case U3 (Figure 4), flow above the recirculating zone remains at the same vertical level above the vortex on the top of the upstream building as it passes over a recirculation region over the downstream

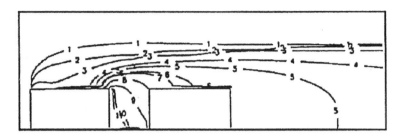

FIGURE 2. Concentration contour levels around canyon in case U1. The normalized χ-contour ($\chi = CU_r L_{sc}/Q$) levels are 1: 1.3×10^{-3}, 2: 1.3×10^{-2}, 3: 1.3×10^{-1}, 4: 1.3, 5: 2.6, 6: 5.3, 7: 6.7, 8: 13.3, 9: 26.7, 10: 53.3, 11: 66.7.

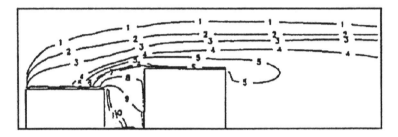

FIGURE 3. Concentration contour around street canyon in case U2. The contour level and views are the same as in Figure 2.

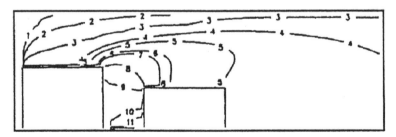

FIGURE 4. Concentration contour around street canyon in case U3. The contour level and views are the same as in Figure 2.

building. The reversed flow on top of the upstream building draws the pollutants from the canyon to the roof of the upstream building in all three cases.

In case U1, with the buildings at an equal height, there was one large vortex generated in the gap between the buildings. In case U2, with the downwind building 1.5 times as tall as the one upwind, the flow passing over the first building is somewhat blocked by the second one. Again a single vortex is formed in the canyon. The size and shape of the vortices inside the canyon for U1 and U2 depend on the relative heights of the buildings at both sides of the canyon. In case U3, the upwind building is 1.5 times higher than the downwind one, and the flow goes downward as it passes. This downward flow has a tendency to form a big wake behind the building. Because there is a second building in this potential wake, the flow hits the second building roof and tends to go back upstream,

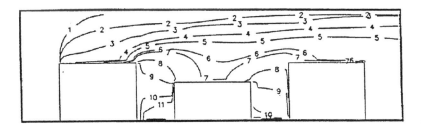

FIGURE 5. Concentration contour around street canyon in case U4. The contour level and views are the same as in Figure 2.

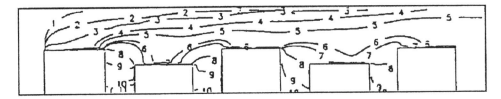

FIGURE 6. Concentration contour around street canyon in case U5. The contour level and views are the same as in Figure 2.

forming two counterdirected recirculating vortices: the clockwise one that is larger and stronger and overlaps the step-down building, and the secondary counterclockwise one at the bottom of the canyon that the canyon itself generates.

The differences in canyon configurations result in different dispersion patterns, and the maximum concentration shifts from the downstream side of the canyon in U3 to the upstream side of the canyon in U1 and U2 (Figure 2 and Figure 3). The concentration on the upstream side of the canyon is twice as high as that on the downstream of the canyon in cases U1 and U2, whereas the concentration downstream of the canyon is more than 2.5 times as high as that on the upstream side of the canyon in U3 (Figure 4).

In cases U1 and U2, the height of the wake behind the downstream building depends mainly on the downstream building height. The length of the wake appears to be proportional to the height of the downstream building. In case U3, the strong reversed flow above the second building deflects the flow above the canyon. The resulting wake size behind this building is double the downwind building height (60 m) and the length of this wake is triple that of case U1. This strong recirculating flow directs the pollutants upstream, and the concentration is smaller inside the wake behind the upstream building.

3.2. *Multiple Canyons*

In urban atmospheric environments, most situations have several canyons in a row. So, we extended our simulations to include cases involving two and four canyons.

When two or more street canyons are involved (cases U4 and U5), we find that the flow and dispersion patterns in the first canyon simply depend on the upstream conditions and the relative heights of the buildings beside the canyon; those inside the following canyons have flow and similar dispersion patterns. Figure 6 presents case U5 with four canyons in the downwind direction. The concentration contour pattern inside the first canyon is very similar to that in case U3; that inside the second canyon is similar to that in the second canyon in the U4 (Figure 5) but has dramatic differences from that in case

U2. The contour pattern inside the fourth canyon still has some of the characteristics of U3. Our results show that the dispersion fields have steady structure after two canyons.

4. Conclusions

Our study was limited to the cases investigated. A different ratio of building height to street canyon width will influence dispersion patterns in the vicinity of the canyons. This study serves as a start for future research on the subject. The canyon geometry has significant influences on the concentration fields in the vicinity of urban street canyons.

Pollutants emitted within a canyon will generally not be transported to the front of the upstream buildings. The recirculating flow above the upstream building depends more strongly on the upstream shear level than the relative canyon configuration. The flow and concentration pattern inside the canyon depends on both the upstream flow conditions (shear level and turbulence level) and the relative heights of the buildings on both sides of the canyon. The dispersion fields behind the downstream building depend on both the flow characteristics within the canyon and the downstream building height.

The high concentration stays on the upstream side of the building, except in the step-down canyon cases U3, U4, and U5.

When there is more than one canyon in the downwind direction, the upstream flow conditions can be very critical to the dispersion fields inside the first canyon, but they are not as important to the concentration fields inside the consecutive canyons. The dispersion fields in the vicinity of the first canyon are very similar to the corresponding cases involving only one canyon and the same upstream boundary conditions. The fields in the vicinity of the following canyons depend only on the relative geometric configuration on two sides of the canyon.

Acknowledgments. This work has been funded in part by the U.S. Environmental Protection Agency under contract 68-D0-0106 to ManTech Environmental Technology, Inc. Mention of trade names or commercial products does not constitute endorsement or recommendation for use. The author is grateful for the help and advice of Dr. Alan Huber in connection with the project.

REFERENCES

Amsden, A. A., and F. H. Harlow, The SMAC Method: A numerical technique for calculating incompressible flows, Report LA-4370, Los Alamos Scientific Laboratory, Los Alamos, NM, 1970.

Gibson, M. M., and B.E. Launder, Ground effects on pressure fluctuations in the atmospheric boundary layer, *J. Fluid Mech.*, **86**, 491–511, 1978.

Trent, S. D., and L. L. Eyler, TEMPEST, a three-dimensional time-dependent computer program for hydrothermal analysis. Volume 1. Numerical methods and input instructions, PNL-4348, Pacific Northwest Laboratory, Battelle, WA, 1989.

Zhang, Y. Q., A. H. Huber, S. P. Arya, and W. H. Snyder, Simulation to determine the effects of incident wind shear and turbulence level on the flow around a building, *J. Wind Engr. Indus. Aerodyn.*, **46 & 47**, 129–134, 1993.

Coupling of an Urban Dispersion Model and an Energy-Budget Model

By Glenn T. Johnson,[1] A. John Arnfield[2] and Jan M. Herbert[3]

[1]Department of Computing, Macquarie University, NSW 2109, Australia

[2]Department of Geography, The Ohio State University, Columbus, OH 43210, USA

[3]School of Land Economy, University of Western Sydney, NSW 2753, Australia.

A numerical model which simulates three-dimensional dispersion of a scalar within an urban area has been developed. The wind field used in the simulation is predicted on the basis of the surface geometry and a given upwind vertical profile by the control-volume technique. The initial application was the prediction of concentration fields associated with carbon monoxide emitted from motor vehicles. The model is now being coupled with an existing building-facet energy-budget model in order to simulate the evolution in time of fields of air temperature within the canyon air space and patterns of the surface energy budget. The current formulation is two-dimensional, for clear skies, and for a dry system. The inputs to the model include the upwind temperature profile, thermal properties of surface materials, temperatures within buildings and solar geometry. Preliminary work is presented, focusing in particular on turbulent heat exchange at the surfaces.

1. Introduction

Population growth and increased industrial activity in recent times have placed enormous pressures on the urban physical environment such that water and air quality have deteriorated to the point of threatening human health and well-being. Careful urban planning and design are recognized as providing some solutions to these problems, but such solutions must be firmly founded on a detailed understanding of the physical basis of urban climates.

The problem is that our understanding of urban climatology is fragmentary and, too often, site-specific. As Mestayer and Anquetin [1995] recently observed,

> A large part of our knowledge comes from incidental results of research and observations in the neighbouring fields of meteorology, dispersion over complex terrains, and wind engineering. The applied character of this last research field is the reason why we can find innumerable studies of the pressure fields on isolated or small groups of buildings, but much fewer complete descriptions of the flows around them, even fewer systematic studies of these flows, and practically no study of the flows and dispersion through large arrays of building like structures.

It has been our aim for some time to take numerical techniques developed in wind engineering and to apply them to systematic studies of urban canyons, the fundamental repeating unit of the urban canopy layer. Our current endeavor is to simulate the energy exchanges at canyon surfaces since the fundamental principles of heat, mass and momentum exchange constitute the physical basis of urban climates. Moreover, the coupling of the urban canopy layer and boundary layer "heat islands" [Oke, 1987] must be, to a large extent, affected by turbulent exchange at the tops of the urban canyons and at roof surfaces. The approach adopted in this investigation provides a means of determining the degree to which this coupling is controlled by meteorological and site conditions and characteristics of urban morphology.

161

2. Model Structure

Our proposed model, the Johnson/Arnfield Canyon Model (JACM), is modular in design and is being built by coupling a two-dimensional version of a dispersion model SCALAR [Johnson and Hunter, 1995] with an energy-balance model ENERBAL. ENER-BAL is an extension of the canyon model described by Mills [1993], which in turn is based on that of Arnfield [1976]. SCALAR is dependent on the output from a three-dimensional wind-flow model CITY [Patterson and Apelt, 1986, 1989] to provide mean steady-state velocity components and momentum eddy diffusivity. A sequence of these steady-state values provides the typical flow regimes of a 24 hour period.

The objective is to make the model as general as possible. Thus it accepts as input latitude of location, canyon orientation, day of month and month of year (specifying solar radiation), and canyon geometry and thermal properties of building fabric and soil (specifying shadowing effects, longwave irradiance and conductive heat flow).

In production mode, we intend to run JACM for two identical days. The first day of the simulation is intended to ensure that unrealistic initial temperature distributions within the building fabric and soil are "forgotten" and replaced by physically consistent fields for the start of the second day.

3. Wind Model

The wind-flow model (CITY), which provides input to SCALAR, is an implementation of a k-ϵ turbulence model [Hunter *et al.*, 1990/1991], which determines a three-dimensional flow between and around a building group from a given upwind vertical profile, the geometry of the buildings and the roughness lengths of the building surfaces. It employs finite differencing using the control-volume technique with hybrid upwinding. The resultant equations are solved by the alternating direction implicit (ADI) method. The model assumptions relate to situations in which vortex flow is established within the canyon. According to DePaul and Sheih [1986], that fact limits the application of CITY (and hence JACM) to situations where the roof-level wind speed is greater than 1.5 to 2.0 m s^{-1}.

A field program [Johnson *et al.*, 1995] that provides data to test the capabilities of CITY has recently been completed. Preliminary analysis of that data indicates that the model is able to predict realistic along-canyon spiral flows.

4. Dispersion Model

SCALAR is a three-dimensional model based on the following form of the atmospheric diffusion equation in which dispersion within, and out of, the canyon is assumed to result from advection and turbulent diffusion:

$$\frac{\partial T}{\partial t} = \frac{\partial}{\partial x}\left[\kappa_H \frac{\partial T}{\partial x}\right] + \frac{\partial}{\partial y}\left[\kappa_H \frac{\partial T}{\partial y}\right] + \frac{\partial}{\partial z}\left[\kappa_H \frac{\partial T}{\partial z}\right]$$
$$-u\frac{\partial T}{\partial x} - v\frac{\partial T}{\partial y} - w\frac{\partial T}{\partial z} + S \tag{4.1}$$

Here $T(x, y, z, t)$ is the air temperature at the point (x, y, z) at time t; $\kappa_H(x, y, z)$ is the eddy diffusivity for heat; $u(x, y, z), v(x, y, z), w(x, y, z)$ are the wind components; and $S(x, y, z, t)$ is the source term. Note that u, v, w and κ are supplied by CITY. Buoyancy is assumed to be a second-order effect and, at this stage, is ignored. The source term enters calculations only in cells near surfaces and in those cases is given by

$$S = \frac{Q_H}{\rho c_p D} \tag{4.2}$$

where Q_H is the turbulent sensible heat flux at the surface, ρ is the air density, c_p is the specific heat of air at constant pressure, and D is the depth of the surface cell measured normal to the surface with which heat is being exchanged.

5. Energy Balance Model

ENERBAL considers each building facet as being broken up into elements and determines the energy budget of each on the basis of

$$Q^* = Q_H + Q_G \tag{5.3}$$

where Q^* is the net radiation and Q_G is the conductive heat flux. Net radiation is composed of direct, diffuse and reflected solar (K) irradiance and received and emitted longwave (L) irradiance

$$Q^* = K \downarrow_S + K \downarrow_B - K \uparrow + L \downarrow_S + L \downarrow_B - L \uparrow \tag{5.4}$$

where the arrows represent directions to ($\downarrow$) and from ($\uparrow$) the surface, and subscripts S and B represent the sky and surrounding building (and ground) sources, respectively. Each of these terms is calculated by using the model of Arnfield [1976], which calculates and retains the canyon and sky view factors for each element.

6. Heat Flux Calculation

ENERBAL relies on a heat transfer function of the form

$$Q_H = h_c(T_e - T_a) \tag{6.5}$$

where T_e is the temperature of an element, T_a is the air temperature adjacent to that element, and h_c is a heat transfer coefficient.

To couple ENERBAL and SCALAR, the assumption is made that a constant flux layer exists in each SCALAR cell adjacent to a surface and that within this cell the air temperature follows a logarithmic profile based on the roughness length (z_0) of the underlying surface. Then

$$Q_H = -\rho c_p \kappa_H \frac{\partial \Theta}{\partial z} \tag{6.6}$$

where Θ is the potential temperature, z is the height above the surface, and κ_H is the eddy diffusivity for heat.

We assume that

$$\kappa_H = \frac{\kappa}{\sigma_t} \tag{6.7}$$

where κ is the momentum eddy diffusivity calculated by CITY and σ_t is a (constant) turbulent Prandtl number.

Now,

$$\frac{\partial \Theta}{\partial z} = \frac{\Theta_*}{kz} \tag{6.8}$$

where Θ_* is the scaling potential temperature and k is the von Karman constant. The preceding equation then gives us

$$\Theta_a - \Theta_s = \frac{\Theta_*}{k} \ln\left(\frac{\delta z}{z_0}\right) \tag{6.9}$$

at a height δz above the surface. Assuming that Θ and T are essentially equal in the region of interest (i.e., the surface barometric pressure is close to 100 kPa), we have

$$Q_H = -\rho c_p \kappa_H \frac{(T_a - T_e)}{\delta z \, \ln(\frac{\delta z}{z_0})} \tag{6.10}$$

and

$$h_c = \frac{\rho c_p \kappa_H}{\delta z \, \ln\left(\frac{\delta z}{z_0}\right)} \tag{6.11}$$

This h_c is used by ENERBAL as the heat transfer coefficient in Eq. (6.5).

T_e is computed numerically at each ENERBAL time step by balancing the surface energy budget, Eqs. (5.3) and (5.4), for each element on the building and ground surfaces. The conductive heat flux is determined in this process from the substrate thermal properties and the subsurface temperature profile. This profile evolves through time in response to surface forcing and conditions placed on the lower boundary temperature. For the ground surfaces, this temperature is constant at about 0.5 m depth. For the building facets, this temperature is the internal temperature of the walls, which is partially controlled by simulated space heating and cooling requirements and represents an anthropogenic impact on the surface energy budget.

7. Module Integration

Given that JACM has been constructed using several preexisting modules which were implemented independently and with different objectives, some care is required in the way in which the modules are linked and some design decisions have been influenced by the existing module structure.

Both CITY and SCALAR were developed as three-dimensional models whereas ENER-BAL was designed to compute energy balances at midcanyon and is essentially two-dimensional. We have decided initially to investigate relatively long canyons in which there is little lateral transfer of heat at midcanyon. Accordingly, although JACM accepts the three-dimensional flow output from CITY, it disperses temperature in a two-dimensional vertical across-canyon slice at midcanyon.

The grid used by SCALAR extends well above the canyon and a variable mesh size is applied to enhance computational efficiency. On the other hand, that used by ENERBAL is related only to surface facets and the substrate beneath, and equal-sized elements are applied to each facet. Rather than force the grid specifications to coincide, we have decided to permit the modules to work with independent grids. When heat fluxes and diffusivities are to be passed between the modules, cubic splines are fitted through the values of the variables along the length of each surface to ensure that accurate interpolation to grid locations is made. Similarly, time steps chosen for one module do not have to match time steps chosen in the other. The only condition required is that they be exact multiples or fractions of each other.

The aim throughout this design phase has been to maintain as much independence for each module as possible, yet at the same time provide seamless integration between them so that the software handles all space/time inconsistencies. This is done to avoid prejudging which numerical techniques should be applied within the modules or even which physical processes should be modeled or parameterized.

8. Current Situation

All of the coding needed to integrate the modules has been completed, and preliminary interface testing has commenced. The next step will be to model a simple test building arrangement and to investigate the physical plausibility of the predicted temperature distribution. After model refinement based on this analysis, JACM will be applied to data recently collected in an east–west asymmetric canyon in Columbus, Ohio, USA [Arnfield and Mills, 1994a, 1994b], as a first step in the model verification process.

Acknowledgments. Work was supported in part by grants from Australian Research Council and Macquarie University, by a research agreement with Digital Equipment Corporation and by a grant from Ohio Supercomputer Center, which provided access to their CRAY Y-MP and Ohio Visualization Laboratory.

REFERENCES

Arnfield, A. J., Numerical modelling of urban surface radiative parameters, *Papers in Climatology: The Cam Allen Memorial Volume*, J. A. Davies, Ed., Department of Geography, McMaster University, Discussion Paper Number 7, 1–28, 1976.

Arnfield, A. J., and G. M. Mills, An analysis of the circulation characteristics and energy budget of a dry, asymmetric, east-west urban canyon. I. Circulation characteristics, *International Journal of Climatology*, **14**, 119–134, 1994a.

Arnfield, A. J., and G. M. Mills, An analysis of the circulation characteristics and energy budget of a dry, asymmetric, east-west urban canyon. II. Energy budget, *International Journal of Climatology*, **14**, 239–261, 1994b.

DePaul, F. T., and C. M. Sheih, Measurement of wind velocities in a street canyon, *Atmospheric Environment*, **20**, 455–459, 1986.

Hunter, L. J., I. D. Watson, and G. T. Johnson, Modelling of airflow regimes in urban canyons, *Energy and Buildings*, **15–16**, 315–324, 1990/1991.

Johnson, G. T., H. A. Cleugh, J. R. Barnett, and L. J. Hunter, Field evaluation of a scalar canyon airflow model, School of MPCE, Macquarie University, Tech. Report No. 95/182c, 1995.

Johnson, G. T., and L. J. Hunter, A numerical study of dispersion of passive scalars in city canyons, *Boundary-Layer Meteorology*, **75**, 235–262, 1995.

Mestayer, P. G., and S. Anquetin, Climatology of cities, in *Diffusion and Transport of Pollutants in Atmospheric Mesoscale Flow Fields*, A. Gyr and F. Rys, Eds., Kluwer Academic, 1995.

Mills, G. M., Simulation of the energy budget of an urban canyon – I. Model structure and sensitivity test, *Atmospheric Environment*, **27B**, No. 2, 157–170, 1993.

Oke, T. R., *Boundary Layer Climates*, 2nd ed., Methuen & Co., 1987.

Patterson, D. A., and C. J. Apelt, Computation of wind flows over three-dimensional buildings, *J. Wind. Eng. Ind. Aero.*, **24**, 192–213, 1986.

Patterson, D. A., and C. J. Apelt, Simulation of wind flows around three-dimensional buildings, *Building and Environment*, **24**, 39–50, 1989.

A Mesoscale Meteorological Model to Predict Windflow in the Valley of Mexico

By M. D. Williams,[1] G. Sosa[2] and V. Mora[2]

[1]Los Alamos National Laboratory, Group A-4, Mail Stop B299
Los Alamos, NM 87545, USA

[2]Instituto Mexicano del Petroleo, Eje Central Lazaro Cardenas 152
07730 México, D.F., México

A three-dimensional, prognostic higher-order turbulence meteorological model (HOTMAC) was modified to include a characteristic urban canopy and urban heat sources for Mexico City. HOTMAC is used to drive a Monte Carlo kernel and nonreactive dispersion code (RAPTAD) also. Moreover, HOTMAC also provides winds and mixing heights for the CIT photochemical model, which is used to predict up-to-date ozone concentrations in Mexico City. The main results of this model's applications are presented in this paper.

1. Introduction

Los Alamos National Laboratory and Instituto Mexicano del Petroleo have embarked on a joint study of options for improving air quality in Mexico City. The intent is to develop a modeling system which can address the behavior of pollutants in the region so that options for improving air quality can be properly evaluated. In February 1991, the project conducted a field program which yielded a variety of data which are being used to evaluate and improve the models. Normally the worst air quality for both primary and photochemical pollutants occurs in the winter in Mexico City.

During the field program, measurements included (1) lidar measurements of aerosol transport and dispersion; (2) aircraft measurements of winds, turbulence, and chemical species aloft; (3) aircraft measurements of earth surface skin temperatures; and (4) tethersonde measurements of wind, temperature and ozone vertical profiles.

The Mexico City Metropolitan Area (MCMA) lies at an elevation of approximately 7500 feet above sea level in a "U" shaped basin which opens to the north. Mountains on the east and southeast sides of the basin form a barrier with a height of approximately 12,000 feet, while two isolated peaks reach elevations in excess of 17,000 feet. The city occupies a major part of the southwest portion of the basin. Upper level winds are provided by rawinsondes at the airport, and low-level winds are measured at several sites within the city. Many of the sites have obstructed upwind fetches for a variety of directions.

A three-dimensional, prognostic higher-order turbulence meteorological model, HOTMAC, was modified to include an urban canopy and urban heat sources. This model is used to drive a Monte Carlo kernel dispersion code, RAPTAD. HOTMAC also provides winds and mixing heights for the CIT photochemical model, which was developed by investigators at the California Institute of Technology and Carnegie Mellon University.

2. Model Formulation

HOTMAC is a three-dimensional time-dependent model developed by T. Yamada [1985]. It uses the hydrostatic approximation and a terrain-following coordinate system. HOTMAC solves conservation relations for the horizontal wind components, potential

TABLE 1. Categories used to determine surface features.

1. Vegetation
2. Mostly bare soil
3. Dark soil
4. Shadow–volcanic–urban
5. Urban lower income
6. Vegetation–foothills–city
7. Water
8. Dark urban material
9. Urban material mixture
10. Urban
11. Vegetation mix
12. Mountain vegetation
13. Mountain vegetation and rock mixture

temperature, moisture, turbulent kinetic energy, and turbulence length scale. HOTMAC describes advection, Coriolis effects, turbulent transfer of heat, momentum, and radiation effects of forest canopies. The explicit form of the equations is described by Yamada [1981] and Mellor and Yamada [1982].

The lower boundary conditions are defined by a surface energy balance and similarity theory. The soil heat flux is obtained by solving a heat conduction equation in the soil which ignores lateral heat transfer. In an urban context the surface energy balance requires an additional term which represents the heat released by man's activities. The additional heat along with differences in thermal and albedo properties between urban and nonurban surfaces produce the urban heat island.

Early in the project, 3 days in which air quality was poor, good, and normal were chosen for detailed modeling. All of the days were in winter of 1987–88. Meteorological inputs were based on the afternoon rawinsonde of the preceding days, which was used to estimate synoptic scale wind and temperature profiles.

We used a nested grid system to model the Valley of Mexico and its surrounding terrain. The outer grid has 6 km spacing and covers the major terrain influences, as shown in Figure 1. The inner grid, also shown in the figure, embraces the city and the immediately adjacent slopes. The inner grid has a size of 2 km.

Initially, the urban canopy was approximated by using the estimated distribution of CO emissions defined on a 1 km grid. The relative CO emissions were used to determine the fraction of the area of a grid cell which was covered by canopy (roof tops), average soil conductivity, average soil heat capacity, and urban heat release intensity. The modeling showed that on days with poor ventilation and consistent upper level winds, the meteorology of the region could be reasonably represented. However, with changing upper-level winds, the model gave poor results.

The original version of the model used only two surface features: water surface and everything else. The model was changed to permit an urban classification, and later it was further altered to accept the results of a satellite-derived surface categorization. The categorization produced 13 categories, which are listed in Table 1.

The satellite data were used to categorize the inner-grid area. Areas outside the inner grid were described as having mountain vegetation for elevations above 2600 m and as having foothill vegetation (category 6) to represent scrub lands for elevations below 2600 m. The satellite data file gave the fractions of each category in the 2 km by 2 km

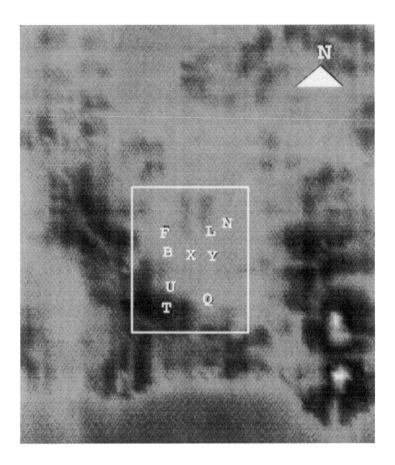

FIGURE 1. Modeling domains for the Valley of Mexico air basin. The entire picture constitutes the outer grid of the meteorological model. The inner grid is delimited by the black box. Shading represents topographic contours.

grid cells, which were 1 km offset in each direction from the meteorological grid. The fractional land coverage data were interpolated to the meteorological grid.

For each classification, an estimate was made of the associated surface characteristics: (1) surface albedo, (2) surface thermal emissitivity, (3) surface daytime Bowen ratio, (4) soil heat capacity, (5) soil density, and (6) soil thermal diffusivity. The land coverage percentages were used with the category values to estimate the appropriate surface characteristics for each grid cell. In the case of Bowen ratio, simple area weighted means, as used for the outer parameters, are not appropriate.

The model moisture treatment was also changed because it was too simple. In the revised treatment moisture could accumulate at night on the surface if the temperature reached the dew point. The next day, the sun's heating would be used to dissipate the surface moisture until it had been evaporated. If the surface was dry and the sun was high, the Bowen ratio, which is the ratio of the sensible heat flux to the latent energy flux, is assumed to be constant. In other words, the energy flux to the atmosphere is apportioned between the heating of the air and evaporation/transpiration of water by a fixed ratio. At other times the evaporation is nil.

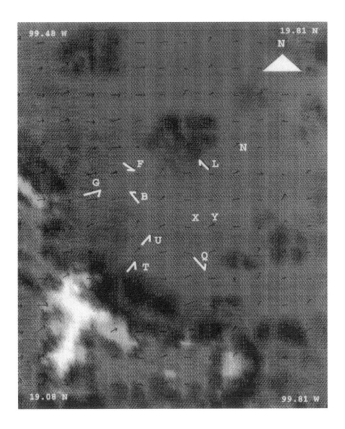

FIGURE 2. Computed winds for 0600 LST, for February 1991.

3. Results and Discussion

The model produces volume-average, ensemble mean vector winds, where these winds are interpolated between grid cells to the monitoring locations in order to make comparisons. The measurements represent scalar-averaged wind speeds and directions. In addition to the uncertainty caused by the averaging techniques, stations may experience local influences on too fine a scale for the model.

Figures 2–4 show model predictions for the early morning, afternoon, and morning transition for February 22. Only the fine grid is shown. The 0600 LST slope winds, as evidenced by stations U and T, which are closest to the mountains, are well represented.

Station B is not well represented, but it generally seems to show anomalous behavior. The city stations such as X and Y show somewhat different behavior, which is not what might be expected because of the likely local effects. There is also some suggestion in the data that there is more wind convergence over the city than the model shows, as indicated by winds from the east on the east side of the city.

The transition to up-slope flows occurs at about 1000 LST, and the model shows a less-developed transition than do the measurements. Figure 4 shows the afternoon flows at 1300 LST, with fully developed slope winds. At 2100 LST strong winds out of the northwest dominate the flow fields, and there is good agreement between the model and the measurements.

This wind is likely the result of a coupled mountain–valley airflow and a maritime flow

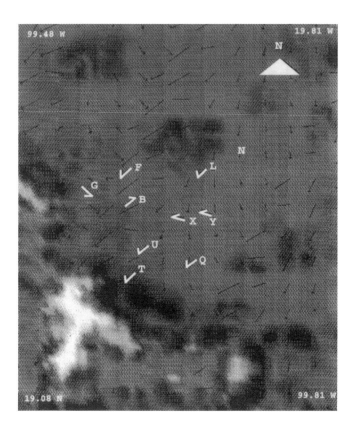

FIGURE 3. Computed winds for 1000 LST, for February 1991.

from the Gulf of Mexico. Some simulations with the Regional Atmospheric Modeling System (RAMS) and a large enough domain to include both oceans indicate that the model is capable of predicting the occurrence of these winds. These three figures demonstrate the variety of wind conditions that can occur in the valley. They also show that the model does a reasonably good job of representing the major features.

4. Lidar-Derived Mixing Heights

For 4 days lidar-derived mixing heights were available in the period of interest. The lidar-derived mixing heights were determined as the height at which 50% of the horizontal area has a signal characteristic of the clean air aloft. There were two sites used in the 4 days. During February 22, the lidar was at the CINVESTAV site, which is a few kilometers north of the city center. On February 26–28, the lidar was at the UNAM site, which is on the southern boundary of the city. The UNAM site is about 100 m above the level of the city, and the mixing height measurements are relative to the height of the site. The actual line of sight is over the city, which is at a lower elevation, so that the mixing heights would be expected to be about 100 m because of line-of-sight restrictions imposed by nearby objects.

The mixing heights were calculated using the HOTMAC model coupled to a random trajectory dispersion model, which is called RAPTAD. Figure 5 displays the mixing

FIGURE 4. Computed winds for 1300 LST, for February 1991.

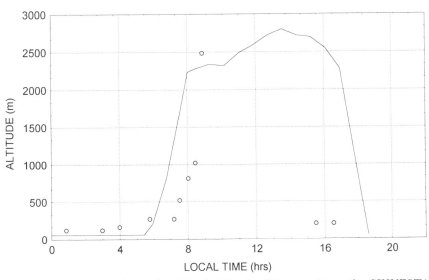

FIGURE 5. Model and experimental mixing heights for February 21, at the CINVESTAV site.

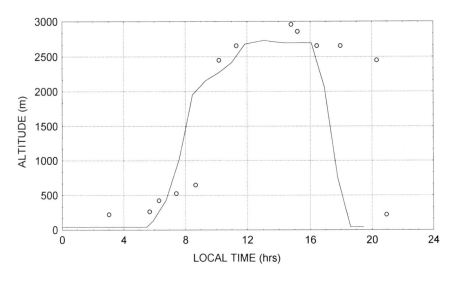

FIGURE 6. Model and experimental mixing heights for February 26, at the UNAM site.

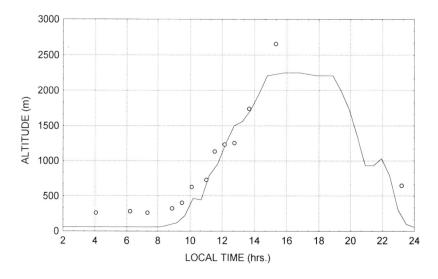

FIGURE 7. Model and experimental mixing heights for February 27, at the UNAM site.

height comparison between the model and measurements for February 22, while Figures 6 and 7 report the comparisons for February 26–27, respectively. The agreement for February 22 is excellent; the principal disagreements are in the night, when the lidar minimum heights may be important, and at one point in the late afternoon when clouds could have influenced the comparison. The agreements for February 26–27 are not as good. In particular, the measurements show a much slower increase in the mixing height in the midmorning.

5. Conclusions

The HOTMAC windflow predictions are consistent with wind measurements, although the results are not good enough, sometimes, in particular places. Generally the model captures the major meteorological features; the winds respond well to major changes in forcing winds. The slope winds develop appropriately and couple well with the large-scale conditions. There are some areas, however, which could be improved. For instance, the model does not have quite as much convergence over the city as it should, and it also appears that the temperatures do not drop as much at night as they should. The rapid rise in mixing heights in the February 25–28 simulations is of concern and is probably related to temperature behavior. The wind speeds seem to be a little low. The light wind speeds may be a function of the way larger-averaged winds are used as input, or they may reflect the fact that the model is currently not using some of the information from soundings in the range from 750 m to 2000 m above the ground.

REFERENCES

Mellor, G. I., and T. Yamada, Development of a turbulence closure model for geophysical fluid problems, *Rev. Geophys. Space Phys.*, **20**, 851–875, 1982.

Yamada, T., A numerical simulation of nocturnal drainage flow, *J. Meteor. Soc. Japan*, **59**, 108–122, 1985.

Yamada, T., Numerical simulation of the night 2 data of the 1980 ASCOT experiments in the California Geysers area, *Archives for Meteorology, Geophysics, and Bioclimatology*, Ser. **A34**, 223–247, 1981.

Some Experiments with a Three-Dimensional Semi-Lagrangian and Semiimplicit Cloud Model

By José A. Vergara

UCAR Visiting Research Program at the U.S. National Meteorological Center, Washington, DC, USA, and Department of Meteorology, University of Maryland, College Park, MD 20742-5121, USA

A three-dimensional cloud model has been developed for the study of the dynamic of convective systems. This model integrates the Euler and microphysical equations using a semi-Lagrangian and semiimplicit scheme. Microphysical processes are included in the model, using warm rain parameterization. The numerical grid permits the use of a very high spatial resolution for the study of cloud evolution. Experiments were performed to assess the role that the vertical temperature and humidity profile play in the formation of clouds. The most important features of this model are its capacity and numerical efficiency for properly simulating the nonhydrostatic and mixing process in the clouds. This model is able to reproduce the many important aspects of convective clouds and tornado evolution. Comparisons with results obtained with traditional cloud models show excellent performance of the model, in terms of both accuracy and computational efficiency.

1. Introduction

The main objective of this paper is to present the development of a new three-dimensional cloud model for the study of moist, precipitating, rotating and fully compressible fluids. This model contains three classes of hydrometeors all treated in a highly parameterized fashion. Dependent variables predicted by the model are the three wind components, as well as pressure, temperature, water vapor, cloud water and rain water. This model solves the compressible equations of motion using a semi-Lagrangian and semiimplicit scheme [Robert, 1969; Robert, 1993; Tanguay *et al.*, 1990; Tapp and White, 1976]. Microphysical processes are included in the model using the warm rain parameterization of Lin *et al.* [1983], Klemp and Wilhelmson [1978], Seman [1991], Vergara and Seman [1993] and Vergara [1996].

2. Model Description

This model solves the compressible equations of motion, using a semi-Lagrangian and semiimplicit scheme that provides numerical efficiency for the sound wave modes and gravity oscillations. For the implementation of semiimplicit and semi-Lagrangian schemes, it is convenient to expand the thermodynamic variables about mean-states profiles as $T = T^* + T'$ and $q = q^* + q'$, where $q = \ln(p/p_0)$. The reference state is used, with an isothermal temperature profile and the corresponding hydrostatic vertical distribution of pressure

$$T^* = \text{constant} \tag{1}$$

$$q^*(z) = -\frac{qz}{RT^*} \tag{2}$$

The basic state for the microphysical model is a dry atmosphere, where $q_v = q'_v$, $q_c = q'_c$ and $q_r = q'_r$. In perturbation form, the model equations in a nonrotating atmosphere can then be written in the following form:

$$\frac{du}{dt} + RT^* \frac{\partial q'}{\partial x} = -RT^* \frac{\partial q'}{\partial x} \tag{3}$$

$$\frac{dv}{dt} + RT^* \frac{\partial q'}{\partial y} = -RT^* \frac{\partial q'}{\partial y} \tag{4}$$

$$\frac{dw}{dt} + RT^* \frac{\partial q'}{\partial z} - g(0.61 q_v - q_c - q_r) = -RT' \frac{\partial q'}{\partial z} \tag{5}$$

$$(1 - \alpha) \left(\frac{dq'}{dt} - g \frac{w}{RT^*} \right) + \text{DIV} = \frac{H}{T} \tag{6}$$

$$\frac{dT'}{dt} - \alpha T^* \frac{dq'}{dx} + \frac{\alpha}{R} gw = \frac{\alpha}{(1 - \alpha)} T' \, \text{DIV} + \frac{[1 - \alpha(T'/T^*)]}{(1 - \alpha)} H \tag{7}$$

$$H = -\frac{L_v}{C_p}(C_N + E_r) \tag{8}$$

Here H represents the heating (condensation/evaporation) and evaporation of rain water; u, v, w are the three wind components; L_v is the latent heat of condensation for water vapor; C_p, the specific heat of air at constant pressure; $\alpha = R/C_p$; and C_N is the condensation/evaporation rate of rain water. The buoyancy term in Eq. (5) arises through linearization of the pressure term [Klemp and Wilhelmson, 1978].

3. Microphysical Processes

The microphysical processes are included in the model using the warm rain parameterization of Lin *et al.* [1983], and three conservation equations are considered here:

$$\frac{dq_v}{dt} = C_N + E_r \tag{9}$$

$$\frac{dq_c}{dt} = -C_N + A_r - C_r \tag{10}$$

$$\frac{dq_r}{dt} = A_r + C_r - E_r - F_r \tag{11}$$

where C_N = condensation of water vapor/evaporation of cloud drops; A_r = autoconversion; C_r = collection of q_c by q_r; E_r = evaporation of rain water; F_r = rain water fallout; and q_v, q_c and q_r are the mixing ratios for water vapor, cloud water and rain water, respectively. The mass-weighted mean terminal fall velocity for the rain (V_t) is based on the relations suggested by Liu and Orville [1969] and Lin *et al.* [1983]. It is expressed in CGS units as

$$V_t = 2115 \frac{\Gamma(4.8)}{6 \lambda^{0.8}_R} \left(\frac{\bar{\rho}}{\rho_0} \right)^{1/2} \tag{12}$$

where λ_R is the slope parameter in rain size distribution

$$\lambda_R = \sqrt{\frac{\pi q_r \eta_{OR}}{\rho \rho_w}} \tag{13}$$

and ρ_w is the density of water (1000 kg m^{-3}). The autoconversion of cloud droplets to form raindrops is parameterized using the relation suggested by Berry [1968] and Lin *et al.* [1983].

$$A_r = \frac{\overline{\rho}(q_c - q_{co})^2}{1.2 \times 10^{-4} + [1.569 \times 10^{-12} N_1/D_0 \rho(q_c - q_{co})]} \tag{14}$$

The autoconversion to rain water (A_r) depends on N_1 (the number concentration of cloud droplets in the assumed Marshall–Palmer distribution) and D_0 (0.15, the dispersion). The autoconversion turns on when the cloud water reaches a threshold of 2 g kg^{-1} [Lin *et al.*, 1983]. The rate of change in mixing ratio of q_r by accretion of cloud water is based on the geometric sweep-out idea and integration over all raindrop sizes for the assumed rain size distribution and is given as

$$C_r = \sqrt{\frac{\rho_O}{\overline{\rho}}} \frac{\pi E_C \eta_{OR} a \Gamma(3 + b) q_c}{4 \lambda_R^{3+b}} \tag{15}$$

The collection of cloud water by falling rain (C_r) is proportional to the amount of cloud water present and the collection efficiency, E_c (here assumed to be 1) [Lin *et al.*, 1983]. The evaporation rate of rain is calculated according to the concept of diffusional growth originally developed by Byers [1965], where

$$E_r = \rho^{-1} \frac{2\pi(S - 1)\eta_{OR}[0.78\lambda_R^{-2} + 0.31 S_c^{1/3}\Gamma((b+5)/2)a^{1/2}\nu^{-1/2}(\rho_O/\overline{\rho})^{1/4}\lambda_R^{-(b+5)/2}]}{[(L_v^2/K_a R_v T^4) + (1/\overline{\rho}q_{vs}\Psi)]} \tag{16}$$

The evaporation of rain water (E_r) is allowed only in unsaturated air $(S < 1)$, Ψ is the diffusivity of water vapor in air, and η_{OR} is the number of raindrops per unit diameter (0.08 cm^{-4}). Empirically we know that atmospheric condensation takes place at very low supersaturation as a result of the abundance of condensation nuclei in the air. Thus, we assume that condensation takes place at a relative humidity of 100%. The condensation/evaporation is solved iteratively [Vergara, 1996]. The first step consists in integrating the numerical model forward one time step without regard for saturation violations. During this step, the dry dynamic and all microphysics processes, except condensation (evaporation) of water vapor (cloud droplets), are activated. The last step is a saturation adjustment of grid points that are either supersaturated or subsaturated with cloud water. Changes in the heating and water vapor are constrained by linearizing Tenten's formula and using conservation relations for thermodynamic energy, water vapor and cloud water [Klemp and Wilhelmson, 1978]. The equations are discretized for integration by the semi-Lagrangian and semiimplicit scheme. The condensation/evaporation term is solved iteratively and the rain water fallout term is computed by the trapezoidal scheme [Vergara, 1994]. Any negative rain water, water vapor or cloud water resulting is adjusted to zero. The boundary conditions applied in this model were assumed to be rigid with free slip conditions. The time filtering and spatial filtering are not necessary for integration of the model.

4. Result

The model has been used for a variety of problems, including cases of hydrodynamic instabilities in rotating and nonrotating cloud and moist processes in a conditional unstable environmental. Here, special attention is given to modeling the convective cloud and tornado simulation. The tests reported here have been carried out in two-dimensional and three-dimensional versions of the model.

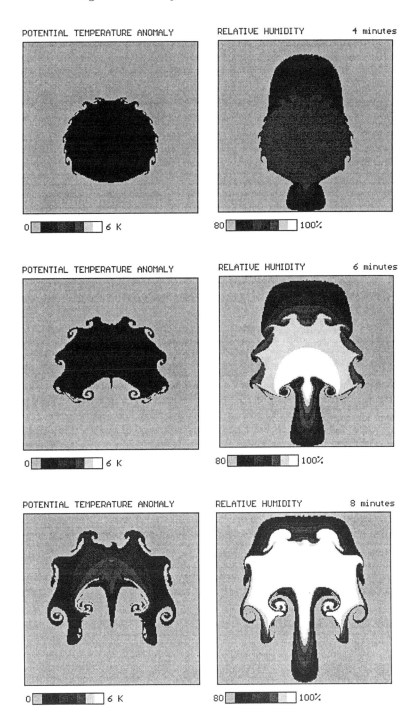

FIGURE 1. Plots of the potential temperature anomaly (right) and relative humidity (left) for three different times (4, 6, 8 minutes) from the 10-meter-resolution experiment.

4.1. *Convective Cloud*

Fifteen minutes of numerical integration was done with the cloud model for the study of bubble convection on a 2000 m (vertical) by 2000 m (horizontal) domain and a grid resolution of 10 m with a time step of 5 s. Note that it is larger than the integration time step (3 s) used by Smolarkiewicz and Grabowski [1995] for solving the same problem with a semi-Lagrangian model. The surface temperature is taken to be 30°C, and then is assumed to decrease dry-adiabatically up to 2 km in height. The relative humidity is specified to be 80%. A circular bubble with a diameter of 1000 m is considered, with the center positioned at 600 m above the surface. The bubble is assigned a uniform potential temperature of 30.5°C. The anomaly temperature and relative humidity predicted by the model at 4, 6, and 8 minutes are shown in Figure 1, which provides an outline of the temperature and relative humidity evolution. At 6 minutes, condensation appears in the bottom of the thermal, and the waves appear on the sides of the cloud; they later develop in the top and bottom parts of the cloud. The cloud water field at 10 minutes (2 and 4 minutes after initial condensation, respectively) is less than 5 g kg^{-1}. Because of the small integration time, no rain is formed in the cloud. Thus, in contrast to Smolarkiewicz and Grabowski [1995], it is not possible to compare the numerical efficiency of the scheme for the rain evolution equation.

4.2. *Tornado*

Most current knowledge of tornado dynamics has been obtained from axisymmetric models with a prescribed initial vertical vorticity and velocity imposed by the exhaust top boundary conditions. One question that remains unanswered is, What is the cloud dynamics/tornado relationship? The model is used to investigate this question. This model permits the use of high spatial resolution for the study of tornado evolution. In this study a grid of 51×51×51 points is employed with a domain of 4×4 km^2 in the horizontal by 7.5 km in the vertical. The integration time step is 4 s. The subgrid scale fluxes of momentum, heat and water are calculated by first-order closure parameterization. Experiments are designed to assess the role of the CAPE in the development of tornadoes. Here the first stage of this tornado modeling approach is presented. The preliminary results show that the model is able to reproduce many important aspects of tornado behavior, including the formation of the funnel and the wall cloud, and the realistic simulation of entrainment/detrainment within the cloud (Figure 2). The maximum horizontal wind speed is 25 m s^{-1} around the core of tornadoes and the maximum and minimum vertical velocities are 35 m s^{-1} and −19 m s^{-1} respectively [Vergara, 1996].

5. Conclusion

In comparisons with results obtained by other authors, the model shows excellent performance in terms of accuracy and computational efficiency. Two features of this model are its capacity and numerical efficiency for properly simulating nonhydrostatic and mixing processes in cloud behavior. However, a much higher resolution will be necessary to solve the complete tornado dynamics. For this purpose, a new version of the model is being developed. I think that this work demonstrates that simulation of a tornado is computationally feasible.

Acknowledgments. The author would like to thank Drs. Eugenia Kalnay and Charles J. Seman and Professor Ferdinand Baer for their comments and support.

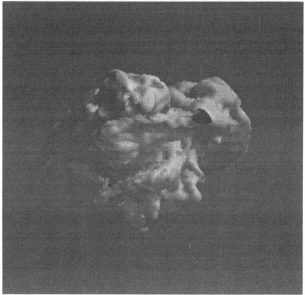

FIGURE 2. Cloud water isosurface (0.01 g kg^{-1}) at 720 and 800 s. The domain is 4×4 km^2 (horizontal) by 7.5 km (vertical). Note the prominent "funnel cloud" in the central part of the domain and intense entrainment on the base of the cloud and detrainment in the top of the cloud.

REFERENCES

Berry, E. X., Modification of the warm rain process, *Preprints 1st Nat. Conf. Weather Modification*, Albany, Am. Meteor. Soc., 81–88, 1968.

Byers, H. R., *Elements of Cloud Physics*, The University of Chicago Press, 1965.

Klemp, J. B., and R. Wilhelmson, The simulation of three-dimensional convective storm dynamics, *J. Atmos. Sci.*, **35**, 1070–1096, 1978.

Lin, Y.-H., R. D. Farley, and H. D. Orville, Bulk parameterization of the snow field in a cloud model, *J. Climate Appl. Meteor.*, **22**, 1065–1092, 1983.

Liu, L. Y., and H. D. Orville, Numerical modeling of precipitation and cloud shadow effects on mountain-induced cumuli, *J. Atmos. Sci.*, **26**, 1283–1298, 1969.

Robert, A. J., The integration of a spectral model of the atmosphere by the implicit method, *Proc. of the WMO-IUGG Symp. on NWP*, Tokyo, Japan Meteor, Agency, VII 19–24, 1969.

Robert, A. J., Bubble convection experiments with a semi-implicit formulation of the Euler equations, *J. Atmos. Sci.*, **50**, 1865–1873, 1993.

Seman, J., Numerical study of nonlinear convective-symmetric instability in a rotating baroclinic atmosphere, Ph. D. Thesis, Dept. of Meteorology, University of Wisconsin, Madison, 1991.

Smolarkiewicz, P. K., and W. W. Grabowski, Semi-Lagrangian/Eulerian cloud model, *Conference on Cloud Physics*, Dallas, Texas, January 15–20, 1995.

Tanguay, M., A. Robert, and R. Laprise, A semi-implicit semi-Lagrangian fully compressible regional forecast model, *Mon. Wea. Rev.*, **118**, 1970–1980, 1990.

Tapp, M., and P. W. White, A non-hydrostatic mesoscale, *Q. J. R. Meteor. Soc.*, **102**, 277–296, 1976.

Vergara, J., A fully compressible cloud model: Preliminary results, *Sixth Conference on Mesoscale Processes*, Portland, Oregon, July 18–22, 1994.

Vergara, J., Numerical simulation of tornado with a fully compressible cloud model: Preliminary result, *18th Conference on Severe Local Storms*, San Francisco, California, February 19–23, 1996.

Vergara, J., and C. Seman, A semi-implicit and semi-Lagrangian fully compressible cloud model, *Fourth International Conference on Southern Hemisphere Meteorology and Oceanography*, Hobart, Australia, 1993.

Large Eddy Modeling of Stratocumulus Clouds

By G. B. Raga

Centro de Ciencias de la Atmósfera, Universidad Nacional Autónoma de México, Ciudad Universitaria, 04510 México D.F., México

The details of large eddy simulation modeling are described. The model utilized in this study was developed at the British Meteorological Office and is based on the original model by Mason [1989]. It has been used to simulate the nocturnal evolution of a cloud-topped marine boundary layer. The initial conditions and the observations to validate the model results were obtained by aircraft during the Atlantic Stratocumulus Transition Experiment. The simulated turbulent structure of the boundary layer is in fairly good agreement with the observations.

1. Introduction

Cloud-topped boundary layers cover a significant fraction of the Earth's surface and are important to both climatology and weather forcasting. The amount and distribution of short- and longwave radiative flux divergence in the boundary layer are altered by the presence of clouds, and these effects are emerging as important aspects of the climate-change problem. Clouds that are limited in their vertical extent by the main subsidence inversion are an intrinsic feature of the cloud-topped boundary layer and consist mainly of three types: (i) shallow cumulus, (ii) stratocumulus and (iii) stratus.

The structure and evolution of the marine cloud-topped boundary layer depend on the interaction of several physical processes, such as surface heating, cloup top entrainment, short- and longwave radiation, drizzle production and large-scale subsidence. Most of these processes occur and interact near the cloud top, where the turbulent boundary layer and the nonturbulent free atmosphere meet. How these processes combine to determine the turbulent transports within the boundary layer is not yet completely understood.

Several studies have concentrated on theoretical aspects of the clear convective boundary layer [Moeng, 1984; Mason, 1989; Moeng and Sullivan, 1994] and have successfully reproduced the main observed features of the flow. The stratus-topped boundary layer has also been the topic of intensive study, and the entrainment rate at the top of the stratus layer was identified as a key parameter in determining the layer evolution [Moeng, 1986; Schumann and Moeng, 1991; Moeng and Schumann, 1991]. The nocturnal evolution of the stratus-topped boundary layer has been studied theoretically by Moeng *et al.* [1992], stressing the importance of radiative cooling in driving the boundary layer circulations.

In this study we use a large eddy simulation (LES) model [Mason, 1989] to simulate the evolution of a stratocumulus layer and to compare the model results directly with the extensive observations obtained by aircraft during the Atlantic Stratocumulus Transition Experiment (ASTEX). This field program was designed to determine the evolution and mesoscale variability of marine stratocumulus clouds and, in particular, to examine cloud top entrainment instability criteria and diurnal decoupling and layer evaporation. The experiment took place during June/July 1992 in the vicinity of the Azores Islands, with the participation of numerous research groups from the United States and Europe. The detailed comparison of the results from an LES model with aircraft data (obtained during ASTEX) has not been previously presented.

2. The LES Model

Large eddy simulation is a technique applied specifically to study turbulence problems. It attempts to resolve explicitly the larger-scale turbulent motions or eddies (which contain most of the turbulent energy and are responsible for most of the turbulent transport), leaving only small-scale eddies to be parameterized. These small eddies do not contribute much to the overall turbulent transport and are mainly responsible for the dissipation of kinetic energy.

The LES model used in this study was developed at the British Meteorological Office, based on the original model by Mason [1989]. It has been successfully applied to the study of numerous turbulent problems in the boundary layer, ranging from the stable [Mason and Derbyshire, 1990] to the neutral [Mason and Thomson, 1987], to the convective cases [MacVean and Mason, 1990].

The basic equations include the three-dimensional Navier–Stokes equations (with the inclusion of rotation and buoyancy terms), the deep anelastic mass continuity equation as well as a thermodynamic equation and continuity equation for water substance. This set of equations is nonhydrostatic but excludes sound waves.

The LES technique consists in applying a filter to the basic equations and decomposing the velocities (and also the scalar variables) into resolved and subfilter fields. The filter operator removes the smallest scales and is not necessarily related to the grid. Use of the subfilter or subgrid model is the procedure by which estimates of the subgrid stress tensor and subgrid scalar fluxes are computed. In LES, it is common to use a simple subgrid model, preferring to devote computational resources to the explicit resolution of turbulence. In a well-resolved simulation, the results should be relatively insensitive to the actual resolution and to the details of the subgrid parameterization. The present model uses a first-order closure (a variant of the Smagorinsky–Lilly model), with eddy diffusivity and eddy viscosity computed from the resolved strain rate, the resolved scalar gradients and a certain prescribed length scale.

2.1. *Cloud Microphysics*

The microphysical processes in a cloud are those which lead to the formation, growth and depletion of the water particles. These particles can be liquid, ice or a combination of both and may have regular or irregular shapes. The model's microphysical scheme divides these particles into several categories commonly used in bulk water schemes: liquid water droplets, rain drops, ice crystals, snow crystals and graupel. The microphysics scheme uses the parameterization of liquid water processes given by Rutledge and Hobbs [1983], while ice processes are parameterized as in Hsie and Orville [1980]. In this study of marine stratocumulus clouds, only warm rain microphysics is active and only two particle categories are included.

Cloud water droplets are assumed to have infinitely small radius, and supersaturation with respect to water is not allowed. Similarly, no liquid water can be present in subsaturated air. Both the condensation and evaporation processes are assumed to be instantaneous. It is not possible for water vapor to condense directly onto rain droplets. The mass mixing ratio is represented by a model variable and the size distribution of the particles is assumed to depend on their mass mixing ratio. The cloud droplets combine with one another through collisions and coalescence, and this process is parameterized by an autoconversion term.

The rain size spectrum is assumed to be an inverse exponential [Kessler, 1969]

$$n(D) = n_0 e^{-\lambda D} \tag{2.1}$$

where the slope parameter (λ) is calculated from the mass mixing ratio and the intercept parameter (n_0) is assumed constant in space and in time for a given simulation. The process of larger droplets sweeping through and collecting smaller cloud droplets is represented by the accretion term in the parameterization scheme. The fall velocities and conversion rates between particle types are expressed in terms of model variables.

2.2. *Radiation*

The model includes a longwave radiative scheme based on the work by Slingo and Wilderspin [1986]; it does not include a shortwave scheme in the present configuration. The lack of a shortwave parameterization is not important for this study, in which the nocturnal evolution of a stratocumulus-capped boundary layer is performed.

Longwave radiation is absorbed and emitted by atmospheric gases, clouds and the earth's surface. The main components of the atmosphere (nitrogen and oxygen) are diatomic and symmetric, and, therefore, their interaction with longwave radiation is negligible. For our purposes, their effect can be neglected and we need to consider only the three most abundant triatomic trace constituents: water (in its three phases), ozone and carbon dioxide. Other "greenhouse" gases may make a significant contribution to anthropogenic warming but are not included in the present radiation scheme. The wavelength dependence of the absorption and emission by these gases is very complex because of the many vibrational and rotational transitions that are excited by longwave radiation. Therefore, the spectral lines have been grouped into six bands in the present parameterization. For water vapor, continuum absorption, which changes slowly with wavelength, is also included.

2.3. *Other Physical Considerations*

This version of the LES allows for superimposed mean vertical motion. A value of large-scale divergence is specified and the mean subsidence is simply computed as a linear function of height and divergence. This superimposed subsidence has the effect of constraining the growth of the boundary layer.

A Newtonian damping layer can be included in the upper part of the domain to absorb the energy radiating upward, before it is reflected off the top boundary. In this layer, all variables (dynamic and thermodynamic) are damped toward their horizontal averages. In this study a 1 km deep damping layer was utilized.

The surface boundary conditions are derived from Monin–Obukhov similarity theory. The surface heat fluxes (both latent and sensible) are specified and the friction velocity is obtained at the first domain level, considering the neutral, stable and unstable cases.

The lateral boundary conditions are essentially cyclic, but with the possibility of an added nonperiodic term, which takes into account the pressure gradients associated with weather systems on scales much larger than the horizontal domain.

2.4. *Numerical Scheme*

It is desirable that the numerical scheme preserve a reasonably high order of accuracy and that it satisfy stability and energy conservation where possible. Most of the dynamic terms are handled with second-order accuracy in space and time. The diffusive terms are treated with only first-order accuracy in time. There is a choice of advection schemes in the model. One is the Piacsek–Williams scheme, which is energy conserving but not positive definite. Where sharp scalar gradients are present (such as in the boundary between cloud and clear air), a more sophisticated algorithm, total variation diminishing (TVD), is preferred. The Piacsek–Williams scheme treats the advection terms as centered in time, while the TVD scheme treats them as forward in time. When the centered difference scheme is used, there is a risk of a potential decoupling of the time levels. To prevent this, a weak time-smoother is applied continuously. This has the undesirable effect of violating energy conservation slightly; the violation is only significant if very precise checks on energy conservation are made.

The grid used has each velocity component staggered in its own direction (Arakawa-C grid). The horizontal spacing is uniform, but vertical spacing can vary, allowing more resolution closer to the surface. The timestep is variable and based on the Courant–Friedicks–Levy (CFL) stability criterion.

A useful feature of the model is the simple way of switching between one-dimensional (1-D), 2-D, and 3-D versions. This allows an easy means of testing a run in 2-D and studying the sensitivity to particular parameters before running the 3-D version.

2.5. *Initial Conditions*

Both the initial conditions and the analyzed observations against which the model results will be validated were obtained by aircraft during the Atlantic Stratocumulus Transition Experiment (ASTEX), in the summer of 1992 in the vicinity of the Azores Islands. The instrumented C-130 of the U.K. Meteorological Research Flight was one of the aircraft that participated in the field project. When sampling a stratocumulus layer the aircraft would typically make five or six flight legs at constant altitude (roughly 40 km in length) in a stack pattern, below, within and above the cloud layer. These stacks were flown in about 1 hour. The nocturnal flight of June 2, during which five vertical stacks were flown, has been analyzed.

The initial thermodynamic sounding for the simulations was based on the observed profile before sunset. The surface sensible and latent heat fluxes needed in the simulations were based on the observations obtained 30 m above the ocean surface. A random initial perturbation in the potential temperature field (maximum amplitude 0.1°C) was introduced in the lower 100 m of the domain, which extended 5 km in the horizontal and 3.4 km in the vertical. The horizontal resolution was 50 m, while the vertical resolution ranged from 25 m in the lower levels to 200 m at the top of the domain.

3. Results

The analyzed observations presented here correspond to the second stack flown by the aircraft, obtained between 1 and 2 hours after sunset, near the surface high-pressure center (1032 mb closed contour). Turbulent fluxes of several thermodynamic variables have been computed at each horizontal level to obtain vertical profiles. Figure 1 shows symbols that correspond to the observed profile of the liquid water flux as a function of the nondimensional height, which takes into account the boundary layer height. The triangles denote the turbulent fluxes, computed as the eddy correlation of the vertical velocity and the liquid water content for the full length of the flight leg (approximately 40 km). The squares correspond to the fluxes computed as the average of four 10 km segments. This was done to eliminate the contribution to the turbulent fluxes of motions on the mesoscale, which are not included in the numerical model. In the lower levels, there is virtually no difference between the two methods of computing the fluxes. In contrast, within the cloud layer there is up to a 10% difference in the magnitude of the fluxes, which corresponds to the mesoscale motions near the inversion.

A large number of 2-D runs were carried out to study the sensitivity of the results to the value selected for the large-scale subsidence and to the presence of the longwave (LW) radiation scheme. The values of divergence ranged from 0 to 5×10^{-5} s^{-1}, with and without the LW scheme. So far, only 3 hours of the 3-D simulation has been analyzed

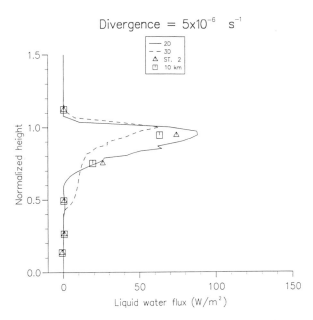

FIGURE 1. Vertical profile of the liquid water flux.

and the results from the third hour are presented here, after the TKE introduced during the spin-up has reached a steady state.

3.1. *Effect of Subsidence*

The presence of superimposed subsidence tends to slow the circulations that develop within the boundary layer. This is evidenced in the domain averaged turbulent kinetic energy (TKE), which is a maximum in the nondivergent case. The cloud depth increased for low divergence (D) values (less than 2.5×10^{-5} s^{-1}) and decreased for higher D values. For the largest D values, total evaporation of the cloud layer occurred after 2 hours of simulation, as a result of the warming and drying effects of the sinking air. The liquid water path (averaged in the horizontal) monotonically decreased with increasing D value. The presence of substratocumulus cumulus clouds was suppressed when a nonzero value of divergence was specified. In contrast, these small cumulus clouds were observed throughout the whole flight. The profile of the longwave radiative flux shows a decrease in the jump at cloud top with increasing D values. This translates into decreased cooling at cloud top and, therefore, decreased circulation because the radiative cooling and its induced entrainment are driving the circulations within the boundary layer.

3.2. *Effect of LW Radiation*

For all values of divergence, the maximum in the TKE occurs sooner in the simulations where LW radiation is included. This is due to the radiatively driven instability that develops near the cloud top. In the simulations without LW radiation, surface heating is responsible for the generation of TKE within the boundary layer. Maximum values of, for example, liquid water content and vertical velocities are about 50% larger in the simulations with LW. The turbulent fluxes are also much larger when the LW is included, and the circulation within the boundary layer, given by the magnitude of up- and downdrafts, is stronger. As we will see later, the presence of LW radiation is crucial to the simulation of a realistic nocturnal cloud-topped boundary layer.

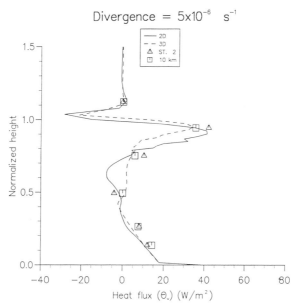

FIGURE 2. Vertical profile of the buoyancy flux.

3.3. 2-D vs. 3-D and Comparison with Observations

The first point to note from the comparison of 2-D and 3-D simulations is that the bulk of the turbulent properties is fairly well reproduced by a 2-D simulation. It is in the small details that the presence of the third dimension makes a difference. The profiles in Figure 2, which correspond to the virtual potential temperature (or buoyancy) flux, are fairly similar, but the small positive flux values at about 0.5 of the normalized height in the 3-D simulation correspond to the presence of small cumulus clouds underneath the main stratocumulus sheet. Therefore, the 3-D simulation is able to reproduce the observed cloud pattern, while this was never found in the 2-D simulations. The boundary layer circulations in the 2-D are much stronger than in the 3-D simulations, and, therefore, favor a better vertical mixing which prevents the formation of the small cumulus clouds.

The simulated curves in Figure 2 compare very favorably with the observations. The low-level positive values reflect the ascending warm plumes due to surface sensible and latent heat fluxes, decreasing linearly to the top of the well-mixed layer. The negative values in the subcloud layer in the 2-D simulations are linked to the cooling by evaporating drizzle. This drizzle develops only in the 2-D simulations, which simulate a stronger circulation than the 3-D case. In contrast, the 3-D curve shows positive values, which, as we have mentioned, correspond to the transport by the cumulus clouds underneath the stratocumulus layer. The negative values at cloud top relate to the entrainment of warm air from above the inversion, while the layer maximum within the cloud layer is due to the LW cooling-induced entrainment.

Figure 3 shows the different terms present in the TKE (calculated from the resolved velocity fields) budget equation for the 3-D simulation. The solid line corresponds to the buoyancy term. There is generation of TKE by buoyancy near the surface, due to the positive latent and sensible heat fluxes. Within the stratocumulus layer, the maximum is due to condensation but mainly to LW cooling induced entrainment, in which cooler parcels are moving downward. There is very little shear production of TKE, given by the short-dash line, except near the surface. The dissipation of TKE (given by the dot–dash

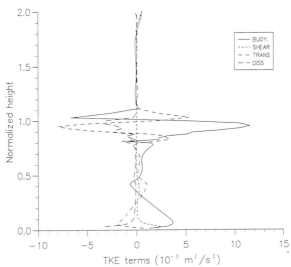

FIGURE 3. Vertical profile of the terms in the TKE budget for the 3-D simulation.

line) is large near the surface and also within the stratocumulus layer. The dashed line corresponds to the combination of vertical advection of TKE and the pressure term, and it shows that the TKE is transported out of the regions where it is produced and into the regions where it is dissipated.

4. Summary

We have studied the effect of large-scale subsidence and LW radiation on the evolution of a nocturnal cloud-topped marine boundary layer using an LES model. The results of the 2-D simulations indicate that the LW helps to strengthen the boundary layer circulations and is responsible for the increased turbulent transports for all values of divergence specified. The domain-averaged TKE decreases with increasing divergence, because the superimposed subsidence tends to suppress the boundary layer circulations. The cloud depth decreases with increasing divergence, leading to total evaporation of the cloud layer in some cases.

The 2-D simulations capture most of the features characteristic of the turbulent transports within the boundary layer, except the presence of the cumulus clouds underneath the stratocumulus layer. This was indeed observed throughout the whole flight and is only simulated in 3-D. The circulations are stronger in 2-D than in 3-D, and this causes a more efficient vertical mixing of the boundary layer, thus preventing the formation of the cumulus clouds.

The model results obtained thus far agree very well with the observations obtained by aircraft during ASTEX. A full night simulation will be performed shortly and the results will provide more insight into the variability of a nocturnal boundary layer in which cumulus and stratocumulus clouds are present.

Acknowledgments. The author gratefully acknowledges Dr. Paul Mason for providing the LES code and Drs. D. Johnson, G. Martin and J. Taylor for providing the ASTEX data and helpful discussions on instrumentation and data quality. Thanks are also due to Drs. J. Pasquier and S. Siems for introducing the author to numerical modeling.

REFERENCES

Hsie, F., and H. Orville, Numerical simulation of the ice-phase convective cloud seeding, *J. Appl. Meteor.*, **19**, 950–966, 1980.

Kessler, E., On the distribution and continuity of water substance, *Meteor. Monogr.*, **32**, Am. Meteor. Soc., 1969.

Mason, P. J., Large-eddy simulation of the convective atmospheric boundary layer, *J. Atmos. Sci.*, **46**, 1492–1516, 1989.

Mason, P. J., and S. H. Derbyshire, Large eddy simulation of the stably stratified atmospheric boundary layer, *Bound. Layer Met.*, **53**, 117–162, 1990.

Mason, P. J., and D. J. Thomson, Large eddy simulation of the neutral static stability planetary boundary layer, *Q. J. R. Meteor. Soc.*, **113**, 413–433, 1987.

Moeng, C.-H., A large eddy simulation model for the study of planeraty boundary layer turbulence, *J. Atmos. Sci.*, **41**, 2052–2062, 1984.

Moeng, C.-H., Large eddy simulation of a stratus-topped boundary layer. Part I. Structure and budgets, *J. Atmos. Sci.*, **43**, 2886–2900, 1986.

Moeng, C.-H., and U. Schumann, Composite structure of plumes in stratus-topped boundary layer, *J. Atmos. Sci.*, **48**, 2280–2291, 1991.

Moeng, C.-H., S. Shen, and D. Randall, Physical processes within the nocturnal stratus-topped boundary layer, *J. Atmos. Sci.*, **49**, 2384–2401, 1992.

Moeng, C.-H., and P. Sullivan, A comparison of shear- and buoyancy-driven planetary boundary layer flows, *J. Atmos. Sci.*, **51**, 999–1022, 1994.

Rutledge, S., and P. V. Hobbs, The mesoscale and microscale structure and organization of clouds and precipitation in midlatitude cyclones. VIII. A model for the seeder-feeder process in warm-frontal rainbands, *J. Atmos. Sci.*, **40**, 1185–1206, 1983.

Schumann, U., and C.-H. Moeng, Plume budgets in clear and cloudy convective boundary layers, *J. Atmos. Sci.*, **48**, 1758–1770, 1991.

Slingo, A., and R. C. Wilderspin, Development of a revised longwave radiation scheme for an atmospheric general circulation model, *Q. J. R. Meteor. Soc.*, **112**, 371–386, 1986.

PART III
Geophysical Data Assimilation

Computational Aspects of Kalman Filtering and Smoothing for Atmospheric Data Assimilation

By Ricardo Todling

USRA/NASA/GSFC, Data Assimilation Office, Code 910.3
Greenbelt, MD 20771, USA

Atmospheric data assimilation is the process of combining model information about the dynamics of the atmosphere with current observations to generate a reliable ongoing estimate of the true atmospheric state. Estimation theory provides the mathematical framework for developing data assimilation schemes. Specifically, for linear systems with known error statistics, the Kalman filter (KF) gives the optimal (minimum variance) solution to the assimilation problem. In practice, however, the equations governing the atmosphere are nonlinear and the error statistics are unknown. Moreover, the dimensionality of the problem makes brute–force algorithms extending the KF for nonlinear dynamics computationally infeasible. Under these circumstances only suboptimal data assimilation schemes are possible, and their performance needs to be evaluated.

An initial phase of this effort has been conducted using a simple linear stable or unstable, shallow–water model, with a full KF as the basis for comparison. Results indicate a systematic path to follow toward improving current data assimilation schemes. In particular, the study of unstable dynamics suggests that only a few eigenmodes of the predictability error covariance matrix, those corresponding to the points where most of the predictability error variance lies, need to be computed. Iterative algorithms can be used for that purpose.

The ultimate goal of this research is to find approximations that are operationally feasible and reliable, not only for the filter problem, but for retrospective analysis (smoothing) as well.

---

1. Introduction

Four-dimensional data assimilation (4-DDA) is the research area of combining model simulations and observational data collected from the atmosphere–ocean–land system to allow future predictions of the interactive behavior of these systems. At present, data assimilation aims to go beyond weather forecasting: the Earth Observing System (EOS) project of NASA, in collaboration with the European and Japanese space agencies (ESA and NASDA), is ultimately expected to contribute to the far more complicated task of climate prediction.

Most major developments in data assimilation have occurred in the field of numerical weather prediction, with important contributions from oceanographers in the last few years. Recently published reviews of data assimilation for the atmosphere by Daley [1991] and by Harms *et al.* [1992], and for the atmosphere and oceans by Ghil and Malanotte-Rizzoli [1991] and by Bennett [1992], have been given elsewhere. Here we are interested in the atmosphere and in the development of data assimilation schemes based on the Kalman filter (KF). The KF cannot be applied operationally for a variety of reasons, for instance, its computational demands. In the past few years there has been an enormous effort to overcome the computational barrier imposed by the KF by proposing feasible and reliable alternative schemes that retain most of the advantages of the KF at a lower cost. A performance evaluation study of many possible alternative schemes

[Todling and Cohn, 1994; Cohn and Todling, 1996] suggests that iterative methods may provide a suitable way of achieving computational feasibility.

To fulfill the goals established by the EOS program, research in this area should lead to applicable algorithms for performing retrospective analysis. This means going beyond filtered analyses by using data to improve upon past analyses and generate reliable model assimilated data sets for climate studies. The fixed-lag Kalman smoother (FLKS) has been proposed by Cohn *et al.* [1994] as an approach for achieving this goal. In its ideal formulation, the FLKS allows for the dynamically consistent transmission of information from new data to past analyses. The FLKS represents an additional burden on the already computationally intensive assimilation methods. The challenge is to develop FLKS-based approximations that are feasible and reliable for dealing with the nonideal problem of retrospective 4-DDA.

The purposes of the present paper are to highlight the computational cost requirements involved in implementation of the KF and FLKS and to present a summary of the alternative schemes with a potential for practical applications. In Section 2, we briefly state the notion of data assimilation and its roots in estimation theory. In Section 3, we tackle computational issues and present alternative schemes to the KF and the FLKS. In Section 4, we summarize the results on the performance evaluation of these alternative schemes; we draw conclusions in Section 5.

2. Data Assimilation: Estimation Approach

A description of the atmosphere and its observation systems can take into account the stochasticity of these systems. A schematic representation is given in Figure 1, where the state of the atmosphere $\mathbf{w}_0$, at a given time, changes according to nonlinear dynamical processes indicated by $\mathbf{\Psi}^t$. These can be forced randomly by the inputs ϵ^t, and deterministically by the inputs $\mathbf{F}$, generating the atmospheric state $\mathbf{w}^t$, at a later time.

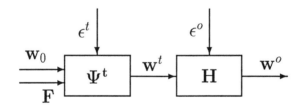

FIGURE 1. Schematic representation of the atmosphere as a stochastic–dynamical system and the observation process.

The observation process, also represented in a very general way in Figure 1, entails looking at the current atmospheric state, $\mathbf{w}^t$, by means of a measuring apparatus $\mathbf{H}$, which is always contaminated by random noise, ϵ^o. The final result of this process is the observed state $\mathbf{w}^o$, representing our rough knowledge of the atmospheric state.

In principle, from the observations $\mathbf{w}^o$, we can produce regularly gridded data $\mathbf{w}^a$ (analysis state), to be subsequently utilized as the initial state of a deterministic (non-stochastic) nonlinear model $\mathbf{\Psi}$, to produce a prediction $\mathbf{w}^f$ (forecast) of the atmospheric state. This forecast is our estimate of the true state $\mathbf{w}^t$ of the atmosphere and we hoped it to be as close as possible to that. The accuracy of the estimate will depend on how well the atmosphere ($\mathbf{\Psi}^t, \epsilon^t, \mathbf{F}$) is represented by the model $\mathbf{\Psi}$, as well as how accurately the initial analysis state is represented by the observations and the "gridding" procedure. In reality, taking observations directly into a model produces very poor quality forecasts given the subjective nature of the analysis procedure, the sparsity/quantity of the observations, and the lack of complete knowledge of atmospheric dynamics.

A systematic way to estimate the state of a system, at a particular time, by combining observations and model predictions objectively is provided by estimation theory. Despite many earlier developments in the field, Kalman [1960] revolutionized the procedure for approaching the estimation problem. Using the concept of state transition from the theory of dynamical systems and concepts of conditional probability distributions, Kalman obtained an algorithm for linear time-discrete systems that combines observations and model simulations in an optimal way to give the best possible estimate of the state of the system. The algorithm was extended to linear time-continuous systems by Kalman and Bucy [1961]. Kalman's original procedure was derived for systems governed by ordinary differential equations (lumped parameter systems; e.g., Gelb [1974]) and later extended to systems governed by partial differential equations (distributed parameter systems; e.g., Stavroulakis [1983]). A glimpse at the importance of Kalman's work and its impact on a variety of fields can be found in Antoulas [1991].

The procedure devised by Kalman aims to filter most of the valuable information contained in noisy observations and combine it with a model forecast based on statistical knowledge of the observation and model error characteristics. The accuracy of the state estimates (analyses) produced by the KF can be further improved upon by combining new observations with past estimates to smooth out those estimates, producing better analyses. The procedures are called Kalman smoothers, among which the FLKS is of special interest to us.

3. Computational Issues

Table 1 lists the equations involved in the KF and FLKS algorithms. These equations are for linear systems; however, the computations involved are close to some extensions to nonlinear dynamics and nonlinear observations [Jaswinski, 1970]. We refer the reader to the literature for a detailed explanation of the quantities listed in the table, e.g., Anderson and Moore [1979], Cohn *et al.* [1994], Ghil *et al.* [1981], and Jaswinski [1970]. The table also lists the computational cost, measured in terms of floating-point operations (FLOPs) – multiplications plus additions – involved in each step of *brute-force* implementations of the KF and the FLKS (see also Mendel [1971]). Here, brute force means that we assume that storage of certain quantities is not a problem; we also assume that the dynamics matrix Ψ is dense; and we take no advantage of the symmetry of the error covariance matrices $P^{f,a}$. Computational stability (e.g., Verhaegen and Van Dooren [1986]) is also not taken into account, as in many cases it may not even be an issue. The brute-force FLKS is $(M + 2)/2$ times as expensive as the KF itself, where M is the number of retrospective analyses calculated at each observation time.

Assuming $n \gg p$, where n is the number of degrees of freedom of the system (dimension of the state vector $w^{f,a}$) and p is the number of observations available at a given time (dimension of the observation vector w^o), a close look at Table 1 shows that equations F2 and S1 account for most of the computational cost for the filter and the smoother, respectively. These equations correspond to the propagation of the error covariance and cross-covariance matrices according to the dynamics of the system. The calculations generated by these equations are order n^3 and thus impractical for atmospheric and oceanic models for which $n \sim 10^6$, or higher. Fortunately, however, it is generally true for these models that the dynamics Ψ is a sparse matrix (e.g., finite difference methods have

TABLE 1. Computational cost requirements for the Kalman filter and smoother.

Brute-force Kalman filter

Eq.	Variable	Equation	Computation	FLOPs
F1	$\mathbf{w}_k^f$	$\mathbf{\Psi}_k \mathbf{w}_{k-1}^a$	$\mathbf{\Psi w}$	$2n^2 - n$
F2	$\mathbf{P}_k^f$	$\mathbf{\Psi}_k \mathbf{P}_{k-1}^a \mathbf{\Psi}_k^T + \mathbf{Q}_{k-1}$	$\mathbf{P\Psi}^T$	$2n^3 - n^2$
			$\mathbf{\Psi}(\mathbf{P\Psi}^T)$	$2n^3 - n^2$
			$(\mathbf{\Psi P\Psi}^T) + \mathbf{Q}$	n^2
F3	$\mathbf{K}_k$	$\mathbf{P}_k^f \mathbf{H}_k^T (\mathbf{H}_k \mathbf{P}_k^f \mathbf{H}_k^T + \mathbf{R}_k)^{-1}$	$\mathbf{HP}$	$2n^2 p - np$
			$(\mathbf{HP})\mathbf{H}^T$	$2np^2 - p^2$
			$(\mathbf{HPH}^T) + \mathbf{R}$	p^2
			$(\mathbf{HPH}^T + \mathbf{R})^{-1}$	$2p^3$
			$(\mathbf{PH}^T)(\mathbf{HPH}^T + \mathbf{R})^{-1}$	$2np^2 - np$
F4	$\mathbf{w}_k^a$	$\mathbf{w}_k^f + \mathbf{K}_k(\mathbf{w}_k^o - \mathbf{H}_k \mathbf{w}_k^f)$	$\mathbf{Hw}^f$	$2np - p$
			$\mathbf{w}^o - \mathbf{Hw}^f$	p
			$\mathbf{K}(\mathbf{w}^o - \mathbf{Hw}^f)$	$2np - n$
			$\mathbf{w}^f + [\mathbf{K}(\mathbf{w}^o - \mathbf{Hw}^f)]$	n
F5	$\mathbf{P}_k^a$	$(\mathbf{I} - \mathbf{K}_k \mathbf{H}_k)\mathbf{P}_k^f$	$\mathbf{K}(\mathbf{HP})$	$2np^2 - p^2$
			$\mathbf{P} - \mathbf{K}(\mathbf{HP})$	n^2

Brute-force fixed-lag Kalman smoother

Eq.	Variable	Equation	Computation	FLOPs
S1	$\mathbf{P}_{k,k-\ell\|k}^{fa}$	$\mathbf{\Psi}_k \mathbf{P}_{k-1,k-\ell\|k-1}^{aa}$	$\mathbf{\Psi P}$	$2n^3 - n^2$
S2	$\mathbf{K}_{k-\ell\|k}$	$(\mathbf{H}_k \mathbf{P}_{k,k-\ell\|k-1}^{fa})^T (\mathbf{H}_k \mathbf{P}_k^f \mathbf{H}_k^T + \mathbf{R}_k)^{-1}$	$\mathbf{HP}$	$2n^2 p - np$
			$(\mathbf{HP})^T(\mathbf{HPH}^T + \mathbf{R})^{-1}$	$2np^2 - np$
S3	$\mathbf{w}_{k-\ell\|k}^a$	$\mathbf{w}_{k-\ell\|k-1}^a + \mathbf{K}_{k-\ell\|k}(\mathbf{w}_k^o - \mathbf{H}_k \mathbf{w}_k^f)$	$\mathbf{K}(\mathbf{w}^o - \mathbf{Hw}^f)$	$2np - n$
S4	$\mathbf{P}_{k-\ell\|k}^a$	$\mathbf{P}_{k-\ell\|k-1}^a - \mathbf{K}_{k-\ell\|k}\mathbf{H}_k \mathbf{P}_{k,k-\ell\|k-1}^{fa}$	$\mathbf{K}(\mathbf{HP})$	$2np^2 - p^2$
			$\mathbf{P} - \mathbf{K}(\mathbf{HP})$	n^2
S5	$\mathbf{P}_{k,k-\ell\|k}^{aa}$	$(\mathbf{I} - \mathbf{K}_k \mathbf{H}_k)\mathbf{P}_{k,k-\ell\|k-1}^{fa}$	$\mathbf{K}(\mathbf{HP})$	$2np^2 - p^2$
			$\mathbf{P} - \mathbf{K}(\mathbf{HP})$	n^2

simple stencils); moreover, these are not addressed as matrices but rather as operators, therefore reducing enormously the storage requirements. With that taken into account, the cost of state propagation as in equation F1, that is, of the application of the dynamics operator to an n-vector quantity, is reduced to order n. With this reasoning, the cost of F2 and S1 is reduced to about n^2. Still, the burden of these computations is forbidden at present computing power.

An intensive area of research is the search for algorithmic simplifications and approxi-

Optimal Interpolation (OI)

Partition forecast error covariance $\mathbf{S}_k^f$:

$$\mathbf{S}_k^f \equiv \begin{bmatrix} \mathbf{S}^{f|uu} & \mathbf{S}^{f|uv} & \mathbf{S}^{f|uh} \\ \mathbf{S}^{f|vu} & \mathbf{S}^{f|vv} & \mathbf{S}^{f|vh} \\ \mathbf{S}^{f|hu} & \mathbf{S}^{f|hv} & \mathbf{S}^{f|hh} \end{bmatrix}_k .$$

Decompose height-height error covariance:

$$\mathbf{S}_k^{f|hh} = (\mathbf{D}_k^{f|h})^{1/2} \mathbf{C}^{hh} (\mathbf{D}_k^{f|h})^{1/2}$$

$\mathbf{D}_k^{f|h}$ - h error variances
$\mathbf{C}^{hh}$ - hh prescribed error correlations.
$\mathbf{D}_k^{f|h}$ is allowed to grow linearly in time:

$$\mathbf{D}_k^{f|h} = \mathbf{D}_{k-r}^{a|h} + \mathbf{D}_{growth}^h,$$

where r is the analyses interval.
Remaining error covariance computed through dynamical balance.

Variance Evolution

Height error variance propagation:

$$\mathbf{D}_k^{p|h} = \mathbf{A}_k \mathbf{D}_k^{a|h},$$

where $\mathbf{A}_k$ represents an advection scheme.

Height-height error covariance:

$$\mathbf{S}_k^{p|hh} = (\mathbf{D}_k^{p|h})^{1/2} \mathbf{C}^{hh} (\mathbf{D}_k^{p|h})^{1/2}.$$

Remaining error covariance computed through dynamical balance.

Simplified Kalman filter

Height-height error covariance propagation:

$$\mathbf{S}_k^{p|hh} = \mathbf{A}_k \mathbf{S}_k^{a|hh} \mathbf{A}_k^T.$$

Remaining error covariance computed through dynamical balance.

Partial Singular Value Decomposition Filter (PSF)

Assume SVD of propagator $\boldsymbol{\Psi}_k$:

$$\boldsymbol{\Psi}_k = \left(\mathbf{U} \mathbf{D} \mathbf{V}^T \right)_k .$$

and leading(L) and trailing(T) partitions:

$$\mathbf{U}_k = [\mathbf{U}_L \, \mathbf{U}_T]_k , \quad \mathbf{V}_k = [\mathbf{V}_L \, \mathbf{V}_T]_k$$

$$\mathbf{D}_k = diag[\mathbf{D}_L \cdot \mathbf{D}_T]_k .$$

Take for $\mathbf{S}_k^p$ the following model:

$$\mathbf{S}_k^p = (\mathbf{S}_L^p + \mathbf{S}_T^p)_k = \left(\tilde{\boldsymbol{\Psi}} \mathbf{S}^a \tilde{\boldsymbol{\Psi}}^T + \mathbf{S}_T^p \right)_k .$$

where

$$\tilde{\boldsymbol{\Psi}}_k = \left(\mathbf{U}_L \mathbf{D}_L \mathbf{V}_L^T \right)_k .$$

$\mathbf{S}_T^p$ is specified by an adaptively tuned co-variance model.
Cost $\sim o(10L)$ forecasts

Partial Eigenvalue Decomposition Filter (PEF)

Assume eigendecomposition of predictability error covariance $\mathbf{S}^p$:

$$\mathbf{S}_k^p = \left(\boldsymbol{\Psi} \mathbf{S}^a \boldsymbol{\Psi}^T \right)_k = \left(\mathbf{W} \hat{\mathbf{S}} \mathbf{W}^T \right)_k .$$

and leading(L) and trailing(T) partitions:

$$\mathbf{W}_k = [\mathbf{W}_L \, \mathbf{W}_T]_k .$$

$$\hat{\mathbf{S}}_k = diag[\hat{\mathbf{S}}_L \cdot \hat{\mathbf{S}}_T]_k .$$

Take for $\mathbf{S}_k^p$ the following model:

$$\mathbf{S}_k^p = (\mathbf{S}_L^p + \mathbf{S}_T^p)_k = \left(\mathbf{W}_L \hat{\mathbf{S}}_L \mathbf{W}_L^T + \mathbf{S}_T^p \right)_k .$$

$\mathbf{S}_T^p$ is specified by an adaptively tuned covariance model.
Cost $\sim o(10L)$ forecasts

FIGURE 2. Description of the suboptimal schemes evaluated in Todling and Cohn [1994] and Cohn and Todling [1996].

mations that would lead to possible applications of a KF-based scheme to routine 4-DDA. All the approximations suggested in the atmosphere/ocean literature concentrate on simplifying F2 in Table 1, the filter problem. Similar investigation for simplifying S2, the retrospective analysis problem, has been started only very recently. The approximations suggested for the filter problem either fall into one of the following categories or represent a combination of them, according to Todling and Cohn [1994]: (i) covariance modeling [Jiang and Ghil, 1995; Gaspari and Cohn, 1996]; (ii) dynamics simplification; (iii) reduced resolution [Hoang *et al.*, 1995; Verlaan and Heemink, 1995]; (iv) local representation [Riedel, 1993; Boggs *et al.*, 1995]; (v) limiting filtering; (vi) a Monte Carlo approach. (The references listed here, for work done in these categories, are meant to be complementary to those cited in Todling and Cohn [1994].)

The panels in Figure 2 briefly describe the approximate schemes evaluated in Todling and Cohn [1994] for stable dynamics, and in Cohn and Todling [1996] for stable and unstable dynamics. All these schemes refer to approximating the forecast error covariance matrix $\mathbf{P}^f$ (F2 in Table 1), which in the figure is denoted by $\mathbf{S}^f$ to emphasize the suboptimal nature of these approximations. The schemes range from a version of the currently operational optimal interpolation (OI; top-left panel), to a slightly improved version of it that allows for advection of the height error variance field through an advection dynamics matrix $\mathbf{A}$ (HVA; top-right panel), to a yet more sophisticated scheme that allows for propagation of the full height–height error covariance field through the simplified dynamics $\mathbf{A}$ (SKF; top-right panel). All these schemes use a methodology to construct dynamically balanced cross-covariance matrices from a height–height error covariance matrix. These approximations have been found to perform well for stable dynamics, but fail in the presence of dynamical instabilities. Schemes more suitable for unstable dynamics, that also work in the stable case, are described in the two panels at the bottom of the figure. Both these approximations are iterative and make no reference to full covariance matrices. The use of a Lanczos-type algorithm is required. The partial singular value decomposition filter (PSF; bottom-left panel) proposes to use only the first L most dominant singular values/vectors of the dynamics (propagator) $\mathbf{\Psi}$ in propagating analysis error covariances; the partial eigendecomposition filter (PEF; bottom-right panel) proposes to use only the first L eigenvalues/vectors of the predictability error covariance matrix $\mathbf{S}^p$. The PSF and PEF are low-rank approximations, and completion of rank, through further approximation of their corresponding trailing error covariance matrix, $\mathbf{S}^p_T$, is important for good performance. An alternative low-rank covariance approximation scheme has been suggested by Courtier [1993] and studied independently by Sheinbaum [this volume] in the context of variational algorithms.

Another approximate scheme, not described in Figure 2 and in principle suitable for unstable dynamics, is the reduced resolution filter (RRF), which simplifies the computational burden by solving F2 in Table 1 for a system with number of degrees of freedom $m < n$, and by interpolating the corresponding gains in F3 to the full resolution n. A reduced resolution fixed-lag smoother (RRFLS) can be devised in a similar way by simplifying S1, S2, S4 and S5 in Table 1.

Although it is not the subject of this paper, the search for approximate schemes based on the KF/FLKS is motivated by more than the computational burden involved in solving the costly equations in Table 1. Other factors that make it worthless to try to solve these equations exactly are as follows:

• The atmosphere and oceans are infinite-dimensional nonlinear systems for which closed solutions to the corresponding filtering and smoothing problems are not known.

• Many observation instruments relate nonlinearly to the variables accounted for in models for the atmosphere and oceans. Again, corresponding filters and smoothers have no exact closed solution in this case.

• Lack of knowledge in the statistics of model and observation errors.

• The assumption $n \gg p$ is no longer true in current Earth science assimilation systems. This being the case, approximations are also required to solve equations F3, F5, S2, S4 and S5 in Table 1.

Even with the promise of computer power increase through massively parallel computing [Hoffmann and Snelling, 1988; Hoffmann and Maretis, 1990; Lyster *et al.*, 1996], the foreseen consequent increase in model resolution and observational data availability already indicates that the next generation of computers will still not resolve the computational issues behind complete error covariance evolution. In any case, the arguments presented mean that whatever assimilation scheme we devise for solving the problem of atmospheric 4-DDA will be suboptimal (approximate). The relevant issue is thus to design feasible and reliable schemes, that is, schemes involving the minimum amount of computations which are stable and accurate.

4. Suboptimal Filter and Smoother Evaluation

Concrete evaluation of approximate schemes for 4-DDA can be conducted only for linear problems since it is only in this case that an exact solution exists. A performance analysis study, of the approximations in Figure 2, has been conducted by using a simple, two-dimensional linear shallow-water model, applied to a domain of about the size of the contiguous United States and discretized on a 25×16 grid, together with a full KF, as the basis for comparison. Results for stable dynamics are discussed in Todling and Cohn [1994] and are summarized in Figure 3 here. The observation network includes radiosondes observing winds and heights every 12 hours and hourly wind profilers. The figure compares the performance of different approximations by looking at the domain-averaged expected error in total energy. The exact result, given by the KF, is shown as the bottom curve (unlabeled), where the effect of error reduction because of the observations is seen hourly as well as at 12 hour intervals. Errors grow between two consecutive observation times due to model imperfections. The OI, HVA and SKF schemes, as described in the top two panels in Figure 2, were evaluated. Positive and significant impact of the hourly wind profilers is noticeable only as the schemes become more sophisticated. The performance of the SKF is almost indistinguishable from the result given by the KF.

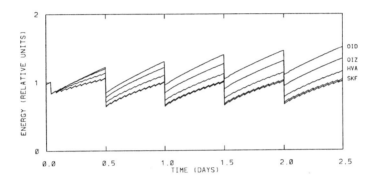

FIGURE 3. Expected root mean square (ERMS) forecast/analysis error in the total energy for a 2.5-day assimilation period, with an imperfect stable model, using the Kalman filter (KF; lowest curve, unlabeled) and four other dynamically balanced approximations suitable to stable dynamics: optimal interpolation (OI) with domain-averaged growth rates (OID); OI with zonally averaged growth rates (OIZ); height variance advection (HVA); and height–height covariance advection (SKF). See Todling and Cohn [1994] for details.

Evaluation of these approximate schemes for a barotropically unstable version of the model indicated that all these schemes performed poorly in this case. As a consequence, Cohn and Todling [1996] have developed and evaluated the performance of three new schemes more suited to unstable dynamics: the RRF, the PSF, and the PEF; the last two schemes are briefly described in the bottom two panels of Figure 2. Figure 4 summarizes the results of these approximations when low resolution and small number of modes are taken into account. The observation network here, and in the experiments that follow, consists of a subset of the radiosonde network used in the experiments displayed in Figure 3. Also, the perfect model assumption is made in these experiments. The RRF is at 13×12 resolution, basically half the resolution of the complete system, while both the PSF and PEF retain only 10 modes. All the approximations diverge at these resolutions.

Increase of resolution and number of modes is enough to prevent the error from growing unboundedly. This is seen in Figure 5, where the RRF is now at 13×16 resolution, and

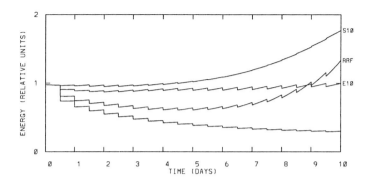

FIGURE 4. ERMS forecast/analysis error in the total energy for a 10-day assimilation period, with a perfect barotropically unstable model, using the KF (lowest unlabeled curve) and three approximations: reduced resolution filter (RRF) on a grid with dimensions 13×12, in contrast to the full resolution of 25×16; partial singular value decomposition filter (S10) including the first 10 singular modes of the dynamics within each 12-hour period; and a partial eigendecomposition filter (E10), including the first 10 eigenmodes of the propagated error covariance structure during each 12-hour period. Error growth between consecutive analysis times is solely due to the presence of unstable modes. Same as Figure 10 in Cohn and Todling [1996].

the PSF and the PEF retain 54 modes. In particular in this case the RRF fully resolves the unstable jet, and the PSF accounts for all singular modes corresponding to singular values larger than or equal to unity. These two factors are required for the stability of the RRF and PSF, respectively (see Cohn and Todling [1996]). The results for the PSF and PEF can be further improved by adaptively estimating the magnitude of a specified trailing error covariance matrix using the method suggested by Dee [1995].

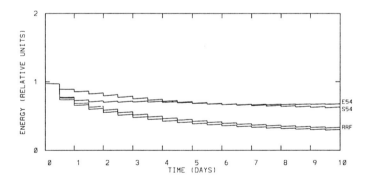

FIGURE 5. Same as in Figure 4, but for RRF at 13×16 resolution, the PSF with 54 singular modes (S54) and the PEF with 54 eigenmodes (E54). The bottom curve is for the KF result. Same as Figure 11 in Cohn and Todling [1996].

The performance of these novel filter approximations can also be tested for the fixed-lag smoother, retrospective analysis problem. Before doing so, we show in Figure 6 the behavior of the FLKS for the unstable system of Figures 4 and 5. The relevant results here are those for the transient part of the assimilation period, before the filter (top curve) almost asymptotes to a steady state. Successive curves in the figure refer to retrospective analyses using data 12, 24, 36 and 48 hours into the future, corresponding to lags $\ell =$

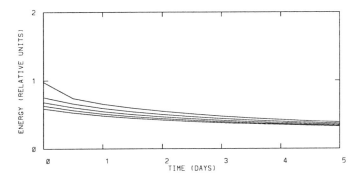

FIGURE 6. ERMS analysis error in total energy for a KF/fixed-lag Kalman smoother (FLKS). The upper curve corresponds to the KF result whereas successively lower curves correspond to retrospective ERMS analysis error including data 12, 24, 36 and 48 hours in the future. Unstable dynamics and the perfect model assumption are used in this case.

1, 2, 3 and 4. Incorporating new data into past analysis reduces the corresponding past analysis errors considerably. In this particular experiment, the initial errors drop to almost half of the filter's first guess error after data 48 hours into the future has been assimilated.

Two questions related to approximating the FLKS arise. The first one is about the behavior of the smoother when only the filter is approximated, with the smoother being computed by using its complete formulation, that is, when F2 is approximated and S1 is used exactly, in Table 1. The second question is about the behavior of the FLKS when both the filter and the smoother procedures are approximated.

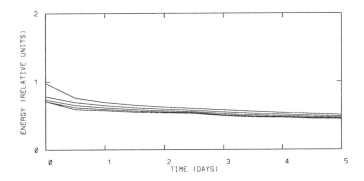

FIGURE 7. As in Figure 6, but using the PEF with 54 modes for the forecast error covariance approximation, and smoother computations without any approximation; that is, the propagated forecast/analysis error cross-covariances $\mathbf{P}^{fa}$ is computed by using the complete dynamics $\mathbf{\Psi}$.

We address the first question in Figure 7, where the filter is approximated by using the PEF algorithm, retaining 54 modes, together with an adaptive scheme to estimate the magnitude of a trailing error covariance matrix $\mathbf{S}_T^p$, used to complete the rank deficiency of the PEF. The filter result has quite good performance, better than that for the experiment displayed in Figure 5 because of the adaptive scheme used now (compare the curve labeled E54 in Figure 5 with the top curve in Figure 7). The performance of the retrospective analysis results is quite reasonable, with less improvement seen for

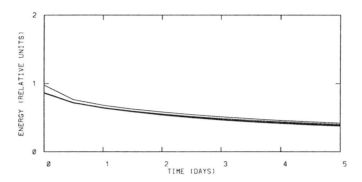

FIGURE 8. As in Figure 6, but using the RRF and a reduced resolution smoother at 13×16 resolution. The top curve corresponds to the filter result, whereas subsequent curves correspond to retrospective analyses including data 12, 24, 36 and 48 hours into the future (all four curves basically coincide). While the filter performs well (compare with top curve in Figure 6), the approximate retrospective analysis results (lower curves) do not compare as favorably with the exact results shown in Figure 6.

lags $\ell = 3$ and 4, when contrasted with the exact solution in Figure 6. Notice that retrospective analysis may allow for recovery from the loss of performance due to use of an approximate filter: at day 2, the lowest curve in Figure 7 (retrospective analysis error for lag $\ell = 4$) reaches an error level comparable to that obtained by the exact filter (KF; top curve in Figure 6).

The second question is addressed in Figure 8, where an RRF and an RRFLS are used, both at 13×16 resolution. According to the results of Figure 5, the RRF is at its best for the filter. The benefit of incorporating future data does not seem significant now, as the retrospective analysis results (lower curves) are almost indistinguishable from the filter result (top curve). The improvement still provided by lag $\ell = 1$ is small compared to that for the cases depicted in Figures 6 and 7.

5. Conclusions

In this paper we have highlighted the main computational difficulties involved in implementing the Kalman filter (KF) and fixed-lag Kalman smoother (FLKS) for atmospheric four-dimensional data assimilation (4-DDA). We have pointed out that the major burden is due to error covariance propagation, in the case of the KF, and to error cross-covariance propagation, in the case of the FLKS.

Concentrating first on the filter problem, we have presented alternative schemes that incorporate dynamics into the filtering procedure at a lower computational cost. We have summarized the performance of these schemes, compared against the exact solution given by the KF, and also against conventional assimilation techniques such as optimal interpolation. Two versions of a simple two-dimensional, linear shallow-water model were used for that purpose. The first version, having stable dynamics, was used to indicate a path of improvement over conventional assimilation schemes. A scheme that advects the complete height–height error covariance matrix coupled with a technique for cross-covariance generation (SKF) had the best performance in this case. The second version, having unstable dynamics, was used to provide a stringent test of simplified dynamics schemes, like the SKF, and to allow for the development of more reliable approximations that remain stable even in the presence of dynamical instabilities. Two iterative methods, the partial singular value decomposition filter (PSF) and the partial eigendecomposition filter (PEF), were shown to provide reliable estimates for a reasonable number of modes calculated, when compared against the KF. A reduced resolution filter (RRF) was also tested and shown to perform quite well whenever fully resolving the system's instability.

Some of these approximate filters were tested in the context of the fixed-lag smoothing retrospective analysis problem. Improvement over filter analyses was reasonable only when the filter calculations were approximated but the smoother calculations were kept exact. When the RRF was used in conjunction with a reduced resolution fixed-lag smoother algorithm, the improvement due to retrospective analysis was not significant.

Approximate schemes like the PSF, PEF and RRF are quite promising for implementation in more realistic data assimilation settings. These schemes require certain flexibilities in the analysis system, such as availability of the analysis error covariance matrix as an operator rather than a full matrix. This is currently under implementation for the physical-space statistical analysis system of the Data Assimilation Office, at Goddard (see Guo and da Silva [this volume]). The performance results discussed here for the retrospective analysis problem leave room for further development of fixed-lag smoother approximations that provide reliable improvement over the filter analysis at reasonable cost.

Acknowledgments. It is a pleasure to thank S. E. Cohn and N. S. Sivakumaran for their collaboration throughout the course of this research. M. Ghil revised part of an earlier version of this manuscript. D. Sorensen provided the iterative eigensystem package ARPACK used in some of the experiments here. The numerical results were obtained on the CRAY C90 through cooperation of the NASA Center for Computational Sciences at the Goddard Space Flight Center. This research was partially supported by a fellowship from the Universities Space Research Association.

REFERENCES

Anderson, B. D. O., and J. B. Moore, *Optimal Filtering*, Prentice-Hall, 1979.

Antoulas, A. C., Ed., *Mathematical System Theory: The Influence of R. E. Kalman*, Springer-Verlag, 1991.

Bennett, A. F., *Inverse Methods in Physical Oceanography*, Cambridge University Press, 1992.

Boggs, D., M. Ghil, and C. Keppenne, A stabilized sparse-matrix U-D square-root implementation of a large-state extended Kalman filter, *Proc. Intl. Symp. on Assimilation of Observations in Meteorology and Oceanography*, Tokyo, Japan, WMO, Vol. 1, 219–224, 1995.

Cohn, S. E., N. S. Sivakumaran, and R. Todling, A fixed-lag Kalman smoother for retrospective data assimilation, *Mon. Wea. Rev.*, **122**, 2838–2867, 1994.

Cohn, S. E., and R. Todling, Approximate Kalman filters for stable and unstable dynamics, *J. Meteor. Soc. Japan*, **74**, 63–75, 1996.

Courtier, P., Introduction to numerical weather prediction data assimilation methods, *Proc. ECMWF Workshop on Developments in the Use of Satellite Data in Numerical Weather Prediction*, 189–207, 1993.

Daley, R., *Atmospheric Data Analysis*, Cambridge University Press, 1991.

Dee, D. P., On-line estimation of error covariance parameters for atmospheric data assimilation, *Mon. Wea. Rev.*, **123**, 1128–1196, 1995.

Gaspari, G., and S. E. Cohn, Construction of correlation functions in two and three dimensions, *Math. Geology*, submitted, 1996.

Gelb, A., Ed., *Applied Optimal Estimation*, MIT Press, 1974.

Ghil, M., S. Cohn, J. Tavantzis, K. Bube, and E. Isaacson, Applications of estimation theory to numerical weather prediction, *Dynamic Meteorology: Data Assimilation Methods*, L. Bengtsson, M. Ghil, and E. Källén, Eds., Springer-Verlag, 139–224, 1981.

Ghil, M., and P. Malanotte-Rizzoli, Data assimilation in meteorology and oceanography, *Advances in Geophysics*, Vol. 33, Academic Press, 141–266, 1991.

Harms, D. E., S. Raman, and R. V. Madala, An examination of four-dimensional data-assimilation techniques for numerical weather prediction, *Bull. Am. Meteor. Soc.*, **73**, 425–440, 1992.

Hoang, H. S., P. De Mey, O. Talagrand, and R. Baraille, Assimilation of altimeter data in a multilayer quasi-geostrophic ocean model by simple nonlinear adaptive filter, *Proc. Intl. Symp. on Assimilation of Observations in Meteorology and Oceanography*, Tokyo, Japan, WMO, 521–526, 1995.

Hoffmann, G.-R., and D. K. Maretis, Eds., *The Dawn of Massively Parallel Processing in Meteorology*, Springer-Verlag, 1990.

Hoffmann, G.-R., and D. F. Snelling, Eds., *Multiprocessing in Meteorological Models*, Springer-Verlag, 1988.

Jazwinski, A. H., *Stochastic Processes and Filtering Theory*, Academic Press, 1970.

Jiang, S., and M. Ghil, Toward monitoring the nonlinear variability of Western boundary currents – assimilation of simulated altimeter data into a wind-driven, double-gyre, shallow-water model, *Proc. Intl. Symp. on Assimilation of Observations in Meteorology and Oceanography*, Tokyo, Japan, WMO, Vol. 2, 533–538, 1995.

Kalman, R. E., A new approach to linear filtering and prediction problems, *Trans. ASME, Ser. D, J. Basic. Eng.*, **82**, 35–45, 1960.

Kalman, R. E., and R. S. Bucy, New results in linear filtering and prediction theory, *Trans. ASME, Ser. D, J. Basic Eng.*, **83**, 95–108, 1961.

Lyster, P. M., S. E. Cohn, R. Ménard, L.-P. Chang, S.-J. Lin, and R. Olsen, Parallel implementation of a Kalman filter for constituent data assimilation, *Mon. Wea. Rev.*, in press, 1996.

Mendel, J. M., Computational requirements for a discrete Kalman filter, *IEEE Trans. Autom. Control*, **AC-16**, 748–758, 1971.

Riedel, K. S., Block diagonally dominant positive definite approximate filters and smoothers, *Automatica*, **29**, 779–783, 1993.

Stavroulakis, P., Ed., *Distributed parameter systems theory, Part II: Estimation*, Hutchinson Ross, 1983.

Todling, R., and S. E. Cohn, Suboptimal schemes for atmospheric data assimilation based on the Kalman filter, *Mon. Wea. Rev.*, **122**, 2530–2557, 1994.

Verhaegen, M., and P. Van Dooren, Numerical aspects of different Kalman filter implementations, *IEEE Trans. Automat. Contr.*, **AC-31**, 907–917, 1986.

Verlaan, M., and A. W. Heemink, Reduced rank square root filters for large scale data assimilation problems, *Proc. Intl. Symp. on Assimilation of Observations in Meteorology and Oceanography*, Tokyo, Japan, WMO, Vol. 1, 247–252, 1995.

Computational Aspects of Goddard's Physical-Space Statistical Analysis System (PSAS)

By Jing Guo[1,2] and Arlindo da Silva[1]

[1]Data Assimilation Office, NASA/Goddard Space Flight Center
Greenbelt, Maryland, USA

[2]General Sciences Corporation, Laurel, Maryland, USA

The Physical-Space Statistical Analysis System (PSAS) is being developed at the Data Assimilation Office (DAO), NASA/Goddard Space Flight Center. PSAS has been designed as an incremental improvement over the current optimal interpolation (OI) based Goddard Earth Observing System Data Assimilation System (GEOS/DAS), and a framework to test advanced forecast error covariance models, in support of suboptimal Kalman filtering research at DAO.

PSAS differs from OI in the numerical method used to solve the analysis equation. By removing the *local* and *data selection* approximations, PSAS includes all observations in a single linear system of equations of order 10^5. A *preconditioned conjugate gradient* algorithm in block form is used to solve this linear system of equations iteratively.

The complexity of the PSAS solution is significantly lower than that of a global OI, but the global solver implemented in PSAS is still computationally expensive. For operational feasibility, several physical approximations have been implemented in PSAS, and the code has been parallelized to utilize a shared memory multiprocessor CRAY C98/6 system.

1. Introduction

Optimal interpolation (OI) has been perhaps the best known and the most commonly used statistical interpolation scheme at many operational numerical weather prediction (NWP) centers until recently. Two common approximations to most OI are (a) *local approximation*, whereby observations only within a given region should affect a *minivolume* of analysis including one or more nearby grid points; and (b) *data selection*, assuming that the localized analysis may be approximated by using only a few selected observations from the given region. With the two approximations, OI analyses are carried out independently for each *minivolume*, neglecting interactions across them. As a result, an OI system is not capable of responding to subtle differences in correlation function models.

As a part of the effort to improve GEOS/DAS in support of the Goddard Earth Observing System, the Physical-Space Statistical Analysis System (PSAS) is under development at DAO, coinciding with efforts by many operational NWP centers around the world to replace OI with more advanced global analysis system. PSAS relaxes the *local approximation* and *data selection* restrictions used in OI and includes all observations in one single-matrix equation to produce a global analysis.

Although the observation and forecast error statistics used in the current PSAS implementation remain similar to those used in the OI system, PSAS has been designed to make fewer assumptions about the structure of the correlation functions, in order to provide a flexible framework for using advanced error covariance schemes in support of suboptimal Kalman filter research activities at DAO.

In this paper, we will first give a brief description of the current GEOS/DAS in Section

203

2. A description of the PSAS implementation will be given in Section 3, including the parallelization schemes used in a *multitasked* version of PSAS along with its predicted performance, given in Section 4. A summary will be given in Section 5.

2. Description of GEOS-1/DAS

GEOS/DAS is one of the primary tools developed at DAO in support of its mission of producing research quality assimilated data sets. Unlike in NWP centers, the success of the DAO's mission will be measured uniquely by the quality and the utility of the assimilated data, rather than by forecasts. With version 1 of the system (GEOS-1/DAS), DAO has been carrying out a multiyear reanalysis project covering the period from 1979 to the present [Schubert *et al.*, 1993]. As of this writing, almost 10 years of assimilated data are available for climate studies and other environmental applications.

The GEOS-1/DAS consists of two major coupled systems: an atmospheric global circulation model (AGCM) with a grid resolution of $2° \times 2.5°$ (latitude by longitude) in horizontal and 20 *sigma*–levels in vertical [Takacs *et al.*, 1994]; and an OI-based analysis system [Baker *et al.*, 1987; Pfaendtner *et al.*, 1995]. The OI system is used to produce *analysis increments*, based on the difference between observations and 6-hour forecasts produced by the AGCM. An empirical latitude- and height-dependent error growth scheme is used to estimate the forecast error variances. The analysis increments are assimilated into the dynamic model by using an incremental analysis update (IAU) technique [Bloom *et al.*, 1991, 1996; da Silva, 1995].

The OI system has the same horizontal $2° \times 2.5°$ grid resolution as the AGCM, but 14 vertical pressure levels at 20, 30, 50, 70, 100, 150, 200, 300, 400, 500, 700, 850, and 1000 hPa. Analyzed variables include geopotential heights and winds with multivariate correlation function models, mixing ratio with a univariate correlation function model, as well as sea-level pressure coupled with surface winds via a frictional-wind balanced correlation model. The sea-level analysis is used to produce synthetic 1000 hPa heights for the upper air analysis.

With the *local approximation* and *data selection* restrictions, the OI system selects a maximum of 75 observations from a cylinder of radius equal to 1600 km, with vertical levels ranging from 1 to 2 mandatory pressure levels above and below the *minivolume*. Grid points are grouped into $\sim 12,000$ *minivolumes*, with the same selected observations used for all grid points in the same *minivolume*.

3. Description of PSAS

PSAS solves the analysis problem in the same physical space as in OI. However, it formulates the analysis equations in a different but mathematically equivalent form. As the result, the PSAS solution has significantly less complexity than a global OI solution (including all observations). Because of the size of the solution, PSAS uses a *preconditioned conjugate gradient* algorithm in block form to solve the equation iteratively, with a series of numerical and physical approximations that simplify the computation.

3.1. *Formulation*

On the basis of the minimum variance principle, an *optimal* estimate of the state of atmosphere from the information contained in observations and model forecasts may be presented as the solution of the equations in the form of OI [Daley, 1991],

$$w_a - w_f = K(w_o - H w_f) \tag{3.1}$$

$$(H P^f H^T + R) K^T = H P^f \tag{3.2}$$

where $w_a \in I\!\!R^n$ is a vector representing the analyzed field, $w_f \in I\!\!R^n$ is the model forecast used as a first guess, and $w_o \in I\!\!R^p$ is the observational vector. The difference $w_a - w_f$ is the *analysis increment*. P^f and R are symmetric positive definite matrices of forecast and observation error covariances, respectively. H is a generalized interpolation operator which transforms model variables into observables. The matrix K is the weights of the *innovation* $w_o - Hw_f$. The size of *analysis increment* n is of the order of 10^6. The size of innovation vector p is of the order of 10^5 for the current meteorological observing system, including conventional meteorological soundings, satellite retrievals, and surface and aircraft measurements. The number p and the number of different data types with distinctive error characteristics will be sure to increase quickly when additional satellite and other air- and surface-based observations become available at the turn of the century from the Earth Observing System.

In an OI system, Eq. 3.2 is solved first to determine the weights K. Eq. 3.1 is then used to derive the *analysis increment*. For a global OI solution with given matrices $HP^fH^T + R$ and HP^f, the number of floating point operations required to solve the n $(p \times p)$ equations in Eq. 3.2 alone is at least $\sim O(np^2)$. For the size of problems expected for the system, such an approach is obviously not feasible.

PSAS rewrites the equations in the form

$$w_a - w_f = P^f H^T x \qquad (3.3)$$
$$(HP^f H^T + R)x = w_0 - Hw_f \qquad (3.4)$$

where x is an intermediate vector of size p. By changing the formulation, the complexity of the solution in PSAS is reduced to $\sim O(p^2)$ for solving Eq. 3.4 and $\sim O(np)$ for deriving the *analysis increment* from Eq. 3.3, a significant simplification from a global OI solution.

3.2. *Computational Considerations*

Some implementation considerations regarding the computational algorithm and efficiency of PSAS are now described.

(a) A *preconditioned conjugate gradient* algorithm [Golub and van Loan, 1989] is used to solve Eq. 3.4 iteratively. The *preconditioner* is a recursive conjugate gradient solver with gradually smaller problems defined by (i) regional (see (c)), (ii) univariate, and (iii) one or more profiles of observations, block diagonal matrices, respectively. The block equations at the lowest level (profiles) are solved directly by using the Cholesky algorithm.

(b) Both matrices $HP^f H^T + R$ in Eq. 3.4 and $P^f H^T$ in Eq. 3.3 are normally too large to be held in memory on current supercomputers and therefore blocks of these matrices have to be regenerated in each iteration, limiting the efficiency of the algorithm. Currently, the time spent on generating the matrices may be well over 95% of the total computing time for a medium sized analysis ($p \sim 50$–80×10^3).

(c) The innovation vector $w_o - Hw_f$ is partitioned into 80 nearly equal area regions (4 spherical triangular regions for each of 20 triangular sides of an icosahedron) according to the observation locations to allow matrix-vector multiplications to be performed in a block form.

(d) Sparsity of the matrices is accomplished by assuming *compactly supported* horizontal correlation functions. Regions separated by more than ~ 6000 km (corresponding to negligible correlation values) are treated as uncorrelated, such that the partial matrix-vector product for the regions can be avoided.

(e) A quasi-equal area grid is used to reduce the cost of computing the *analysis increment* from Eq. 3.3. In an equal-interval grid, single locations at the poles are represented by many grid points with the same latitude but different longitude values. In a quasi-equal area grid, however, different longitude grid intervals are used at different latitudes, to keep the grid point density ($\propto$ area^{-1}) about the same over the globe. Therefore, the redundancy in an equal-interval grid over higher latitude regions is removed. The results are further interpolated to the required equal-latitude–longitude-interval grid by one-dimensional *cubic-spline* interpolations.

(f) Some adjustable physical approximations have also been introduced to limit the computation cost without degrading the quality of analyses, including *superobing* – a process averaging dense observations within a user specified range to limit the size of the innovation vector; relaxing the convergence criterion for the iterative solution to prevent excessive iterations within the significant accuracy of the solution; and reducing the spatial extent of the *compactly supported* correlation functions to create extra sparsity.

4. Multitasked PSAS

In order to take the advantage of a shared memory multiprocessor CRAY C98/6 system, PSAS has been parallelized by utilizing the *multitasking* technology of the system. For this application, the parallelism is clearly identifiable at the levels of block matrix formation and block matrix-vector multiplication in the algorithm. However, severe memory contention problems would have been present if multiple processors had been programmed to update the same block of the product vector concurrently. Recognizing that the matrix $P^f H^T$ in Eq. 3.3 is rectangular, whereas the matrix $H P^f H^T + R$ in Eq. 3.4 is symmetric, computations for the two equations are parallelized with two different partition schemes, both schemes designed to prevent any two processors from updating the same memory section within a parallelized program segment.

4.1. *Rectangular Matrix-Vector Multiplication*

Eq. 3.3 may be perfectly parallelized by dividing the *analysis increment* vector into blocks according to variable types and the vertical levels of its elements

$$
Cx = \begin{bmatrix} C_1 \\ C_2 \\ \vdots \\ C_m \end{bmatrix} x \quad \Longrightarrow \quad \begin{pmatrix} C_1 x \\ C_2 x \\ \vdots \\ C_m x \end{pmatrix}
$$

where $C = P^f H^T$, m is the size of the *analysis increment* vector Cx in blocks, and C_i is a block of rows of the matrix with a unique variable type and vertical level combination. The tall parenthesis pair following "$\Longrightarrow$" delimits a program segment with parallel partitions. A partition is represented by a single expression component $C_i x$.

Since the sparsity of the matrix depends on several factors, the computational load among parallelized partitions may not be well balanced. This appears not to be a problem for the actual *extent of the parallelism* (the number of parallelized partitions, which is $m \sim 60$–75), compared to the number of processors (≤ 6) on our CRAY C98/6. A reasonable load balance seems to be achieved dynamically during the computation. A larger *extent of parallelism* (m) with smaller *granularity* (amount of work for a partition) may be obtained by further dividing C into smaller blocks according to the regional indices of its rows.

4.2. *Symmetric Matrix-Vector Multiplication*

Eq. 3.4 is solved iteratively with each iteration involving matrix-vector multiplications with a symmetric matrix $H P^f H^T + R$ (or any preconditioner matrix) and a vector x.

Multitasking of the multiplication is implemented with a rearranged order of addition operations.

The matrix $C = HP^f H^T + R$ is symmetric. It is divided into blocks $\{C_{ij}\}$ symmetrically corresponding to the block partition of the vector x, which is partitioned to $\{x_i\}$ according to the regional indices and the variable types of its elements. Every C_{ij} has a symmetric image $C_{ji} = C_{ij}^T$. Since most computations will be spent on regenerating the matrix, it is highly desirable to compute $C_{ij}^T x_i$ and $C_{ij} x_j$ at the same time when the block matrix C_{ij} is created. Therefore, the multiplication is split into three multiplications of partial matrices represented in skewed forms:

$$
Cx = \begin{bmatrix} C_{11} & C_{12} & \cdots & C_{1q} \\ C_{12}^T & C_{22} & \cdots & C_{2q} \\ \vdots & \vdots & \ddots & \vdots \\ C_{1q}^T & C_{2q}^T & \cdots & C_{qq} \end{bmatrix} \begin{bmatrix} x_1 \\ x_2 \\ \vdots \\ x_q \end{bmatrix}
$$

$$
= \begin{bmatrix} & C_{11}x_1 & +C_{12}x_2 + \ldots + C_{1(q-1)}x_{q-1} + C_{1q}x_q \\ C_{12}^T x_1 + & C_{22}x_2 & +C_{23}x_3 + \ldots + C_{2q}x_q \\ & \vdots & \\ C_{1q}^T x_1 + \ldots + C_{(q-1)q}^T x_{q-1} + & C_{qq}x_q & \end{bmatrix}
$$

$$
= C_l x + C_d x + C_u x
$$

where q is the number of blocks in vector x. Product $C_d x$ involves all diagonal matrix blocks $\{C_{ii}\}$; $C_u x$ involves all matrix blocks in the upper triangle $\{C_{ij}; i < j\}$; and $C_l x = C_u^T x$ involves all matrix blocks in the lower triangle $\{C_{ji} = C_{ij}^T; i < j\}$.

The partial products are parallelized with the following schemes:

$$
C_d x \implies \begin{pmatrix} C_{11}x_1 \\ C_{22}x_2 \\ \vdots \\ C_{qq}x_q \end{pmatrix}
$$

and

$$
C_u x \implies \begin{pmatrix} C_{12}x_2 \\ C_{23}x_3 \\ \vdots \\ C_{(q-2)(q-1)}x_{q-1} \\ C_{(q-1)q}x_q \\ 0 \end{pmatrix} + \begin{pmatrix} C_{13}x_3 \\ C_{24}x_4 \\ \vdots \\ C_{(q-2)q}x_q \\ 0 \\ 0 \end{pmatrix} + \cdots + \begin{pmatrix} C_{1q}x_q \\ 0 \\ \vdots \\ 0 \\ 0 \\ 0 \end{pmatrix}
$$

As in Section 4.1, a pair of tall parentheses after "$\implies$" delimits a parallelized program segment. A single-expression component represents a partition. A similar scheme is used for $C_l x = C_u^T x$. Instead of being performed separately, $C_l x$ is computed along with $C_u x$ by creating a second vector storage and linking partitions to their transposed matrix blocks (e.g., both $C_{12}x_2$ and $C_{21}x_1 = C_{12}^T x_1$ are computed in the same partition with results stored in two separate vectors). With this partition scheme, no two partitions (therefore, no two processors) will write their outputs to the same memory location inside a parallelized program segment.

At the far right of the expression for the $C_u x$ scheme, there is only one partition $C_{1q}x_q$ to be computed. However, in a real case with a proper regional order, the summation will

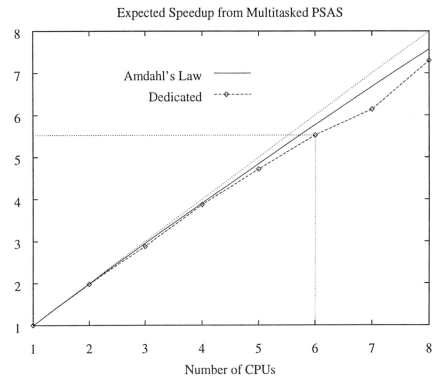

FIGURE 1. Expected *speedup* according to *ATExpert* (Cray Research, Inc.), for one analysis experiment. The program is 99.2% parallel. The dotted diagonal line is the theoretical *speedup* limit of any parallelized code. Amdahl's law represents an idealized *speedup* with the realized parallelization percentage. For 6 CPUs, 0.25 CPUs of overhead and a 5.52 *speedup* on a *dedicated* system are predicted.

be terminated long before it reaches the far right end because of the sparsity structure of matrix C created by using *compactly supported* horizontal correlation functions.

The actual load balance among partitions in each parallelized program segment is irregular and difficult to estimate, because of the irregular regional data distribution and the complex sparsity structure of the matrix. However, the leading terms in the partition expression of parallelized $C_u x$ dominate the total computation cost and have large *extents of the parallelism* (number of blocks $q \sim 80$–560). Therefore, a dynamic load balance can be achieved, while the inefficiency of the trailing terms may be either prevented or ignored.

4.3. *Predicted Multitasked Performance*

The predicted performance of the *multitasked* version of PSAS has been examined with a system utility. The result shown in Figure 1 indicates that with the partition schemes described in this section, the PSAS computation can be well parallelized on a multiprocessor CRAY C98/6 system with very small overhead. For real applications with different situations from case to case, the performance may vary. However, variations among the predicted *speedup* and overhead factors are small, and the larger the size of the analysis equation, the better the predicted performance.

5. Summary

In this paper, we have given a brief description of the ongoing development of PSAS for the Goddard Earth Observing System Data Assimilation System and outlined the parallelization strategy on a shared memory multiprocessor CRAY C98/6 system with its predicted performance.

The PSAS solution, while mathematically equivalent, is different from OI in its numerical approaches and approximations. Preliminary experiments made to assess the effects of relaxing OI *local* and *data selection* approximations have shown various distinctive spectral differences in the *analysis increments* [da Silva *et al.*, 1995]. Having a different approach to solving the analysis equations, PSAS also significantly simplifies the complexity of a global solution compared to using the OI formulation. Although its computation and memory costs are still high, PSAS can be effectively parallelized for a shared memory multiprocessor CRAY C98/6 system and is operationally feasible with other proper approximations. A version of PSAS for massive parallelized processor systems is also under study in cooperation with a group of scientists led by Dr. R. Ferraro at NASA/JPL.

Acknowledgments. We would like to dedicate this paper to the memory of Dr. James Pfaendtner, the chief designer and developer of the original version of PSAS, on which this work is based. We would like also to thank Meta Sienkiewicz, Steve Cohn, and Dave Lamich for fruitful discussions and their contributions to the development of PSAS.

REFERENCES

Baker, W. E., S. Bloom, J. Woollen, M. Nestler, E. Brin, T. Schlatter, and T. Branstator, Experiments with a three-dimensional statistical objective analysis scheme using FGGE data, *Mon. Wea. Rev.*, **115**, 272–296, 1987.

Bloom, S. S., L. L. Takacs, and E. Brin, A scheme to incorporate analysis increments gradually in the GLA assimilation system, *Ninth Conf. on Numerical Weather Prediction*, Denver, CO, Am. Meteor. Soc., 110–112, 1991.

Bloom, S. S., L. L. Takacs, A. M. da Silva, and D. Ledvina, Data assimilation using incremental analysis updates, *Mon. Wea. Rev.*, **124**, 1256–1271, 1996.

Daley, R., *Atmospheric Data Analysis.* Cambridge University Press, 1991.

Golub, G. H. and C. F. van Loan, *Matrix Computations*, 2nd ed., The Johns Hopkins University Press, 1989.

Pfaendtner, J., S. Bloom, D. Lamich, M. Seablom, M. Sienkiewicz, J. Stobie, and A. da Silva, Documentation of the Goddard Earth Observing System (GEOS) Data Assimilation System – Version 1, *NASA Tech. Memo. 104606*, Vol. 4, 1995.

Schubert, S. D., R. B. Rood, and J. Pfaendtner, An assimilated data set for earth science applications, *Bull. Am. Meteor. Soc.*, **74**, 2331–2342, 1993.

da Silva, Filtering Properties of Incremental Analysis Updates (IAU), *International Symposium on Assimilation of Observations*, Tokyo, Japan, March 13–17, 1995.

da Silva, A., J. Pfaendtner, J. Guo, M. Sienkiewicz, and S. E. Cohn, Assessing the effects of data selection with DAO's Physical-Space Statistical Analysis System, *International Symposium on Assimilation of Observations*, Tokyo, Japan, March 13–17, 273–278, 1995.

Takacs, L., A. Molod, and T. Wang, Documentation of the Goddard Earth Observing System (GEOS) General Circulation Model – Version 1, *NASA Tech. Memo. 104606*, Vol. 1, 1994.

A Study on the Influence of the Pacific and Atlantic SST on the Northeast Brazil Monthly Precipitation Using Singular Value Decomposition

By Cíntia Bertacchi Uvo,[1] Carlos Alberto Repelli,[2] Stephen Zebiak[3] and Yohanan Kushnir[3]

[1]Centro de Previsão de Tempo e Estudos Climáticos, Instituto Nacional de Pesquisas Espaciais, São José dos Campos, SP, Brazil

[2]Foundation for Meteorology and Water Resources – FUNCEME, Fortaleza, CE, Brazil

[3]Lamont-Doherty Earth Observatory, Columbia University, Palisades, New York, USA

The objective of this work is to diagnose the spatial relationship between the monthly sea surface temperature (SST) and the precipitation over Northeast Brazil. The methodology employed is based on multivariate analysis known as singular value decomposition (SVD). The results obtained in this work are the first step toward intraseasonal precipitation forecasting for the Brazilian semiarid region. They show clearly the importance of both the Pacific and the Atlantic SST conditions in driving the precipitation of the region and also show the different effects of the SST patterns on different parts of the semiarid region.

1. Introduction

The Political Northeast Region of Brazil (hereafter referred to as the Northeast) is a densely populated region located approximately between 1°S and 18°S and 35°W and 47°W. The northern part of the Northeast (hereafter referred to as Nordeste) is known for its semiarid climate with its rainy season concentrated between February and May and very large interannual variability.

Many authors have identified relationships between the precipitation over the Northeast and El Niño–Southern Oscillation (ENSO) events: Atlantic Ocean sea surface temperature, trade winds, and sea level pressure; the position of the Intertropical Convergence Zone (ITCZ) over the Atlantic Ocean; cold fronts and other effects [Moura and Shukla, 1981; Ropelewski and Halpert, 1987; Uvo, 1989]. Nordeste exhibits not only high variability in the total amount of precipitation from year to year, but high spatial and temporal variability in the precipitation within its rainy season. Late February/March is the period when the ITCZ over the Tropical Atlantic Ocean reaches its southernmost position, beginning what is called the principal rainy season over Nordeste. The return of the ITCZ to its more northern position is what determines the end of the Nordeste principal rainy season. The direct cause of a good (precipitation above normal) or a bad (precipitation below normal) rainy season is the ITCZ position during April and May [Uvo, 1989]. In a normal year, the ITCZ starts its return to the northern positions in mid-April. During a dry year the ITCZ either doesn't reach positions south of the equator at all, isn't active enough to cause precipitation over the region, or moves northward early.

Different methodologies have been developed to forecast the Nordeste's rainy season and have been used during the last several years for operational purposes at the National Institute of Spatial Research (INPE) in São José dos Campos, SP, Brazil, and at the

Fundação Cearense de Meteorologia e Recursos Hídricos (FUNCEME) in Fortaleza, CE, Brazil. Those methodologies are based on empirical atmospheric and oceanic parameters, on results from coupled models, and on statistical models developed specifically for forecasting the Nordeste rainy season, including those of Hastenrath *et al.* [1993] and Ward and Folland [1991]. These methodologies have been formulated in terms of the precipitation of the entire rainy season, without consideration of the variations within the rainy season. The constancy of the rainfall during the rainy season is, however, important for the development of the local agriculture and the phenomenon called "Veranico" – the suppression of the precipitation during a period greater than 10 days – which can be disastrous for local crops. Improving the understanding of intraseasonal variability in the Nordeste precipitation and assessing the potential of monthly precipitation forecasts were the primary motivations for this work.

2. Data Sources

For this study we have utilized monthly precipitation anomalies normalized by standard deviation for a network of 105 rain gauge stations well distributed over the Northeast Brazil Region (NEB), and sea surface temperature (SST) for the Tropical Region for the period between 1946 and 1985. The precipitation data set was obtained from the Superintendjncia do Desenvolvimento do Nordeste (SUDENE) and FUNCEME. The SST data set was obtained from an analysis of the Comprehensive Ocean Atmosphere Data Set (COADS). The data were analyzed on a $1° \times 1°$ latitude $\times$ longitude grid and interpolated to the Gaussian grid of the GFDL R15 spectral model, which has a resolution of approximately $4.5°$ latitude $\times 7.5°$ longitude [Lau *et al.*, 1990]. Two different latitude ranges were selected: $42.75°$S to $42.75°$N and $24.75°$S to $24.75°$N.

3. Methodology

The techniques employed for this study are based on a multivariate analysis procedure known as singular value decomposition (SVD), which utilizes the cross-covariance matrix between two data sets [Bretherton *et al.*, 1992]. Its use allows one to isolate pairs of the spatial pattern that explain the maximum squared temporal covariance between two physical variable fields. It is based upon a fundamental matrix operation: That is, it is a generalization of the diagonalization process used in principal component analysis (PCA) for square matrices.

The theory of SVD is very well discussed and compared with other multivariate techniques in Bretherton *et al.* [1992] and Wallace *et al.* [1992]. The SVD of the cross-covariance matrix of two fields gives two matrices of singular vectors and one set of singular values. A singular vector pair describes spatial patterns for each field which have overall covariance given by the singular value. As described in Bretherton *et al.* [1992], the set of singular values can be used to find the squared covariance fraction (SCF) explained by each pair of singular vectors, in the same way that the eigenvalues of one covariance matrix are useful to measure the percentage of variance assumed by the principal components. In other words, the SCF is useful to compare the relative importance of each expansion mode. It can become undetermined when the squared covariance of the fields is close to zero. Also, a high SCF value is not a guarantee that the mode is statistically significant, as shown by Monte Carlo simulations in Bretherton *et al.* [1992]. On the basis of these facts, Wallace *et al.* [1993] suggest the standardized squared covariance. The normalized squared convariance ranges from 0, when two fields

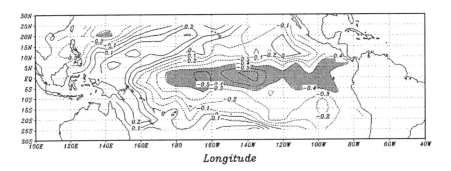

FIGURE 1. Heterogeneous correlation map for the first mode in the SVD expansion for Feb–Mar–Apr–May Pacific SST field. Shaded regions indicate significance at a level around 95% based on Student's t distribution.

are unrelated, to 1 if the variations at each grid point in the first field are perfectly correlated with the variations at all grid points in the second field [Wallace *et al.*, 1993].

From the singular vector pairs of a two-vector set, one can obtain the temporal expansion series of each field projecting the data onto the singular vector. The temporal field can be reconstructed by the linear combination of its spatial vectors with the respective coefficient, over all modes. As described in Bretherton *et al.* [1992], this results in the so-called heterogeneous covariance maps of the left and right fields. This represents the correlation coefficients between the expansion coefficients of one field and the grid points of the other. In our application, the patterns shown by the heterogeneous correlation maps for the kth SVD expansion mode indicate how well the pattern of the precipitation (SST) anomalies relates to the kth expansion coefficient of SST (precipitation). The heterogeneous maps of each field are mutually orthogonal in the space domain. The correlation coefficient between two temporal expansion coefficients for the first modes indicates how strong the relationship between the fields is.

The SVD technique was applied to the covariance matrix between monthly normalized precipitation anomalies over Northeast Brazil and tropical SST, for the period between December and June. Results for individual months, as well as the combined February–May period, were obtained.

4. Results

SVD analyses between monthly SST and precipitation over Northeast Brazil were computed for diagnostic purposes; thus the analyses were done with both simultaneous precipitation and SST data.

First, analyses were made of global SST between 40°N and 40°S and the precipitation in the whole Northeast. Second, SVD was done for SST for the Tropical Pacific and Atlantic (25°N to 25°S) separately.

The following results are based on SVD analyses for the monthly precipitation over Nordeste from January to May and the simultaneous SST fields over the Tropical Pacific and Atlantic Oceans.

During January the rainy season is not yet established over Nordeste and the precipitation that can occur over the region is mostly due to cold fronts or their remnants. The relationship found between Pacific SST and the precipitation for January was as follows: SST in the El Niño region (1+2 and 3) over the Pacific relates positively to precipitation

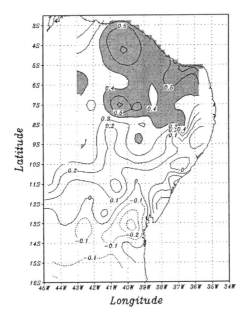

FIGURE 2. Heterogeneous correlation map for the first mode in the SVD expansion for Feb–Mar–Apr–May precipitation field. Shaded regions indicate significance at a level around 95% based on Student's t distribution.

over some areas in the southern part of the Northeast (nearly at the end of its rainy season in this month) and negatively to precipitation over small areas in Nordeste (mostly in Piaui State).

The heterogeneous maps for the Tropical Atlantic Ocean SST and Northeast precipitation did not show a strong relationship in January. Only a very small region in the Central Atlantic was significantly positively correlated to the precipitation over the central and southern parts of Northeast. During February, when the rainy season starts in Nordeste, some interesting correlations begin to emerge. Pacific Ocean SSTs do not appear to exert a large influence over the precipitation in this month, except for a small region along the equator between 160°W and the dateline that is significantly positively correlated with the precipitation over the southern part of Northeast (mostly south of 12°S). On the other hand, the Atlantic Ocean shows a strong relationship with the precipitation over Nordeste in February. The central region of the Atlantic, between 5°N and 10°S, is related positively to the precipitation over north-Nordeste; that is, warm SST anomalies in the Central Atlantic are related to an earlier start of the rainy season in Nordeste, mostly in the region that is north and west of Ceara State, west of Piaui, and part of Rio Grande do Norte State, where the presence of the ITCZ is one of the most important causes of precipitation.

In April, the Pacific equatorial band between 120°W and the dateline presents the highest negative correlation with the precipitation over Nordeste (mostly north of 10°S and west of 39°W). Over the Atlantic, for the first time, the so-called dipole [Moura and Shukla, 1981] feature is observed and is highly correlated to the precipitation in most parts of Nordeste, north of 10°S. This map presents the highest normalized squared covariance value observed for the heterogeneous maps.

For May, the region with highest correlation over the Pacific is displaced to the east,

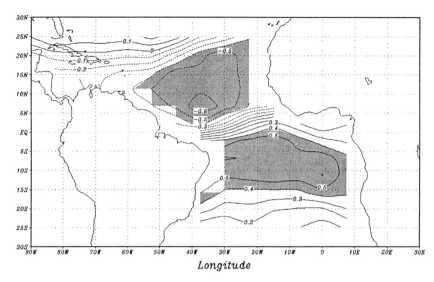

Longitude

FIGURE 3. Heterogeneous correlation map for the first mode in the SVD expansion for Feb–Mar–Apr–May Atlantic SST field. Shaded regions indicate significance at a level around 95% based on Student's t distribution.

from 150°W to the South American coast, along the equator. This region is mostly related to the Nordeste precipitation north of 6°S. The dipole pattern appears again over the Atlantic, related also to the precipitation over Nordeste, mostly north of 6°S, and also to a small area in the northeastern part of Bahia State.

The heterogeneous fields for the average February–May SST and precipitation show clearly the different influences of Pacific and Atlantic SST over the Nordeste rainy season. For the Pacific (Figure 1), the highest correlations were observed in the equatorial band from 170°E to the South American coast, correlated negatively with the precipitation over the northern and western parts of Nordeste (Figure 2).

For the Atlantic, the dipole pattern is evident (Figure 3). The north basin of the Atlantic Ocean is negatively correlated with Nordeste precipitation (mostly north of 10°S) while the south basin showed a positive correlation with the precipitation over the same area. Features over the Atlantic Ocean seem to explain more of the precipitation variance than Pacific patterns (Figure 4). In summary, the influence of the Pacific Ocean on the Nordeste rainy season is more evident from March on. Atlantic SSTs show a relatively strong relationship, explaining a higher part of the covariance. SST anomalies over the Atlantic seem well related to the start of the rainy season and to its duration. It is worth emphasizing that the quality of the rainy season is primarily related to how long the ITCZ remains over its southernmost position.

The three regions most correlated with the Northeast seasonal precipitation in the equatorial Pacific, in the North Atlantic, and in the South Atlantic have also been determined with a simple correlation analysis. For each of the three regions an SST index was calculated, defined as the spatial average over each region of the monthly average normalized SST anomaly.

From the comparison of these monthly indices with a similar index for the Northeast precipitation, it is apparent that extremely dry (wet) years occur with anomalous warm (cold) equatorial Pacific SST together with anomalously warm (cold) North Atlantic SST and anomalously cold (warm) South Atlantic SST. Any situation between these

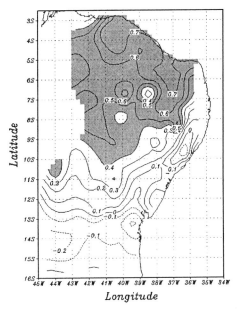

FIGURE 4. Heterogeneous correlation map for the first mode in the SVD expansion for Feb–Mar–Apr–May precipitation field. Shaded regions indicate significance at a level around 95% based on Student's t distribution.

two extremes depends on the relative strength of the Pacific and Atlantic SST patterns. This characteristic was observed to be valid either for the averaged rainy season or for individual months within the rainy season.

5. Conclusions

We have utilized singular value decomposition (SVD) to analyze the relationships between monthly normalized tropical SST anomalies and monthly normalized precipitation anomalies over Northeast Brazil, from December to June, considering simultaneous relationships. The results showed different influences of Atlantic and Pacific SSTs on Nordeste precipitation during different months of the rainy season. In February, when the rainy season starts, no pattern over the Pacific was found to influence the precipitation over Nordeste significantly. However, Atlantic Ocean SST anomalies seem to play a more important role: Positive anomalies over the central Atlantic in February are significantly related to the precipitation in the northern part of Nordeste. Presumably, positive SST anomalies in the central Atlantic can influence the southward displacement of the ITCZ, generating an earlier start of the rainy season.

March is the month when the ITCZ cloud band and tropical trough are normally at about their southernmost position. The proximity of the cloud band and the tropical trough to Nordeste increases the precipitation over the region. This occurs virtually every year regardless of SST anomalies in the Atlantic basin. Consequently, during this month, the Atlantic SST did not present any significant relationship with the precipitation over Nordeste. On the other hand, SST anomalies in the equatorial East Pacific do show some relationship with precipitation over parts of Nordeste during March.

Precipitation during April and May is very important in determining the quality of the rainy season, because a good rainy season means precipitation above normal. The proximity of the ITCZ to the vicinity of Nordeste during these 2 months, and consequently the amount of precipitation over the region, is what results in a good rainy season. Our analyses showed that during the 2 months, both the dipole over the Atlantic Ocean and the El Niño over the Pacific have a large influence on the ITCZ position, but the Atlantic pattern presented the stronger relationship in both months.

The results obtained in this work are a first step toward intraseasonal precipitation forecasting for the Brazilian semiarid region. They show clearly the importance of both the Pacific and the Atlantic SST conditions in driving the precipitation of the region and also indicate that different parts of the semiarid region are affected differently by the SST patterns. Further development of intraseasonal forecasts will be especially useful for agricultural planning in this region where irrigation is almost nonexistent.

REFERENCES

Bretherton, C. S., C. Smith, and J. M. Wallace, An intercomparison of methods for finding coupled patterns on climate data, *J. Climate*, 5, 541–560, 1992.

Hastenrath, S., and L. Greischar, Further work on the prediction of Northeast Brazil rainfall anomalies, *J. Climate*, 6, 743–758, 1993.

Lau, N.-C., and M. J. Nath, A general circulation model study of the atmospheric response to extratropical SST anomalies observed in 1950–79, *J. Climate*, 3, 965–989, 1990.

Moura, A. D., and J. Shukla, On the dynamics of droughts in Northeast Brazil: Observations, theory, and numerical experiments with a general circulation model, *J. Atmos. Sci.*, 38, 2653–2675, 1981.

Ropelewski, C. F., and M. S. Halpert, Global and regional scale precipitation patterns associated with the El Niño–Southern Oscillation, *Mon. Wea. Rev.*, 115, 1606–1626, 1987.

Uvo, C. R. B., The Intertropical Convergence Zone and its relationship with the precipitation over north-northeast region of Brazil, INPE-4887-TDL/378, 1989, available from Instituto Nacional de Pesquisas Espaciais, 12200 São José dos Campos, SP, Brasil.

Wallace, J. M., C. Smith, and S. Bretherton, Singular value decomposition of wintertime sea surface temperature and 500mb heights anomalies, *J. Climate*, 5, 561–576, 1992.

Wallace, J. M., Y. Zhang, and K. H. Lau, Structure and seasonality of interannual and interdecadal variability of the geopotential height and temperature fields in the Northern Hemisphere troposphere, *J. Climate*, 6, 2063–2082, 1993.

Ward, M. N., and C. K. Folland, Prediction of seasonal rainfall in the North Nordeste of Brazil using eigenvectors of sea-surface temperature, *Int. J. Climatol.*, 11, 711–743, 1991.

Numerically Efficient Methods Applicable to the Eigenvalue Problems Arising in Linear Stability Analysis

By Marco De la Cruz-Heredia and G. W. Kent Moore

Atmospheric Physics Department, University of Toronto
60 St. George Street, Toronto, Ontario, M5S 1A7, Canada

A persistent problem in atmospheric physics consists of solving the large eigenvalue problems that arise when one attempts to determine the stability of a given basic state to small-amplitude perturbations. Even a problem as simple as determining the stability of a one-dimensional basic state with a constant vertical wind shear leads to a generalized eigenvalue problem of the form $Ax = cBx$ where A and B are non-Hermitian. The QR algorithm is generally applicable but its usefulness decreases rapidly as the complexity of the system is increased, either by assuming a more general set of governing equations or by considering a more complicated basic state. However, high-resolution solutions of these systems can be obtained by exploiting the sparse nature of the eigenvalue problem. In particular, the application of inverse vector iteration coupled with an LU algorithm within a sparse storage scheme leads to a dramatic reduction in the storage and computational requirements.

In view of this, we have approached the instability problem by taking full advantage of its sparseness. In this paper, we will describe the solution of some simple cases that will serve to illustrate the advantages that our approach has over conventional methods.

---

1. Introduction

The basic equations that govern the dynamics of the atmosphere represent a complex set of nonlinear equations, namely, the three momentum equations, the continuity equation, and a thermodynamic equation together with an appropriate equation of state. Simplification of this set of primitive equations is achieved through scaling and linearization about a certain basic state. One can then apply the normal modes method to obtain a system which, together with suitable boundary conditions, leads to an eigenvalue problem.

Among the most popular techniques for solving the instability problem are the "shooting method" and the "matrix method." The former uses relatively small amounts of computer memory but relies on a rootfinder and a good initial guess. Solving the eigensystem using the QR algorithm (regarding which full details can be found in Jennings and McKeown [1992]) requires no guess and produces all the eigenvalues and eigenvectors, but, since the computation time grows as the cube of the order of the matrix, it soon becomes impractical as the grid resolution is increased. Even on a fast computer, it is the large storage requirements which eventually limit the use of matrix methods in solving the instability problem.

In order to avoid the disadvantages of the shooting method and to eliminate the huge memory requirements of the conventional matrix approach, efficient storage methods coupled with sparse techniques have been applied to this problem. Using the QR algorithm, we solve over a coarse grid to obtain an approximate eigenvalue spectrum. We can then choose an eigenvalue (in our case the fastest growing mode) and use it as a first approximation to initiate an inverse vector iteration. Since it preserves the band structures of the original matrices, very large systems can be treated. Following a continuation procedure, we can then afford to increase the grid resolution and repeatedly solve until changes in eigenvalues and eigenvectors are within a certain tolerance.

In the next section we will briefly describe the instability problems we have tackled using the approach. A description of the inverse vector iteration and continuation procedures follows. The implementation of these algorithms to the resulting eigensystems is discussed in Section 4. The final section contains the results obtained.

2. Basic Instability Theory

Given an initial state, perturbations in the atmosphere may grow or decay, depending on their ability to extract energy from the basic flow. In particular, if a disturbance grows in time by transfer of potential energy from the mean flow, the wave instability is said to be baroclinic. This type of instability is associated with vertical shear in the wind; that is, the mean zonal wind velocity is a function of height $\overline{U} = \overline{U}(z)$. Barotropic instability, on the other hand, results from the presence of horizontal shears in a jetlike current, $\overline{U} = \overline{U}(y)$, and requires the conversion of kinetic energy from the flow into the perturbation to grow. Although both types of instabilities are normally present in the atmosphere, it is customary to assume that one is absent so as to reduce the problem to a one-dimensional case.

Assuming a Boussinesq, adiabatic, inviscid and hydrostatic flow linearized around the basic state it is possible to write down the governing equations the following way

$$\frac{\partial \tilde{u}}{\partial t} + \overline{U}\frac{\partial \tilde{u}}{\partial x} + \tilde{v}\frac{\partial \overline{U}}{\partial y} + \tilde{w}\frac{\partial \overline{U}}{\partial z} - f\tilde{v} = -\frac{\partial \tilde{\phi}}{\partial x} \tag{2.1}$$

$$\frac{\partial \tilde{v}}{\partial t} + \overline{U}\frac{\partial \tilde{v}}{\partial x} + f\tilde{u} = -\frac{\partial \tilde{\phi}}{\partial y} \tag{2.2}$$

$$\frac{\partial \tilde{u}}{\partial x} + \frac{\partial \tilde{v}}{\partial y} + \frac{\partial \tilde{w}}{\partial z} = 0 \tag{2.3}$$

$$\frac{\tilde{\theta}}{\theta_0}g = \frac{\partial \tilde{\phi}}{\partial z} \tag{2.4}$$

$$\frac{\partial \tilde{\theta}}{\partial t} + \overline{U}\frac{\partial \tilde{\theta}}{\partial x} - \frac{\theta_0 f}{g}\frac{\partial \overline{U}}{\partial z}\tilde{v} + \frac{\partial \overline{\theta}}{\partial z}\tilde{w} = 0 \tag{2.5}$$

Here $\tilde{u}$, $\tilde{v}$, $\tilde{w}$ represent the x, y, z perturbation velocities; $\tilde{\theta}$ is the potential temperature and $\tilde{\phi}$ is the geopotential; f and θ_0 are the Coriolis parameter and a reference temperature, respectively.

For the baroclinic case we assume that $\overline{U} = \Lambda z$ is independent of y. Substituting perturbations of the form

$$\widetilde{\Psi} = \Psi(z)e^{i(kx-\omega t)} \tag{2.6}$$

we can obtain a single equation for the vertical velocity

$$\left[k^2(\overline{U} - c)^2 - f^2\right]w_{zz} + \frac{2f^2\Lambda}{\overline{U} - c}w_z + k^2 N^2(z)w = 0 \tag{2.7}$$

where $c \equiv (\omega/k)$. N is called the Brunt–Väisälä frequency and is related to the vertical gradient of the mean potential temperature. It is in general a function of height, and we shall assume it has the form

$$N^2(z) = \left(a \tanh \frac{z - d}{\Delta d} + b\right) \tag{2.8}$$

where a, d, Δd and b are constants. Together with the rigid-lid boundary conditions

$$w = 0 \quad \text{at} \quad z = 0, H \tag{2.9}$$

Eq. 2.7 can be written as the following eigenvalue problem:

$$A_0 w = c A_1 w + c^2 A_2 w + c^3 A_3 w \tag{2.10}$$

where

$$A_0 = \Lambda z D^{(2)} - 2\Lambda D^{(1)} - \frac{k^2 N^2(z)}{f^2} \Lambda z - \frac{k^2}{f^2} (\Lambda z)^3 D^{(2)}$$

$$A_1 = D^{(2)} - \frac{k^2 N^2(z)}{f^2} - 3\frac{k^2}{f^2} (\Lambda z)^2 D^{(2)}$$

$$A_2 = 3\frac{k^2}{f^2} \Lambda z D^{(2)} \qquad A_3 = \frac{k^2}{f^2} D^{(2)}$$

$D^{(1)}$ and $D^{(2)}$ represent the first and second finite difference operators. With the preceding notation it is easy to see that the resulting matrices are sparse since, for example, using a three-point central difference formula will yield four tridiagonal matrices. This is essentially the same problem solved by Nakamura [1988] using a shooting method technique.

A similar approach is used when formulating the barotropic problem; i.e., we now consider a mean zonal profile with no vertical shear $\overline{U} = (1 - \cos y)/2$, so that upon substituting solutions of the form

$$\widetilde{\Psi} = \Psi(z) e^{i(kx + mz - \omega t)} \tag{2.11}$$

into the primitive equations we can reduce the system to an equation in the meridional (y direction) velocity v only. Note that in this case we assume that the vertical domain is finite, and hence there is a vertical wavenumber dependence through m. This assumption is normally made under the so-called quasi-geostrophic (QG) theory, which deals with a simplified version of the primitive equations (PEs). It then turns out that m enters the eigenvalue problem in a trivial way, which effectively renders irrelevant the fact whether or not the domain is bounded. The novelty in our approach consists in taking into account the presence of "lids," but now using the full PE. This is clearly a more realistic model, and one that follows naturally from the boundary conditions we use in the baroclinic case, e.g., a bottom boundary (the earth) and a lid (the tropopause). Assuming rigid walls at the horizontal boundaries we again obtain a generalized eigensystem, but now of fifth order

$$A_0 v = c A_1 v + c^2 A_2 v + c^3 A_3 v + c^4 A_4 v + c^5 A_5 v \tag{2.12}$$

The full form of the matrices is rather complicated. Their general features, however, are similar to those appearing in the baroclinic case, e.g., matrix A_3 is

$$m^2 \left(2k^2 N^2 - 10 k^2 m^2 \overline{U}^2 - f m^2 (\overline{U}_y - f) - N^2 D^{(2)} \right)$$

and thus is also sparse. For simplicity, we will assume that N is constant for the barotropic case.

3. Method of Solution

The problem at hand is now to solve the eigensystems 2.10 and 2.12. We are interested in finding the most unstable modes and hence we seek the eigenvalues with the largest positive imaginary part; i.e., given $c = c_r + ic_i$ the exponent in Eq. 2.6 will be $e^{ik(x-c_r t)} e^{c_i t}$ and the perturbation will grow in time.

Let us start with the simpler baroclinic problem. By introducing the vector

$$r = (w, \ cw, \ c^2 w)^T$$

we can rewrite Eq. 2.10 in the following way

$$Ar = cBr \tag{3.1}$$

where

$$A = \begin{bmatrix} A_0 & -A_1 & -A_2 \\ 0 & A_2 & 0 \\ 0 & 0 & A_3 \end{bmatrix} \tag{3.2}$$

and

$$B = \begin{bmatrix} 0 & 0 & A_3 \\ A_2 & 0 & 0 \\ 0 & A_3 & 0 \end{bmatrix} \tag{3.3}$$

Following the same procedure, but now introducing the vector

$$r = (w, \ cw, \ c^2 w, \ c^3 w, \ c^4 w)^T$$

we find that the barotropic problem (Eq. 2.12) reduces to the form 3.1. However, in this case the matrices involved are much larger since A and B are "composed" by a combination of five "submatrices."

Both A and B are nonsymmetric, real matrices with only a few diagonals. A five-point central difference is used in all cases. In order to take advantage of the sparsity of the system, inverse vector iteration was the algorithm chosen to solve Eq. 3.1. Since this procedure preserves the structure of the matrices, sparse storage can be used throughout the calculation, and high-resolution grids can be achieved. The algorithm is described in Kerner [1986] and will now be briefly reviewed.

Given an initial guess c_0 close to the eigenvalue we seek, $c = c_0 + \Delta c_0$, we can rewrite Eq. 3.1 as

$$(A - c_0 B)r = \Delta c_0 \, Br \tag{3.4}$$

A second sequence is required since the eigenvalues will in general be complex, and hence the inverse vector iteration procedure is defined by

$$\tilde{r}_0 = r_0 \tag{3.5}$$

$$(A - c_0 B)r_i = \Delta c_{i-1} \, Br_{i-1} \tag{3.6}$$

$$(A - c_0 B)^* \tilde{r}_i = \overline{\Delta c_{i-1}} \, B\tilde{r}_{i-1} \tag{3.7}$$

where $i = 1, 2, \ldots$. The asterisk and overbar denote the Hermitian and complex conjugate, respectively. Corrections to the initial eigenvalue are then obtained by using the generalized Rayleigh quotient

$$\Delta c_i = \frac{\tilde{r}_i^* \tilde{A} r}{\tilde{r}_i^* B r} \tag{3.8}$$

where $\tilde{A} = A - c_0 B$. In order to solve the system of equations more efficiently we first carry out the LU decomposition of matrix $\tilde{A}$. Note, however, that because of the large bandwidth of this matrix, sparsity is lost when carrying out the factorization. To circumvent this problem, a simple reordering of the equations was performed, so that the original eigenvector

$$r = (w_1, w_2, w_3, \ldots, cw_1, cw_2, \ldots, c^2 w_1, c^2 w_2, \ldots)^T \tag{3.9}$$

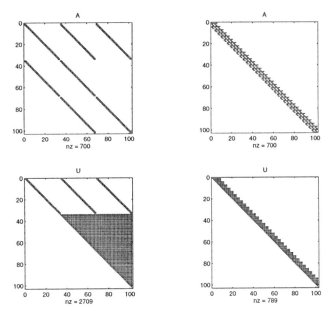

FIGURE 1. Matrices $\tilde{A}$ and U before (left) and after (right) reordering (baroclinic case).

now reads

$$r = (w_1,\ cw_1,\ c^2 w_1,\ w_2,\ cw_2,\ c^2 w_2,\ \dots)^T \tag{3.10}$$

where w_i is the ith element of the eigenvector (in this case the vertical wind profile). Using this scheme the number of nonzero elements in both lower and upper matrices remains similar to that of matrix $\tilde{A}$. Figure 1 shows a plot of the structure of $\tilde{A}$ and U before and after reordering.

The continuation procedure consists in repeating the process by using increasingly finer meshes until the change in eigenvalue is less than a given tolerance. The eigenvalue and eigenvector of the previous step (the coarser grid) are used as initial guess; the resizing of the eigenvector is carried out by interpolation.

4. Implementation

The language chosen to carry out the calculations is MATLAB. Its built-in sparse matrix storage class and robust sparse factorization routines make it an ideal choice in which to program the preceding algorithms. Although "external" programs ("m-files") are notably slower than "core" routines, this poses no problems since most of the time is spent performing the LU decomposition. Hence only two programs are required, one for the inverse iteration and the other for the continuation procedure.

An outline of the code of the inverse iteration is as follows:

```
function [eigvec,lambda] = ivi(A,B,lambda,eigvec,N,TOL)

% Purpose:
% This program solves the generalized eigenvalue problem
%                          _       _
%                          A x = lambda B x
```

```
%
% using inverse vector iteration.

  lambda0 = lambda;
  y = eigvec;
  errold = inf;
  Deltalambda = inf;
  A = A-lambda*B;
  [L,U] = lu(A);
  while (k < N)
    Deltalambdaold = Deltalambda;
    eigvecold = eigvec;
    Deltalambda = (y'*A*eigvec)/(y'*B*eigvec);
    eigvec = U\(L\(Deltalambda*B*eigvec));
    y = L'\(U'\(conj(Deltalambda)*B*y));
    lambda = lambda0 + Deltalambda;
    err = Deltalambda - Deltalambdaold;
    err = abs(err/lambda0);
    errold = err;
    if (err < TOL)
        fprintf ('ivi: converged after %g iterations \n',k);
    end
    k = k+1;
  end
```

The actual code is somewhat more complicated since extensive error checking is done throughout. Also, eigenvectors are standarized at each iteration, so that the largest component is $1 + i0$. This is done to prevent overflow in subsequent calculations. The convergence criteria are also more stringent, since the eigenvector and both the real and imaginary parts of the eigenvalue must converge separately. However, the preceding outline basically represents how the calculation is carried out. Note, in particular, that nowhere in the code is the storage scheme mentioned explicitly. Whether sparse LU should be carried out or not depends solely on the function's input.

Although apparently simpler, the routine that performs the continuation procedure is somewhat more involved. The reason is that increasing the number of points within the domain increases the the dimension of the systems, and care must be taken regarding the correct resizing of matrices and vectors. For example, in order to use the eigenvector that results from an iteration in the continuation procedure as a guess in the next iteration, an interpolation has to be carried out (in our case using interp1, a linear one-dimensional (1-D) built-in interpolation routine). Instead of interpolating over the complete eigenvector, doing so only over the values that span the wind velocity per se and ignoring the elements which are multiplied by a power of the eigenvalue is clearly much more efficient. However, we must remember that the eigenvectors are no longer in the sequence shown in Eq. 3.9, but are in that given by 3.10. Thus, the interpolation procedure consists of four steps: reordering the eigenvector, extracting the relevant elements and interpolating, rebuilding the full eigenvector by multiplying the elements by the appropiate power of c, and reordering again so as to minimize the bandwidth and continue with the next iteration. Although straightforward, these steps contribute to complicating an otherwise simple program.

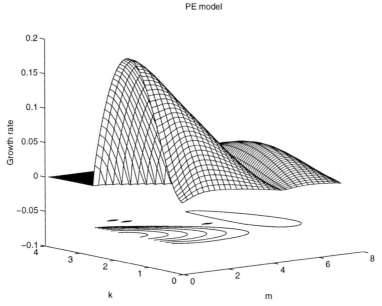

FIGURE 2. Growth rates in the barotropic case.

5. Results and Conclusions

Regarding the initial guess for the first iteration, it is found that a coarse grid solved by using QR is adequate. In the baroclinic case, the number of interior points was initially set to 10 (system size = 30). Specifying a tolerance of 0.001, convergence was typically reached at around 140 grid points and eventually reached sizes of up to 296 grid points (system size = 888). The resulting curve agrees with the results obtained by Nakamura [1988].

In the barotropic problem, 30 points was used initially (system size = 150), and convergence was normally at around 100 points, although grid sizes of up to 507 points (system size = 2535) were sometimes necessary to achieve the required tolerance.

A great advantage of using an iterative technique for calculating eigenvalues is that it is possible to use the results at one point in wavelength space as an initial guess to commence iteration on a nearby point. By "point in wavelength space" is meant that we are normally interested in analyzing the growth rate $\sigma \equiv kc_i$ as a function of wavelength k. Hence, instead of solving the full eigensystem at point k we simply use the result at point $k - \Delta k$ to initiate the inverse vector iteration procedure.

This is especially useful in the barotropic case when we analyze the growth rate dependence on both k and m. Figure 2 shows the regions of instability in the $m - k$ space. Two regions can be readily identified, one at low values of m and another, less unstable region at higher values of m (the vertical wavelength approaches infinity as $m \to 0$, and $m = 6$ is about 10 km, the approximate height of the troposphere). Only the first "hump," the one at low values of m, is found within the QG formulation. The important fact to notice, however, is that although these longwave modes grow faster than the shallow ones, they are located in a region which is *not* physically realistic; i.e., the atmosphere is finite in extent and hence the shortwave modes are the ones which are relevant when describing the atmospheric dynamics.

The graph was made by iterating along curves of constant m. It took 862 cpu-minutes on a MIPS R4400/4010 at 174 MHz (an "Indy") to calculate the 2400 points that were used to interpolate the surface, or about 20 seconds per point on average. For comparison, solving for one typical point using the full system takes 200 seconds. Storage requirements were very low, since, when $n = 2500$, $\tilde{A}$ occupied about 620 kB, compared to the 95 Mb needed to store a full (complex) matrix of this size.

In the future, it is the intention of the authors eventually to apply these techniques to 2-D problems, where vertical and horizontal shears are present simultaneously within the mean flow.

Acknowledgments. The authors are greatly indebted to Mr. Yasuhiro Yamazaki for his invaluable suggestions regarding the numerical techniques. De la Cruz wishes to acknowledge the financial support he received from CONACYT during the course of this work. Support was also received from the Natural Sciences and Engineering Research Council of Canada.

REFERENCES

Gary, J., and R. Helgason, A matrix method for ordinary differential eigenvalue problems, *J. Comput. Phys.*, **5**, 169–187, 1970.

Gilbert, J. R., C. Moler, and R. Schreiber, Sparse matrices in MATLAB: Design and implementation, *SIAM J. Matrix Anal. Appl.*, **13**, 333–356, 1992.

Jennings A., and J. J. McKeown, *Matrix Computation*, 2nd ed., John Wiley & Sons, 1992.

Kerner, W., Computing the complex eigenvalue spectrum for resistive magnetohydrodynamics, Proceedings of the IBM Europe Institute Workshop on Large Scale Eigenvalue Problems, Oberlech, Austria, July 8–12, 1985, J. Cullum and R. A. Willoughby, Eds., 241–265, 1986.

Nakamura, N., Scale selection of baroclinic instability: Effects of stratification and non-geostrophy, *J. Atmos. Sci.*, **45**, 3253–3267, 1988.

Nash J. C., *Compact numerical methods for computers: Linear algebra and function minimisation*, 2nd ed., Adam Hilger, 1990.

Pedlosky J., *Geophysical fluid dynamics*, 2nd ed., Springer-Verlag, 1987.

Use of Canonical Correlation Analysis to Predict the Spatial Rainfall Variability over Northeast Brazil

By Carlos Alberto Repelli and José Maria Brabo Alves

Ceará State Foundation for Meteorology and Water Resources – FUNCEME
Caixa Postal D–3221, 60.325–002 Fortaleza, CE, Brazil

This paper presents a model for prediction of the spatial variability of precipitation over the semiarid region of Northeast Brazil. The model is based on canonical correlation techniques and uses the sea surface temperature field as predictor. Lead/lag January sea surface temperature versus February through May precipitation is discussed. In general, the model was able to capture the tendency of the rainy season to be above normal, below normal or average for all cases considered.

1. Introduction

The Brazilian Northeast (hereafter referred to as the Northeast) is located approximately between 1°S and 18°S and 35°W to 47°W. The so-called semiarid region (SAR) is one subregion of the Northeast, and its climate is defined specifically by the precipitation regime. The circulation mechanisms of SAR rainfall anomalies can be understood largely as modulation of the average annual cycle. Several studies have indicated that the patterns of the sea surface temperatures in the Pacific and Atlantic during the beginning of the year play an important role in SAR precipitation variability. Nobre and Oliveira [1984], Ropelewski and Halpert [1987], and others have shown that the El Niño phenomenon over the Pacific can produce a reduction in the SAR precipitation during its rainy season. Alves and Repelli [1992] analyzed the influences of the El Niño over subregions of the SAR. In the same way the variability of the sea surface temperature (SST) over the Atlantic Ocean can contribute to the quality of the SAR rainy season [Hastenrath and Heller, 1977; Markham and McLain, 1977; Ward and Folland, 1991; Moura and Shukla, 1981; Mechoso et al., 1990]; and more recently Uvo et al. [1996], using singular value decomposition (SVD) techniques, have deeply analyzed the spatial influence of the Pacific and Atlantic SST on SAR rainfall variability.

The objective predictability of the SAR precipitation anomalies has attracted several researchers [Hastenrath et al., 1984; Ward and Folland, 1991, and others]. These methodologies have been formulated in terms of the precipitation in the entire rainy season, without consideration of the variations within the rainy season, and the techniques used up to now are based on linear multiregressional models and linear discriminant analysis [Hastenrath and Greischar, 1993; Ward and Folland, 1991].

The purpose of this paper is to present a model based on canonical correlation analysis (CCA) techniques in order to predict the spatial variability of the precipitation over the SAR. The predictor used is the SST field. In this study the lead/lag January SST versus February–March–April–May precipitation is analyzed; different lags will be applied soon, following those suggested by Uvo et al. [1996].

2. Data Source

The precipitation data set was obtained from the Superintendência do Desenvolvimento do Nordeste (SUDENE) and Fundação Cearense de Meteorologia e Recursos Hídricos (FUNCEME); a total of 55 stations are used. The SST data sets used to build the model (1946–1985) were obtained from the analysis of Pan and Oort [1990], which in turn was based on the Comprehensive Ocean Atmosphere Data Set (COADS) analysis. The blended interpolated data set obtained from the Climate Analysis Center was used to test the model (1986–1992). The spatial resolution of the SST data set is 4.5° latitude × 7.5° longitude and both SST and precipitation data are monthly-mean values.

3. Methodology

The technique employed in this work is based on a multivariate procedure known as canonical correlation analysis (CCA). It consists in finding a linear combination, called the first canonical variable, from two data sets (Y predictor and Z predictant, for example) that maximizes the correlation between them. The first canonical variates are two series of values (or scores) of the linear combinations of the original sets of variables. There are as many scores in each series as there are observations for the original variables. The analysis continues by finding a second linear combination from each set, uncorrelated with the first pair, that produces the next highest correlation coefficient, and so on. The form of the canonical vectors is best depicted by the canonical structure patterns, which are the correlations of each of the original variables with the canonical variates. CCA can be applied to statistical modeling, since it is at the top of the hierarchy of regression modeling approaches [Barnett and Preisendorfer, 1987; Barnston and Ropelewski, 1992].

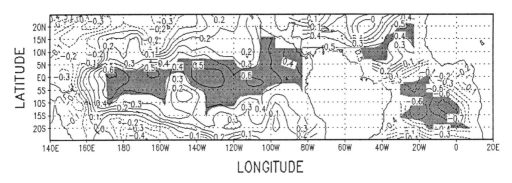

FIGURE 1. First canonical map for January SST. Shaded regions indicate significance at a level of 95% based on Student's t distribution.

The basic methodology employed in this paper is based on Barnett and Preisendorfer [1987]. The reader is referred to that paper for a thorough discussion of the method, interpretation and validation of the model. A model is built with the January SST anomaly field as a predictant and the precipitation field for February–March–April–May as a predictor. Both fields are prefiltered with an empirical orthogonal function (EOF) [Horel, 1981] in order to eliminate the noise contained by the original data. The EOFs retained by prefiltering together contain at least 80% of the total variance in the predictor and predictant. The significance test of the skill of the model is made by using a "cross-validation" procedure a priori [Stone, 1974; Efron, 1983]. It consists of isolating the first pair (year) of predictor–predictant, building the model, and so making the prediction for that year. The procedure is repeated for the second pair, the third, and so on. Cross-validation is meaningful only if the predictors are serially uncorrelated. Thus, in the current context, the predictant field (precipitation) should not be correlated from one year to the next. The predicted and observed data fields were next decomposed into terciles for one experiment of validation. Observed and predicted data were sorted separately for the construction of the three categories (above, normal and below). The percentage of correct forecasts at a particular location (station) is called "local skill."

The significance of the local skill is determined in the normal manner from a binomial distribution. The percentage of stations over SAR that showed significant local skill was used to estimate the "global skill." Similarly, a significance was obtained from the binomial distribution after accounting for the correlation between adjacent stations. To determine whether the correlation coefficients calculated for each map differ significantly from what is expected as a result of chance, a test of the null hypothesis based on Student's t distribution was performed. Shaded regions on Figure 1 indicate significance greater than or equal to 95%.

4. Results

Among the six eigenvalues obtained by CCA using SST over the Pacific and Atlantic versus precipitation field, the first one alone can explain almost 80% of the total variance of the system, and the two first modes are significant according to a Monte Carlo test. The first mode of the temporal canonical components presents correlation coefficients higher than 0.8, meaning a good relationship between the temporal variability of SST and precipitation fields. The very wet years (1964, 1967, 1974 and 1985, for example) have positive peaks coincident with high negative peaks for SST. The opposite situation (very dry years like 1951, 1953, 1958, 1970 and 1983, for example) occurs as well.

The canonical maps of SST and precipitation represent the spatial correlation between the SST (precipitation) field and the precipitation (SST) canonical component. Figures 1 and 2 show the first mode of both fields. A positive correlation over the Tropical Pacific Ocean, with a maximum located around 130°W, can be seen (Figure 1). This is related to the El Niño phenomenon, which corresponds negatively with precipitation variability. This suggests that the maximum influence is over the northern part of the SAR (in agreement with Ropelewski and Halpert [1987] and Alves and Repelli [1992]). On the other hand, the North Atlantic presents a positive significative region related negatively to the precipitation field and a negative region to the southern part. It indicates the presence of the so-called Atlantic-dipole configuration [Moura and Shukla, 1981] and has a strong influence on SAR precipitation. It can also be seen that the Southern Atlantic appears to have a slightly stronger influence on SAR precipitation variability than the Pacific does, since it has the highest correlation values. These results agree with very recent work by Uvo *et al.* [1996] using SVD techniques.

For the categories "above" and "below," the years of "extreme" events considered can hold most of the stations. The number of coincidences occurring in each tercile for the total period and all stations is 323 for the "below," 298 for "normal" and 366 for "above" categories. Figure 3 shows local skill as a percentage calculated by cross-validation at levels of 90%, 95% and 99% significance. Except in the Southern part, most of the region has a skill level higher than 40%, and the North Region has values above 60% of local skill at 99% significance. The global skill of the model is equal to 56.4% with a significance level of 100%. The average significance level through all the stations is equal to 44.9% and there are 31 stations with a 95% level of significance among the correct terciles verified. Some forecasts for precipitation field expressed in categories were made

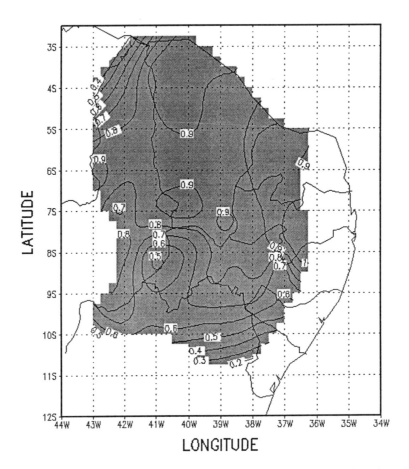

FIGURE 2. First canonical map for Feb–Mar–Apr–May SST. Shaded regions indicate significance at a level of 95% based on Student's t distribution.

using the blended interpolated SST field for years 1986 through 1992 (maps not shown). It could be noted that the stations located over the regions with best skill in general could be more correctly predicted. For those years with extreme climatic events (like 1986, 1989, 1987) the model also responds better.

5. Conclusions

In general, the model is able to catch the tendency of the quality of the rainy season for all tested years, and the results obtained encourage continuing this research. This method of establishing of "categories" is more valid for skill calculations. In the near future the authors intend to continue this work, taking into account the following points related to the model: make new statistical tests in order to get a better understanding of the model confidence; test the model by using preranking of inputs/outputs; use SST from the Pacific Region predicted by the Cane–Zebiak model and the persistence for the Atlantic Ocean; combine CCA with linear discriminant analysis; apply the CCA technique using a different predictor parameter (alone and combined with SST); apply

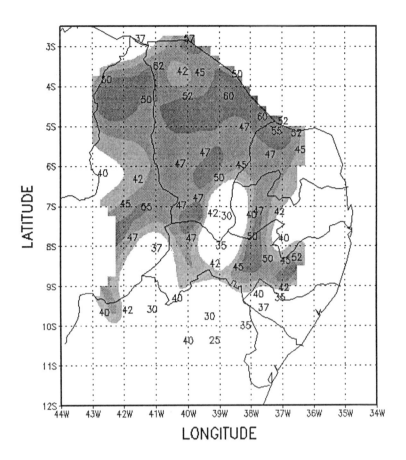

FIGURE 3. Local skill (percentage) calculated from cross-validation using three categories. Shaded regions indicate significance at levels of 90%, 95% and 99% from lighter to darker, respectively.

the model for other lead/lags suggested by the SVD analysis to predict intraseasonal precipitation variability over the region.

REFERENCES

Alves, J. M. B., and C. A. Repelli, Variabilidade pluviométrica no setor norte do Nordeste e os eventos El-Niño Oscilação Sul, *Revista Brasileira de Meteorologia*, 1992.

Barnett, T. P., and R. Preisendorfer, Origins and levels of monthly and seasonal forecast skill for United States surface air temperatures determined by canonical correlation analysis, *Mon. Wea. Rev.*, **115**, 1825–1849, 1987.

Barnston, A. G., and C. F. Ropelewski, Prediction of ENSO episodes using canonical correlation analysis, *J. Climate*, **5**, 1316–1343, 1992.

Efron, B., Estimating the error rate of a prediction rule: Improvement on cross validation, *J. Am. Stat. Assoc.*, **78**, 316–331, 1983.

Hastenrath, S., and L. Greischar, Further work on the prediction of Northeast Brazil rainfall anomalies, *J. Climate*, **6**, 743–758, 1993.

Hastenrath, S., and L. Heller, Dynamics of climate hazards in Northeast Brazil, *Q. J. R. Meteor. Soc.*, **103**, 77–92, 1977.

Hastenrath, S., M. C. Wu, and P. S. Chu, Towards the monitoring and prediction of Northeast Brazil droughts, *Q. J. R. Meteor. Soc.*, **110**, 411–425. 1984.

Horel, H. D., A rotated principal component analysis of the interannual variability of the Northern Hemisphere 500 mb height field, *Mon. Wea. Rev.*, **109** (10), 2080–2092, 1981.

Markham, C. G., and D. R. McLain, Sea surface temperatures related to rain in Ceará, northeastern Brazil, *Nature*, **265**, 320–323, 1977.

Mechoso, C. R., S. W. Lyons, and J. A. Spahr, The impact of sea surface temperature anomalies on the rainfall over Northeast Brazil, *J. Climate*, **3**, 812–826, 1990.

Moura, A. D., and J. Shukla, On the dynamics of droughts in Northeast Brazil: Observations, theory, and numerical experiments with a general circulation model, *J. Atmos. Sci.*, **38**, 2653–2675, 1981.

Nobre, C. A., and A. S. Oliveira, Precipitation and circulation anomalies in South American and 1982–83 El Niño/Southern Oscillation episode, Extend Abstracts of the Second International Conference on Southern Hemisphere Meteorology, December 1–15, 1986, Wellington, New Zealand, 442–445, 1984.

Pan, Y. H., and A. H. Oort, Correlation analyses between sea surface temperature anomalies in the eastern equatorial Pacific and the world ocean, *Climate Dynamics*, **4**, 191–205, 1990.

Ropelewski, C. F., and M. S. Halpert, Global and regional scale precipitation patterns associated with the El Niño–Southern Oscillation, *Mon. Wea. Rev.*, **115**, 1606–1626, 1987.

Stone, M., Cross validatory choice and assessments of physical predictions, *J. R. Stat. Soc.*, **B36**, 111–147, 1974.

Uvo, C. R. B., C. A. Repelli, S. Zebiak, and Y. Kushnir, A study on the influence of the Pacific and Atlantic SST on the Northeast Brazil monthly precipitation using singular value decomposition (SVD), *J. Climate*, 1996, in press.

Ward, M. N., and C. K. Folland, Prediction of seasonal rainfall in the North Nordeste of Brazil using eigenvectors of sea-surface temperature, *Int. J. Climatol.*, **11**, 711–743, 1991.

Identification of the ITCZ Axis by Computational Techniques

By Celeste C. Ferreira and Fernando A. A. Pimentel

Foundation for Meteorology and Water Resources – FUNCEME
Caixa Postal D – 3221, 60.325-002 Fortaleza, CE, Brazil

A nonsubjective method was developed to localize the axis of the convective cloud band that characterizes the Intertropical Convergence Zone (ITCZ) by the thermal infrared images of the METEOSAT geostationary satellite. The identification of the position of the ITZC axis in the maximum convective cloud band is obtained by analysis of the temperature of cloud tops. Two methodologies can be used to localize (latitude, longitude) the ITZC axis: the centroid method, which determines an equivalent value of the center of mass, and the method of minimum temperature, which considers the minimum temperature of the convective cloud tops. Thus we are able to draw a curved line showing the ITCZ axis.

1. Introduction

In Northeastern Brazil (NEB) precipitation is concentrated in just one period of the year. The better the localization of the ITZC center, the more precise will be the assessment of the consequences of the system. Therefore, monitoring of the ITCZ axis position is fundamental to the estimation of the quality of the rainy season. In the past, only subjective methods have been used to localize the ITZC axis. But subjective methods depend on the experience and knowledge of the one who makes the analysis, and, in spite of good technicians, the results are variable. Then, it becomes essential to look for a method that gives the precise position of that weather system. The method discussed in this paper permits the determination of a curved line that localizes the axis of the convective clouds that correspond to the ITZC. The entire process is done by computer.

2. Methodology

The area of the study was the tropical Atlantic Ocean between 10°N and 10°S, and 0°W and 60°W. A Sun-A4/SPARC computer with 32 MBytes of memory was used. The METEOSAT geostationary satellite provided hourly (24 per day) thermal infrared images in the infrared window region between 10.5 and 12.5 μm. The period from 0900 UTC to 2100 UTC was chosen because the convection is more active in that period. The spatial sampling at the subsatellite point corresponds to 5 km $\times$ 5 km in the infrared channel.

Each hourly satellite image was transformed to binary images and added over 5 day periods (pentads). The binary images were obtained by using brightness radiance fields made by selecting the coldest brightness temperature of the top cloud (in terms of raw count). The best threshold chosen to identify the convection had a temperature equal to or less than -40°C (233 K). Those temperatures are related to the top of convective clouds. To visualize the satellite images, those temperatures are transformed into gray levels.

The purpose in considering pentads is to eliminate ITCZ short time oscillations, while the binary image serves only to determine the specific cloudiness (temperature) in the threshold chosen.

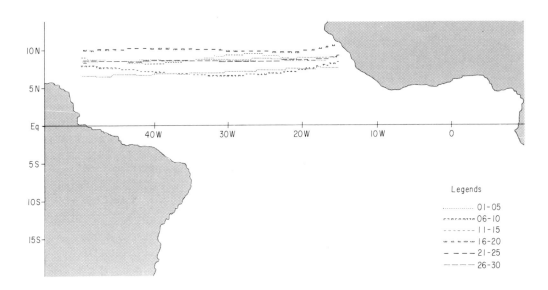

FIGURE 1. Pentad positions showing the variation of the ITZC axis in August 1994.

When the pentad image is ready, the determination of the curved line position of the ITCZ axis begins. The algorithm used in the determination of the ITCZ axis consists of the following process: (1) definition of the area through a box that contains the convective cloud band and (2) application of the chosen method, centroid or minimum temperature, to define the points where the curve will pass.

The centroid method determines the equivalent value of the center of mass in each column of the box chosen. The magnitude used for a weight was the freezing point of water minus the temperature of the pixel (Kelvin).

To control the limitations imposed by the precision of the computers used, for each pixel the sum of the products described previously was calculated. The presence of warm points in the box can cause difficulties in the evaluation. Trying to minimize that problem, we go through the column in the box from end to end in both directions, up and down, until the first pixel with temperature below $-40°C$ (233 K) is found.

The method of minimum temperatures takes into consideration the minimum temperature point in the column within the chosen area.

After the determination of the points, a curved line is fitted that corresponds to the graphic representation of a polynomial function of degree 5 at most. To verify the dispersion of the points, two curves are drawn above and below the main curve. Those curves are obtained from the same fitting as the central line, but now the points analyzed are the pixels above and below (in accordance with the upper or lower band) the central curve.

In case the convective cloudiness is broken or has a gap, it is possible to divide the region into smaller boxes. The points are determined for each box in accordance with the method chosen, and the curve is drawn from all of the point groups, thereby increasing precision.

From that calculation three files are recorded: one with the position of the curved line, another with the raster image of the axis-line, and a third with a raw image. From the

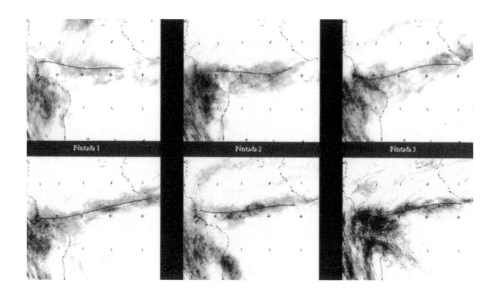

FIGURE 2. METEOSAT4 binary images showing the ITCZ axis in the cloud cover for six pentads in February 1994.

file of the axis position it is possible to produce graphics that show the ITCZ variation (1) at their pentad positions (Figure 1) and (2) at definite points (longitude, for example).

The file that contains the raw image is used to compute the convective cloudiness axis. The raster file is an image file for printing. It contains the image documentation that will allow the printer to identify the different gray tones (Figure 2).

The analysis to determine the ITZC axis can be done for any period for which satellite images are available.

3. Conclusions

It is verified that the curved line fits the center of the convective cloud band (i.e., the ITCZ axis) perfectly in the computational method.

Comparing the subjective method by visualization and the computational method, the curves coincide in areas with a good definition of clouds. In the regions where the definition is not good, the limited human eye cannot draw the ITZC axis as well in the subjective method as does the computational method.

In accordance with that result, and because of the need for precision in the localization of the ITZC axis, the computational method has been used in FUNCEME.

Acknowledgments. The authors would like to thank Mr. Boanerges Soares Matos for his help in reviewing the text.

REFERENCES

Citeau, J., The watch of ITZC migration over tropical Atlantic as an indicator in drought forecast over Sahelian area, *Ocean–atmosphere Newsletter*, **45**, 1–3, 1988.
Hastenrath, S., *Climate and Circulation of the Tropics*. D. Reidel, 1985.

Purdom, J. F. W., F. Weng, and T. H. V. Haar, The diurnal variation of convective cloudiness along the inter-tropical convergence zone (ITZC). Preprints, 4th Conf. on Satellite Meteorology and Oceanography, Am. Meteor. Soc., May 16–19, San Diego, CA, 1989.

Uvo, C. R. B., A Zona de Convergência Intertropical (ZCIT) e a sua relação com a precipitação da Região Norte do Nordeste do Brasil. Dissertação de Mestrado em Meteorologia, INPE-4887-TDL.378, INPE, Sao José dos Campos, Brasil, 1989.

Variational Assimilation of Acoustic Tomography Data and Point Observations: Some Comparisons and Suggestions to Perform Error Analysis

By Julio Sheinbaum

Centro de Investigación Científica y de Educación Superior de Ensenada (CICESE),
Ensenada, BC, México

A simple advection–diffusion model is used to generate synthetic acoustic tomography data and local temperature observations, which are then assimilated by using a variational "adjoint" method. It is found that with the same number of acoustic rays and point observations, the initial temperature field is well reproduced whether acoustic data or local temperature data are used.

To estimate the "goodness of fit" of the analyses, the analysis error covariance should be calculated. Since the computational cost of this calculation is extremely high, approximate methods must be designed to extract this error information. Usually, a great deal of information is contained in just a few of the eigenvectors and eigenvalues of error covariance matrices. Here, Arnoldi's technique is used to calculate a few of the eigenpairs of the Hessian matrix (inverse error covariance matrix) without the need to calculate the full matrix explicitly. The calculations require the product of the matrix (Hessian) with a vector, which is here approximated by a finite difference procedure. The method can be applied to large-scale models at a relatively low computational cost.

1. Introduction

Acoustic tomography was first proposed as a viable method to observe the ocean by Munk and Wunsch [1979]. Since then, several experiments [e.g., Ocean Tomography Group (OTG), 1982; Cornuelle *et al.*, 1985; Howe *et al.*, 1987] have been performed to assess its capabilities. These experiments focused on the study of mesoscale variability in particularly well observed areas of the ocean. More recent experiments [Spiesberger and Metzger, 1992; Baggeroer and Munk, 1992] have shown its viability with long-range (greater than 1000 km) networks of sources and receivers and have therefore proved its potential for studying and monitoring several mean properties of the gyre-scale circulation.

However, it is still a matter of debate whether acoustic tomography data can be as useful as point observations for the purpose of reconstructing the physical fields of interest. Some success and experience has been gained with static inversions of acoustic data, but there is almost no experience in using these observations in a data assimilation framework.

To investigate how to assimilate acoustic tomography data, a numerical advection–diffusion model is used as a constraint to assimilate data from different times using the adjoint technique. A pattern of ray paths inspired by the Acoustic Thermometry of Ocean Climate Project (ATOC) [Hyde, 1993] Baseline Network is used to reconstruct a temperature anomaly represented by a Gaussian eddy. The results are compared to those of experiments in which point observations are assimilated instead of acoustic data, with the number of point observations equal to the number of rays used in the tomography experiments. The information content may not be the same for a single travel time and a point temperature observation. So, using the same number of tomographic measurements and point observations does not guarantee that both data sets contain the same information. Ideally they will provide independent information which, together with other a priori knowledge, will be sufficient to permit reconstruction of the whole temperature field.

Observation errors are introduced in the experiments by adding random noise to the model-generated observations whether they be tomographic or local temperature measurements. Sheinbaum [1995] discusses the more general problem in which model errors are also present. The reader is strongly encouraged to read that reference since what we present here are some new results that follow from that paper.

Knowledge of the "true fields" is one of the advantages of using synthetic data, which are very useful to gauge the success of the analyses. However, in a real data experiment those true fields will never be known. Thus, error and resolution analysis is more complicated and requires computation of the analysis error covariance matrix and the resolution matrix. A spectral decomposition of these matrices can give a great deal of information about the structure of the analysis error [Rodgers, 1990]. In fact, many times much of this information is contained in just a few of the eigenvalue–eigenvector pairs (e.g., the low or high end of the spectrum). We compute eigenpairs of the Hessian matrix without constructing the full matrix explicitly. Using finite differences of gradients, we calculate the projection of the Hessian along a given direction and use that to obtain some of its eigenvalues and eigenvectors using Lanczos's or Arnoldi's method.

2. Adjoint Assimilation: Numerical Experiments

Acoustic tomography data consist of path integrals of sound slowness (reciprocal of sound speed) along ray trajectories. In general, the sound speed is a complicated function of temperature, salinity and depth; a general formula can be found, for example, in Clay and Medwin [1977]. In our experiments we assume, for simplicity, just a linear relation between sound speed and temperature, given by

$$c = c_r + \alpha(T - T_b) \tag{2.1}$$

where T_b is a constant background temperature and c_r a constant reference sound speed. We have chosen the relevant parameters so that the maximum deviation from the reference sound speed corresponds to $\alpha \, \Delta T \leq c_r/3$, where $\Delta T = T - T_b$; this is a larger signal than expected in practical applications where $(\alpha \, \Delta T/c_r) \approx 10^{-2}$ [Munk and Wunsch, 1979]. Experiments with anomalies of this order and with a realistic signal to noise ratio show no essential differences from the experiments reported here.

Using (2.1), the travel-time data can be expressed in terms of temperature as

$$\tau_i = \int_{\Gamma_i} (c_r + \alpha \, \Delta T)^{-1} \, ds \tag{2.2}$$

where τ_j is the travel time along the Γ_j ray.

The numerical model used in these experiments solves the two-dimensional advection–diffusion equation

$$\frac{\partial T}{\partial t} + u \frac{\partial T}{\partial x} + v \frac{\partial T}{\partial y} - \nu \nabla^2 T = 0 \tag{2.3}$$

where u and v are zonal and meridional velocities, respectively, and ν is an eddy diffusion coefficient. Notice that we have eliminated the vertical coordinate z. This is obviously done for the sake of simplicity since in this work we don't want to deal with baroclinic temperature and velocity fields. Consistent with this simplification, the rays in (2.2) will be horizontal. The velocity field in (2.3) is time-independent and is obtained from the classical Stommel solution for a barotropic gyre with bottom friction. The stream function for this classical problem is [see, for example, Hendershott, 1986]

$$\Psi = A \left(1 - \frac{x}{L} - \exp\left(-\frac{\beta x}{\varepsilon}\right) \right) \pi \sin\left(\pi \frac{y}{L}\right) \tag{2.4}$$

where x, y (longitude and latitude, respectively) belong to an $L \times L$ square box, ε is the bottom friction coefficient, β is the derivative of the Coriolis parameter with respect to y, and A is the amplitude of the stream function. The model (2.3) is nondimensionalized using L as length scale and a characteristic velocity U. A nondimensional eddy coefficient of $(\nu/UL) = 4 \times 10^{-2}$ is used in the numerical experiments. If, for example, $L = 1 \times 10^8$ cm, $\nu = 1 \times 10^7$ cm^{2}s^{-1}, the characteristic velocity U is about 10 cm s^{-1}. This leads to a diffusion time scale, $t_d = R^2/\nu$ where R is a characteristic length-scale (say, the radius) of the feature to be advected by the flow (see later discussion). This time scale is about an order of magnitude larger than the gyre circulation time $t_c = L/U$.

To generate the data, a Gaussian anomaly of warm water given by the expression

$$T_0(x, y) = T_b + T_A \exp\left(\frac{-r^2}{(L/3)^2}\right) \tag{2.5}$$

with T_b a background constant temperature, T_A the amplitude of the anomaly, and r the distance to its center (in units such that $\alpha T_A = c_r/3$). The maximum size of the anomaly is $\Delta T = T_0 - T_b = 1.0$. We take $T_b = 0.5$ in the experiments reported although its actual value is not important. This is the initial condition used to integrate (2.3) using the velocities derived from (2.4).

The model is run from $t = 0$ to $t = t_c$ ($t_c = 400$ time steps), and either tomography data or point observations are extracted every $t_c/8$. For the tomography data experiments, the travel times are calculated by using a simple trapezoidal rule to calculate the integrals in (2.2). This constitutes our "true" ocean state, and the synthetic tomography data and the local temperature observations extracted from this model are used in the data assimilation experiments. Random Gaussian noise of standard deviation $\sqrt{\sigma_d} = 0.1$ is added to the model-generated data whether they be tomographic or point observations to simulate observation errors.

Figure 1 panels (a) to (c) show the initial temperature anomaly and two different snapshots of its time evolution ($t = t_c/2$, $t = t_c$, respectively). Panel (d) shows the stream function from which the advective velocities are derived. The acoustic travel time measurements are computed along the 16 rays shown in Figure 1(a). They are shown only once, but are extracted starting at $t = 0$ and then every one-eighth of t_c along the same trajectories. The ray coverage is not uniform; it was chosen to mimic some of the rays expected from the ATOC baseline network for the North Pacific Ocean, with a source located off Hawaii and several receiving stations distributed along the boundaries of the domain; we refer to them as "ATOC" rays hereafter. The geometry of the rays does not change even though the temperature changes; this result is also obtained when linear assumptions are made in the static inverse procedures. Also shown in Figure 1(a) is the uniform network of 16 sensors which will take point observations of temperature every one-eighth of t_c starting at $t = 0$, for the experiments in which these data will be assimilated.

To apply the adjoint method, a cost function CF must be defined. It consists of two terms: The first term measures the misfit between data and their model counterparts, weighted by a matrix which in a statistical framework is given by the inverse of the

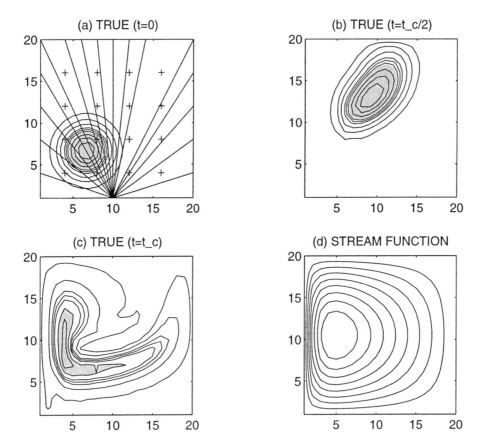

FIGURE 1. Plots of normalized temperature anomaly $(T - T_b)/T_A$ for the true ocean state from which the travel time data are generated and then assimilated by the numerical model. The ray trajectories and point observation positions are shown only in panel (a), but measurements are computed every one-eighth of t_c, the eddy's turnover time. Panel (a) also shows the true initial conditions. Panels (b) and (c) show two different snapshots ($t_c/2$ and t_c, respectively) of the time evolution of the eddy which is advected by the velocity field derived from the Stommel–gyre stream function shown in panel (d). Random noise is added to these observations, simulating observation error.

error covariance matrix of the observations; the second term weights a background field, usually used as the first guess estimate of the solution. An explanation of the need for this term is given in Sheinbaum [1995].

Since two kinds of data are generated and then used for assimilation, i.e., the travel time measurements from the ATOC network, and local temperature observations, two cost functions are defined. The first, for the tomographic observations, is given by

$$CFT = C1T + C2 \tag{2.6}$$

$$C1T = \frac{1}{2} \sum_t \sum_{i,j} \Delta\tau_i \, \mathbf{W}_{i,j}^{-1} \, \Delta\tau_j \tag{2.7}$$

$$C2 = \frac{1}{2}(T_0 - T_b)^t \mathbf{B}^{-1}(T_0 - T_b) \tag{2.8}$$

where $\Delta\tau_i = \tau_t^{\text{obs}}(\Gamma_i) - \tau_t^{\text{mod}}(\Gamma_i)$ is the difference between "observed" travel time along ray Γ_i and its model counterpart (both at time t), respectively, and given in this case by (2.2), but depending on which model was used to generate them.

The second cost function, for which the measurements are point observations of the temperature field taken at different times, has the form

$$CFP = C1P + C2 \tag{2.9}$$

$$C1P = \frac{1}{2} \sum_t \sum_{i,j}^{N_{\text{obs}}} \Delta T_i \, \mathbf{W}_{i,j}^{-1} \, \Delta T_j \tag{2.10}$$

$$C2 = \frac{1}{2}(T_0 - T_b)^t \mathbf{B}^{-1}(T_0 - T_b) \tag{2.11}$$

where $\Delta T_k = T_{\text{obs}}(k) - \mathbf{M}_{kj} T_{\text{mod}}(j)$ is the difference between the observed and modeled temperature field T at the time observations are available, with $\mathbf{M}$ a matrix that interpolates from the model grid to the observation position k. In the experiments reported here, $M_{kj} = \delta_{kj}$ (δ_{kj} is Kronecker's delta) because the observation positions coincide with the model grid.

For both cost functions, the independent or control variables are the values of the initial temperature conditions $T_0(x, y) = T(x, y, t = 0)$ on the model grid, from which the numerical model is integrated. The weighting matrix $\mathbf{W}$ or observation error covariance matrix contains only "measurement noise" and will be taken to be $\mathbf{W}_{ij} = \sigma_d \delta_{ij}$ ($\sigma_d = 0.01$, no correlation between measurements) since the random noise added to the observations has a variance equal to $\sigma_d = 0.01$ (whether they be travel times or temperature observations).

The second term $C2$ weights the background field using $\mathbf{B}$, which is the background error covariance matrix, whose structure is discussed later. Notice that the same term $C2$ is used in both cost functions (2.6) and (2.9).

The background error covariance matrix $\mathbf{B}$ is given by

$$B_{ij} = \sigma_b \exp\left(\frac{-r_{ij}^2}{2b^2}\right) \tag{2.12}$$

where r_{ij} is the distance between grid point i and grid point j, b is the correlation length scale, and the constant of proportionality σ_b is the expected error variance. It is chosen for its smoothing properties rather than as an exact representation of the background error covariance.

In experiment TOMO1 we seek to minimize (2.6) with respect to the initial temperature field by using the numerical model as constraint. Travel times are computed along the ATOC rays of Figure 1(a). A Lagrange function is constructed using the adjoint variables or Lagrange multipliers

$$J = CF + \sum_{x,y,t} \lambda^t \left(\frac{\partial T}{\partial t} + u\frac{\partial T}{\partial x} + v\frac{\partial T}{\partial y} - \nu\nabla^2 T\right) \tag{2.13}$$

where $\lambda(x, y, t)$ are the Lagrange multipliers. Following the general procedure of the variational method [see Sheinbaum and Anderson, 1990a] the forward and adjoint (Lagrange multiplier) equations for this problem are solved iteratively by using the LBFGS method [Liu and Nocedal, 1989]. In most of the numerical experiments performed in this paper, a limit of 20 iterations of the minimization routine is imposed. This is done because an operational implementation of the method is not likely to proceed until full convergence (because of the high computational cost of each iteration) but also because a substantial reduction of the cost function to a value even smaller than the limit imposed

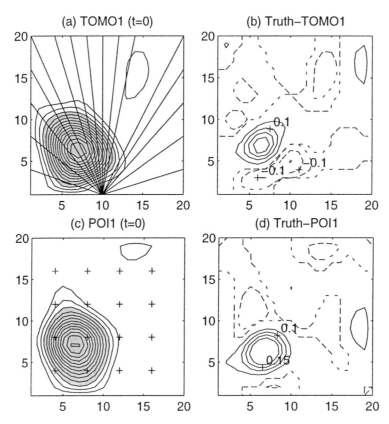

FIGURE 2. Plots of $(T - T_b)/T_A$ of the initial conditions analyses from experiments TOMO1 (panel a) and POI1 (panel c). The first guess temperature condition for the experiments was a flat field of $T_0 = T_b$ everywhere. In experiment TOMO1, travel times were computed along the rays shown in Figure 1(a). In experiment POI1, the point observations at the locations shown in Figure 1(a) are assimilated. Panel (b) shows the difference between the true field (Figure 1(a)) and the analysis for experiment from TOMO1 (panel (a)), whereas panel (d) shows the difference between the true field (also Figure 1(a)) and the analysis of experiment POI1 (panel (c)). There are differences in amplitude and position of the anomaly, but both data permit very good reconstruction of the anomaly.

by the data errors is achieved in about 20 iterations. A first guess (FG) of $T_0 = T_b$ (no anomaly) is used to start the iterations and assimilate the tomography data.

For comparison with the tomography assimilation experiment (TOMO1), experiment POI1 is carried out; in it, point observations of temperature are assimilated. The location of the observations is shown in Figure 1(a). Since we use the same number of observations, whether tomographic travel times or point observations, as in experiment TOMO1, in experiment POI1 there are 16 point observations every one-eighth of the total integration time and the variational-adjoint method used to minimize cost function CFP (2.9) starting from a FG of $T = T_b$. The iteration process also stopped after 20 iterations of the optimization routine.

Figure 2 shows results of these experiments. Panel (a) shows the analysis from experiment TOMO1 at $t = 0$. Panel (b) shows the difference between the analysis shown in panel (a) and the "true" field, shown in Figure 1(a). One can see that the anomaly is fairly well reproduced, although it is tilted toward the direction of the rays. Its amplitude is less than the truth by about 25% and it is slightly misplaced, as can be seen in panel (b). Here one can also see that some small-amplitude features appear in the analysis in addition to the main anomaly; they form as a result of the noisy data. Panels (c) and (d) are similar plots but for experiment POI1. In panel (c) one can see that the position and form of the anomaly are slightly better than the analysis from TOMO1 (panel (a)). However, the amplitude is also underestimated (see panel (d)). Differences between both analyses are relatively small so one may conclude that in this experimental setup tomographic measurements and point observations give similar results. The goodness of fit of these analyses has been estimated here, using the fact that the true fields are known in these synthetic data experiments, but in a real situation, the true fields will never be available. In that case expensive error and resolution analysis must be carried out.

3. Analysis of Error and Resolution

Error and resolution analysis requires the calculation of the Hessian or second derivative matrix of the cost function with respect to the control or independent variables. The Hessian is the inverse of the error covariance matrix of the solution [Thacker, 1989; Tarantola, 1987]. In experiments TOMO1 and POI1, the control variables are the initial temperature conditions on a finite difference grid of 20×20 grid points. So, the Hessian is a matrix of 400×400 whose calculation is feasible in this case. However, when large-scale numerical models are used for data assimilation (10^6 or more degrees of freedom), such precise calculation is impossible. We propose here some practical and feasible methods to calculate a few eigenvectors and eigenvalues of the analysis error covariance matrix (inverse Hessian). Actually, we work with the "curvature" matrix, since this is what needs to be done on spaces where the scalar product is not Euclidean, as is the case here [Tarantola, 1987; Sheinbaum, 1995]. The method requires calculation of some extra gradients (i.e., integration of the forward and adjoint equations), but there is no need to compute or calculate large matrices, as explained later. For lack of space, results are shown only for experiment TOMO1.

The cost function (e.g., Eq. (2.6)) is an implicit function of the model initial conditions (or control parameters). Writing it explicitly as a function of these control variables and making a Taylor series expansion about the minimum, i.e., the analysis T_{0A}, yields

$$CF(T_0) = CF(T_{0A}) + (T_0 - T_{0A})^t \nabla_{T_0} CF(T_{0A})$$
$$+ (T_0 - T_{0A})^t \mathbf{H}(T_0 - T_{0A}) + \cdots$$

where the second term on the right-hand side vanishes since the gradient is evaluated at the minimum T_{0A}, and $\mathbf{H}$ is the Hessian or second derivative matrix. The gradient of cost function (2.6), when it is written explicitly as a function of the control variables, has the form

$$\nabla_{T_0} CF(T_0) = \mathbf{J}^t \mathbf{W}^{-1}(\tau_{\text{mod}}(T_0) - \tau_{\text{obs}})$$
$$+ \mathbf{B}^{-1}(T_0 - T_b) \tag{3.14}$$

where

$$J^{ij} = \frac{\partial \tau^i}{\partial T_{0j}}$$

is the Jacobian or sensitivity matrix. Here, τ^i is the ith travel time measurement and T_{0j} is the jth element of the initial conditions vector. Therefore, the second derivative of the cost function with respect to the control variables is given by

$$\mathbf{H} = \nabla^2_{T_0} = \mathbf{J}^t \mathbf{W}^{-1} \mathbf{J} + \mathbf{B}^{-1} \tag{3.15}$$

where higher-order terms involving derivatives of the Jacobian have been neglected. Such terms exist when the observations are nonlinearly related to the model parameters.

When performing error and resolution analysis it is very useful to calculate the spectral decomposition of the Hessian (or its inverse, the analysis error covariance) or the resolution matrices. In the case of the Hessian, the eigenvector associated with the eigenvalue of largest (smallest) magnitude is the best (worst) determined direction in parameter space [Thacker, 1989; Tziperman and Thacker, 1989; Tarantola, 1987; Rodgers, 1990]. Spectral analysis of the resolution matrix can give us information about which parameters, or linear combinations thereof, are determined by the data and which by the prior information [Rodgers, 1990]. To calculate some of these eigenvectors, we use iterative methods, such as Lanczos's or Arnoldi's method [Golub and Van Loan, 1989], which can determine the eigenvalues and eigenvectors of a matrix knowing only the product of the matrix and a vector. The product of the Hessian, $\mathbf{H}$, and a vector, d, is approximated by

$$\mathbf{H}d = \frac{\nabla CF(T_{0A} + \varepsilon d) - \nabla CF(T_{0A})}{\varepsilon} \tag{3.16}$$

which requires only one extra calculation of the adjoint integration for each iteration of the eigenvalue solver. These calculations, although expensive, are feasible even for large-scale numerical models.

Care should be taken when a prior error covariance matrix is used in the cost function, as is the case in our experiments, because "distances" in parameter space should be normalized by such a matrix. The "principal axes" or the best (or worst) determined directions in parameter space are found by solving the generalized eigenvalue problem

$$\mathbf{B}\mathbf{H}d = \mu d \tag{3.17}$$

where $\mathbf{B}\mathbf{H}$ is the "curvature" matrix [Tarantola, 1987]. This matrix–vector product can also be calculated using (3.16). Both $\mathbf{H}$ and $\mathbf{B}$ are symmetric positive-definite matrices, but their product is in general not symmetric. This extra complication can be handled efficiently by, for example, Arnoldi's method, as implemented by Sorensen [1992].

Figure 3, first row, shows the first four eigenvectors of (3.17) with the largest eigenvalues (in decreasing order) for experiment TOMO1. These eigenvectors were calculated exactly; that is, the full curvature matrix was computed as

$$\mathbf{B}\mathbf{H} = \mathbf{I} + \mathbf{B}\mathbf{J}^t \mathbf{W}^{-1} \mathbf{J} \tag{3.18}$$

with the Jacobian matrices calculated as explained in Sheinbaum [1995]. Then the eigenvectors and eigenvalues of this matrix were computed by standard routines. The second row of Figure 3 shows the same eigenvectors but calculated by the finite difference method (Eqs. (3.16) and (3.17)) and Arnoldi's technique. One can see that they are very similar, which gives us confidence in the approximate calculation using the finite difference technique. These eigenvectors are the four best determined directions in parameter space.

As tends to happen in inverse problems, large scales are determined better than the small scales, and that is why going down through the spectrum the eigenvectors gain more and more structure. At the lower end of the spectrum, one finds the worse determined directions, which are associated with the smallest eigenvalues. These are difficult and expensive to obtain since many iterations of Arnoldi's method are required to get them, probably because of the flatness of the low end of the spectrum (350 eigenvalues are

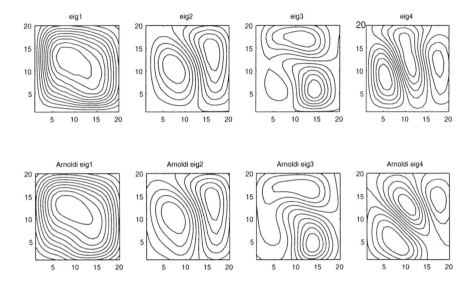

FIGURE 3. Spectral analysis of the curvature matrix of experiment TOMO1. Top row: first four eigenvectors (of largest eigenvalue) of the curvature matrix (see text) calculated exactly, i.e., computing the full curvature matrix and the computing its whole spectrum. Bottom row: the same eigenvectors but computed by the finite difference method and Arnoldi's technique. They are very similar. These are the best determined directions in parameter space. The lower end of the spectrum is more difficult to extract.

very close to a value of 1, as shown later). Besides, Sorensen [1992] also reports better convergence when the algorithms are used to find the largest eigenpairs.

Figure 4 is a plot of the 50 largest eigenvalues of the curvature matrix (calculated exactly). Most of the other 350 are very close to 1. What this figure shows is that the data are telling us something only about 25 or so directions in parameter space (the ones associated with eigenvalues significantly different from 1) and the rest are determined by prior information. The good results of the analysis (see Figure 2(a)), together with the form of the spectrum just described, may be an indication that the real degrees of freedom or control variables may be less than the 400 point values of the temperature initial conditions on the model grid. This is a matter for future research. Results for experiment POI1 are similar but are not shown for lack of space.

4. Summary and Conclusions

Although the numerical model used in this study has limitations, it does illustrate the possibility of assimilating indirect observations like acoustic tomography on the same grounds as direct point measurements. In fact, results indicate that tomography data can be as useful as point observations for the purpose of field reconstruction. The results presented here are complementary to those in Sheinbaum [1995]. Particular attention is given to developing techniques for error analysis on large-scale assimilation problems.

In experiment TOMO1 acoustic travel times are assimilated, whereas point observations are used in experiment POI1 to obtain the analyses. Figure 1 shows the true field from which data are generated and Figure 2 shows the analyses from both experiments. They do indicate that both data are equally useful to obtain the analyses.

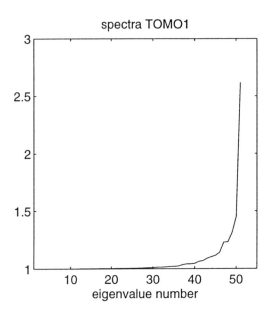

FIGURE 4. The largest 50 eigenvalues of the curvature matrix for experiment TOMO1. The other 350 are very close to a value of 1. This shows that just about 25 directions in parameter space are determined from data (those whose eigenvalue is significantly different from 1). The rest is determined by prior information.

To gauge its results, any least-squares fitting method requires an error estimate so that the precision of the analyses can be assessed. When large-scale assimilation problems are tackled, error analysis is overlooked because of the high computational cost it entails. We suggest here a method to calculate a few eigenvalues and eigenvectors of the analysis error covariance matrix (or the "curvature" matrix if the scalar product in parameter space is not Euclidean) at a relatively low computational cost. The method is based on computing the product of the Hessian matrix with a vector, using finite differences of the gradient vectors at two different points in parameter space. The results of these calculations are then used in an off-the-shelf eigenpackage routine such as Arnoldi's [Sorensen, 1992], Lanczos's, or the Power method [Golub and Van Loan, 1989]. We compare the results of calculations using this technique to those obtained in an exact calculation where the Hessian and Jacobian matrices are computed explicitly and their whole spectrum is calculated. This is shown in Figure 3. It is clear that the best determined directions in parameter space (largest eigenvalues) can be determined very well by using this technique. The lower end of the spectrum is more difficult to extract.

Techniques have been suggested to perform error analysis in Kalman filters or variational methods in an approximate way (see, for example, the paper by Todling in this volume). The method suggested here could in principle be applied to calculate matrix projections on either variational or Kalman assimilation schemes.

Acknowledgments. I am indebted to Prof. David Anderson, who originally suggested the problem of assimilating tomography data, for his continuous advice and encouragement. I should also like to thank Prof. Jorge Nocedal for providing his minimization algorithms and for suggesting the use of Eq. (3.16) to calculate eigenvalue–eigenvector pairs of the Hessian matrix. Thanks go also to Prof. D. C. Sorensen for allowing me to use his Arnoldi's method code and advising me on its implementation. Calculations were carried out at the CRAY Y-MP of the National Center for Supercomputer Applications and at UNAM's CRAY Y-MP. Support for this work came from CONACYT grants 1000-T9111 and 1002-T9111 and through CICESE's normal funding.

REFERENCES

Anderson, D. L. T., and J. Willebrand, Eds., *Oceanic Circulation Models: Combining Data and Dynamics*, Kluwer Academic, 1989.

Baggeroer, A., and W. Munk, Heard Island feasibility test, *Physics Today*, **45**(9), 22–30, 1992.

Clay, B. D., and H. Medwin, *Acoustical Oceanography*, John Wiley & Sons, 1977.

Cornuelle, B. D., C. Wunsch, D. Behringer, T. Birdsall, R. Heinmiller, R. Knox, M. Metzger, W. Munk, J. Spiesberger, R. Spindel, D. Webb, and P. Worcester, Tomographic maps of the ocean mesoscale. Part I. Pure acoustics, *J. Phys. Ocean.*, **15**, 133–152, 1985.

Daley, R., *Atmospheric Data Analysis*, Cambridge Atmospheric and Space Science Series, Cambridge University Press, 1991.

Ghil, M., and P. Malanotte-Rizzoli, Data assimilation in meteorology and oceanography, *Advances in Geophysics*, **33**, 141–266, 1991.

Gill, P. E., W. Murray, and M. H. Wright, *Practical Optimization*, Academic Press, 1981.

Golub, G. H., and C. F. Van Loan, *Matrix Computations*, 2nd ed., Johns Hopkins University Press, 1989.

Hendershot, M. C., Single layer models of the general circulation. *General Circulation of the Ocean*, Abarbanel, H. D. I., and W. R. Young, Eds., Springer-Verlag, 1986.

Howe, B. M., P. Worcester, and R. C. Spindel, Ocean acoustic tomography: Mesoscale velocity, *J. Geophys. Res.*, **92**, 3785–3805, 1987.

Hyde, D. W., ATOC network definition. Oceans '93, Engineering in Harmony with the Ocean, Proceedings, Vol. 1, Victoria, British Columbia, October 18–21, 1993.

Liu, D. C., and J. Nocedal, On limited memory BFGS method for large-scale optimization, *Mathematical Programming*, **45**, 503–528, 1989.

Munk, W., and C. Wunsch, Ocean acoustic tomography: A scheme for large-scale monitoring, *Deep Sea Research*, **26**, 123–161, 1979.

Ocean Tomography Group (OTG), A demonstration of ocean acoustic tomography, *Nature*, **299**, 121–125, 1982.

Rodgers, C. D., Characterization and error analysis of profiles retrieved from remote sounding measurements, *J. Geophys. Res.*, **95**, 5587–5595, 1990.

Sheinbaum, J., *Assimilation of Oceanographic Data in Numerical Models*, Ph.D. thesis, University of Oxford, 1989.

Sheinbaum, J., Field reconstruction from simulated acoustic tomography data and point observations using variational data assimilation: A comparative study, *J. Geophys. Res.*, **100**, No. C10, 20745–20761, 1995.

Sheinbaum, J., and D. L. T. Anderson, Variational assimilation of XBT data. I., *J. Phys. Ocean.*, **20**, 672–688, 1990a.

Sheinbaum, J., and D. L. T. Anderson, Variational assimilation of XBT data. II. Sensitivity studies and use of smoothing constraints, *J. Phys. Ocean.*, **20**, 689–704, 1990b.

Sorensen, D., Implicit application of polynomial filters in a K-step Arnoldi method, *SIAM J. Matrix Anal. Appl.*, **13**, 357–385, 1992.

Spiesberger, J. L., and Metzger K., Basin-scale tomography: A new tool for studying weather and climate, *J. Geophys. Res.*, **96**, C3, 4869–4889, 1991.

Tarantola, A., *Inverse Problem Theory: Methods for Data Fitting and Model Parameter Estimation*, Elsevier, 1987.

Thacker, W. C., The Role of the Hessian matrix in fitting models to measurements, *J. Geophys. Res.*, **94**, 6177–6196, 1989.

Thacker, W. C., and Long R. B., Fitting dynamics to data, *J. Geophys. Res.*, **93**, 1227–1240, 1988.

Tziperman, E., and W. C. Thacker, An optimal control-adjoint equations approach to studying the oceanic general circulation, *J. Phys. Ocean.*, **19**, 1471–1485, 1989.

PART IV
Methods and Applications in Geophysics

Do Man-Made Obstacles Produce Dynamical P-Wave Localization in Mexico City Earthquakes?

By G. Báez,[1] J. Flores[1,2] and T. H. Seligman[1,3]

[1]Instituto de Física, Universidad Nacional Autónoma de México (UNAM), Apdo. Postal 20-364, 01000 México, D.F., México

[2]Centro Universitario de Comunicación de la Ciencia, UNAM, Zona Cultural, Ciudad Universitaria, Coyoacán, 04510 México, D.F., México

[3]Laboratorio de Cuernavaca, Instituto de Física, UNAM, Apdo. Postal 139-B, 62190 Cuernavaca, Morelos, México

To explain the damage distribution always observed in Mexico City earthquakes, we have developed a model in which the SP phase plays a crucial role, since it may produce a resonant response. To see whether man-made obstacles buried in the lake-bed region could alter these resonances, we have studied the effect of a set of randomly distributed obstacles. We consider waves obeying Neumann boundary conditions in a bounded two-dimensional region and analyze the localization behavior of the low-frequency states as a function of the density of obstacles. We have studied approximately 350 states and densities of obstacles up to the percolation limit and have found that dynamical localization plays no significant role.

1. Introduction

We have recently proposed [Flores et al., 1987; Mateos et al., 1993a] a model to explain several facts always observed in the seismic response of the closed basin where Mexico City lies: first, a concentration of damage in the former lake bed; second, a peculiar distribution of damage, high- and low-damage areas alternating; and third, a selectivity for buildings of a certain height. Our model takes into account the geological structure of the basin, which consists of three types of soil: the rigid rocky mountains surrounding the basin, semirigid sediments extending down several kilometers, and on top of these sediments, a shallow layer of very soft waterlogged clays, which correspond to the ancient lake beds. It is in the lake-bed region where the strongest motions are observed and, in fact, where all damage was concentrated during the Michoacan earthquake of September 19, 1985. According to our calculations [Mateos et al., 1993b] when an SV wave incides on very soft layers an evanescent P wave appears; it is called the SP phase [Aki, 1988]. This wave has a characteristic frequency, related to a vertical resonance of the S wave, and moves horizontally as a longitudinal wave, thereby contributing to the horizontal motion only. This SP phase moves with the P-wave velocity, which in the soft clays in Mexico City is equal to 1500 m/s – curiously enough, the same value as the speed of sound in water. Since the lake beds have a typical length of 10,000 m, if the earthquake lasts long enough, the horizontal P wave has sufficient time to bounce back and forth on the lake-bed boundaries, establishing interference patterns and eventually exciting the normal modes of this two-dimensional (2-D) system.

At the time of the Michoacan earthquake, there was a lot of discussion whether subsoil constructions, like large building foundations or subway stations, could have had an effect on damage distribution. On the other hand, one could always ask whether a set of obstacles adequately distributed could also alter the resonance pattern and reduce the

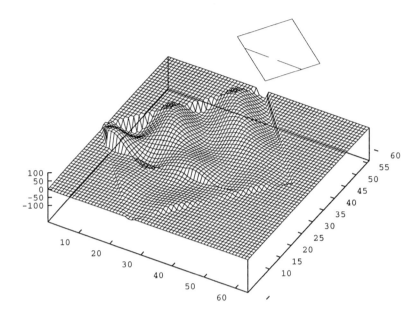

FIGURE 1. The 26th eigenmode (Neumann conditions) of a rectangle with a barrier and a gate of size a, as shown in the inset. For a wavelength $\lambda \gg a$, the eigenfunction barely penetrates.

large response which caused so many buildings to fall down. In order to answer these questions we propose in what follows a set of calculations to study the localization of waves in two-dimensional systems such as the lake-bed region of Mexico City.

Wave localization by obstacles has been studied in many instances. In disordered solids, for example, it has been extensively analyzed in one-dimensional arrays [Ziman, 1979]. Recently, it has also been observed experimentally for electromagnetic waves in two-dimensional systems [Dalichaouch *et al.*, 1991]. Many other examples can be of interest in this context: array of pillars in shallow waters and surface waves, the effect of obstacles in quasi-two-dimensional liquids or gases under pressure waves, and of course the effect of underground constructions in shallow waterlogged clays in the former lake beds where seismic P waves might play a major role in determining earthquake damage [Flores *et al.*, 1987], as mentioned. In many of these cases, Neumann boundary conditions rule; in contrast with the Dirichlet case, they have not been studied in this respect.

We can distinguish two types of wave localization, dynamical [Ziman, 1979] and geometrical. By the latter we mean the simple effect whereby a given chain of obstacles may divide the area available to the waves into disjoint parts, thus localizing the wave functions. This geometrical localization clearly occurs near the percolation limit, but with Dirichlet conditions, or any other boundary conditions that limit the intensity near the obstacle, this type of localization can occur at a much lower density of obstacles. Whenever a chain of obstacles is formed whose elements are separated by open spaces significantly smaller than the wavelength λ, this situation appears, while dynamical localization can be observed only at rather high frequencies.

The preceding conclusions may not be true for waves obeying Neumann boundary conditions, since now the intensity near the obstacles is not forced to be small. To illustrate this, we present in Figures 1 and 2 the effect of opening a gate of size a in a barrier dividing a rectangular region into two parts. For $\lambda \gg a$, we show in Figure 1

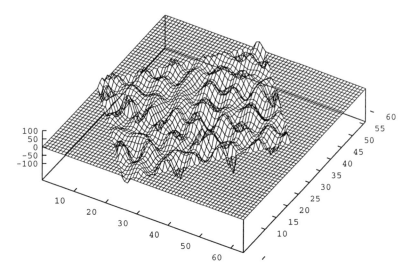

FIGURE 2. The same as in Figure 1, but for mode 140, for which $\lambda \approx a$. The eigenfunction penetrates completely.

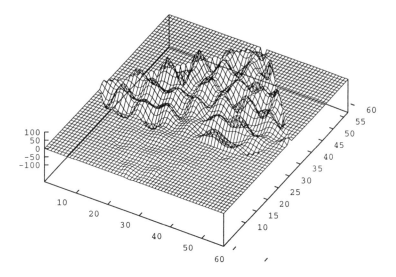

FIGURE 3. The 120th eigenmode (Dirichlet conditions) for the same geometry of Figure 1. The values of λ and a are similar to those of Figure 2, but the function does not penetrate.

that the eigenfunction barely penetrates, whereas for $\lambda \approx a$ we have essentially complete penetration, as will be seen in Figure 2. This shows that it is worthwhile discussing the localization of two-dimensional waves obeying Neumann boundary conditions, since they can differ substantially from other cases under study. In particular, the situation with Dirichlet boundary conditions is such that the wave penetrates only slightly even for $\lambda \approx a$. This can be seen by comparing Figures 3 and 2, which are computed for similar values of a and λ: Geometrical localization in the Dirichlet case occurs more readily.

In what follows we shall consider a two-dimensional region of arbitrary shape and evaluate the wave localization as the number of randomly distributed obstacles is increased. Neumann boundary conditions are assumed to hold at all the obstacles and boundaries. We shall discuss in Section 2 how the calculation is done, and present in Section 3 our results and conclusions, applying them in particular to the effect of underground constructions in the lake-bed region over which downtown Mexico City is built.

2. Numerical Evaluation of Wave Localization

Using the finite element method we shall obtain the normal modes of the Helmholtz equation which obey the Neumann boundary condition, both at the boundary ∂R of a given two-dimensional region R and at a set of N identical obstacles located at random sites; the boundary ∂R has an arbitrary shape. The wave function $\Psi(x, y)$ can be expressed in terms of an orthogonal set of n wave functions φ_i as the combination

$$\Psi(x, y) = \sum_{i=1}^{n} c_i \varphi_i(x, y) \tag{2.1}$$

where the $\varphi_i(x, y)$ are localized around the point (x_i, y_i).

The participation ratio, defined as [Leyvraz *et al.*, 1991]

$$\alpha = \frac{1}{\sum_{i=1}^{n} c_i^4} \tag{2.2}$$

provides us with information regarding the localization properties of the state $\Psi(x, y)$ given in (2.1). Indeed, when all states φ_i equally participate in Ψ so c_i is independent of i, Ψ is completely delocalized and $\alpha = 1$. When only one of the states φ_i participates, so Ψ is extremely localized, $\alpha = 1/n$.

We now indicate how to proceed in order to calculate numerically the coefficients c_i and therefore the localization parameter α using the finite element method. We start with a variational principle associated with the Helmholtz differential equation

$$\nabla^2 \Psi + \lambda \Psi = 0 \tag{2.3}$$

valid inside a two-dimensional region R with arbitrary boundary ∂R and Neumann conditions on ∂R. We define the functional [Lanczos, 1952]

$$F = \frac{1}{2} \int_R (\nabla \Psi)^2 \, dS \tag{2.4}$$

where dS is the element of area. We require that Ψ satisfies the normalization condition

$$\int_R \Psi^2 \, dS - 1 = 0 \tag{2.5}$$

To solve the Helmholtz equation, we need to find an extremum of the following expression:

$$H = \int_R (\nabla \Psi)^2 \, dS - \lambda \left(\int_R \Psi^2 \, dS - 1 \right) \tag{2.6}$$

Taking the variation $\delta H = 0$, interchanging the variation with the gradient, and using the identity

$$\nabla \Psi \cdot \nabla(\delta \Psi) = \nabla \cdot (\delta \Psi \, \nabla \Psi) - \nabla^2 \Psi \, \delta \Psi$$

together with Gauss's theorem, we obtain

$$-\int_R (\nabla^2 \Psi + \lambda \Psi) \delta \Psi \, dS + \oint_{\partial R} (\hat{n} \cdot \nabla \Psi) \delta \Psi \, dS + \delta \lambda \left[\int_R \Psi^2 \, dS - 1 \right] = 0 \tag{2.7}$$

Since $\delta \lambda$ is arbitrary, the normalization condition follows. We then have two cases:

the Neumann and the Dirichlet boundary conditions. For an arbitrary variation $\delta\Psi$ including the boundary points, this equation can be fulfilled only if both integrands vanish independently. The first integral yields as desired Eq. (2.3) and the second integral, the condition

$$\hat{n} \cdot \nabla\Psi \big|_{\partial R} = 0 \tag{2.8}$$

As we can see, this is the natural condition in the functional (2.4) because nothing is actually required of the wave function Ψ.

Summarizing, if we minimize the functional H we not only solve the Helmholtz equation with fixed normalization for the eigenfunctions but we simultaneously fulfill homogeneous Neumann conditions on the boundary ∂R.

On the other hand, we also have the possibility to impose Dirichlet boundary conditions. If $\delta\Psi = 0$ on ∂R we impose Dirichlet boundary conditions and for arbitrary variation $\delta\Psi$, the second integral must be zero. The mathematical details of this method are discussed in the thesis by Méndez [1992].

The implementation of the numerical algorithm starts by discretizing the region R. In general, this region is immersed in a lattice with basic cells of a given shape, the nodes adjusted to the boundary ∂R so the cells adjacent to it are deformed. The number of nodes m used in discretizing the region R is proportional to its area. The next step is to label the m lattice nodes and localize on top of them a basis wave function ϕ_i. Méndez [1992] used a hexagonal basic cell and a nonorthogonal set of m functions ϕ_i. Each ϕ_i is a pyramid of unit height. We now define an "element" as each of the triangles forming this hexagonal cell. The typical pyramid for the inside will consist of six sides, while near the border our algorithm may produce grid points with different numbers of triangles touching them. With Neumann conditions, the algorithm defines additional grid points on the border, and the functions pertaining to these points are truncated pyramids, i.e., pyramids which lack some of their faces. The number m of trial functions will be exactly equal to the total number of grid points for Neumann conditions.

For Dirichlet boundary conditions, the only difference is that there are no truncated pyramids at the border since the wave function there is already zero.

The only variational parameters left are the coefficients c_i' of the expansion in the nonorthogonal basis $\{\phi_i\}$

$$\Psi = \sum_{i=1}^{m} c_i' \phi_i \tag{2.9}$$

Once we introduce the explicit form of the elements ϕ_i in the functional H, the problem reduces to solving the $m \times m$ linear equations system

$$A\mathbf{z} = \lambda B\mathbf{z} \tag{2.10}$$

where A and B are matrices whose elements are given by

$$A_{ij} = \int_R \nabla\phi_i \cdot \nabla\phi_j \, dS \tag{2.11}$$

and

$$B_{ij} = \int_R \phi_i \phi_j \, dS \tag{2.12}$$

Both integrals extend only over the triangles that touch both nodes i and j since either of the two functions is zero in all other triangles; thus, both matrices are extremely sparse. The vector $\mathbf{z}$ has as its components the variational parameters $\{c_i'\}$. We solve the system using a standard routine of EISPACK.

Finally, we change to an orthogonal basis $\{\varphi_i\}$ in order to apply the localization

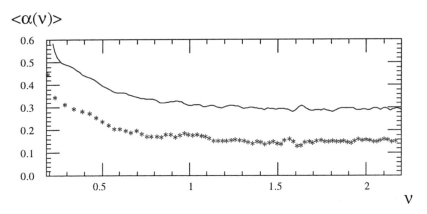

FIGURE 4. The average participation ratio $\langle \alpha \rangle$ as a function of frequency (in arbitrary units). The solid line is for 10% occupation of the rectangle and the dotted one for 30% occupation.

criterion correctly. We define the function φ_i such that

$$\varphi_i = \frac{1}{3}[c_j' \phi_j + c_k' \phi_k + c_l' \phi_l] \tag{2.13}$$

within the element i formed by the points (x_j, y_j), (x_k, y_k) and (x_l, y_l), and $\varphi_i = 0$ otherwise. Note that $\varphi_j(x_j, y_j) = 1$ and i denotes an arbitrary labeling of the finite triangular elements.

The computing code was implemented for a CRAY Y-MP4/464 belonging to UNAM and was checked in several instances. First, we computed hundreds of eigenvalues and eigenstates for integrable systems, such as the rectangle, the circle and an annular 2-D region, for which exact solutions are known. Second, we computed the eigenvalues for the lake-bed boundaries of the ancient Tenochtitlan lake (see Figure 7(a)), which we had previously obtained by the finite difference method [Mateos *et al.*, 1993a]. In all cases, the results were satisfactory, with a typical relative error of 0.5%. In order to obtain 400 states with good resolution, 2500 grid points, and therefore 2500 × 2500 matrices, are needed. Both eigenvalues and eigenfunctions are obtained in this case with a central processing unit (CPU) time of 16.8 min. An alternative code using inverse-power iterations [Méndez, 1992], which is efficient when only a few states are needed, takes much longer (84.5 min) when a large number, such as 400 states, is computed.

3. Results and Discussion

We present the results of wave localization by a random set of obstacles in two cases: first, when the region empty of obstacles corresponds to an integrable problem and, second, when it has a boundary such that the problem is chaotic [Berry, 1983]; the Valley of Mexico lake beds correspond to a high degree to the latter.

We consider a rectangle of size 30 × 37 and include within it a set of p square obstacles of unit length, distributed at random. Ensembles of 10 such systems are chosen. In Figure 4 we show the participation ratio $\langle \alpha \rangle$ of the eigenstates as a function of the frequency ν. The average indicated by $\langle \, \rangle$ is taken over the ensemble as well as over a range of 10 eigenfunctions around the frequency ν. The last average was performed in order to improve statistics without increasing the ensemble size. Local ergodicity, which is implicitly assumed, usually holds. Two values of p are shown, corresponding, respectively,

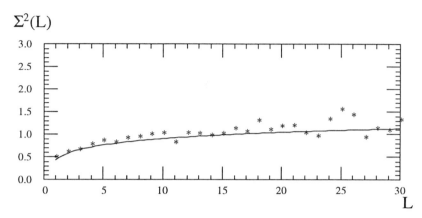

FIGURE 5. The number variance $\sum^2$ as a function of the interval of size L for 10% obstacle occupation of the rectangle. The solid line corresponds to the GOE.

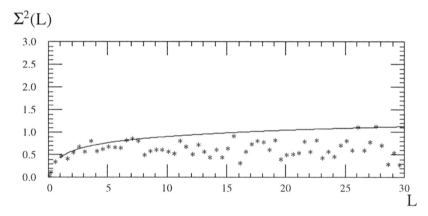

FIGURE 6. The number variance $\sum^2$ as in Figure 5 for Tenochtitlan Lake without obstacles.

to cases in which the obstacles cover 10% and 30% of the total area. When $\nu = 0$, $\alpha = 1$ because then the eigenfunction is a constant, but when ν grows, $\langle \alpha \rangle$ quickly drops to the values 0.30 and 0.18 for the 10% and the 30% cases, respectively, as shown in Figure 4. The first value is close to the one typical of eigenvectors obtained from diagonalizing a matrix whose elements are Gaussianly distributed random numbers, which constitute what is called the Gaussian orthogonal ensemble (GOE) of random matrices [Brody *et al.*, 1981]. The value for GOE is $\alpha = 1/3$ [Ulla and Porter, 1965]. Note that the spectral properties are also close to GOE for the 10% case, as can be seen in Figure 5, where the number variance $\sum^2$ [Brody *et al.*, 1981] is shown for the 10% case. Other statistics, not displayed here, confirm this conclusion. For the 30% case, $\langle \alpha \rangle$ is much smaller than the GOE value (Figure 4) and the spectral statistics no longer fit GOE. Thus for p near the percolation limit, that is, for a large number of obstacles, we find geometrical localization, while dynamical localization is not in evidence for the frequency range we considered.

We now present the results of wave localization with Neumann boundary conditions starting with a boundary leading to a nonintegrable region, which presumably shows a

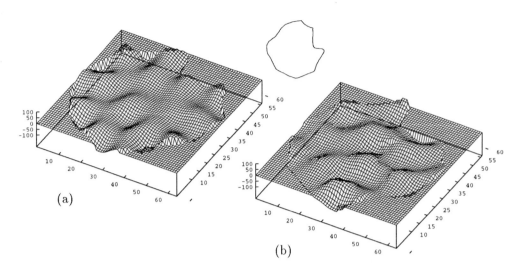

FIGURE 7. Typical wave functions near 0.5 Hz for the Tenochtitlan lake bed (see the inset). In (a) there are no obstacles and (b) corresponds to 10% obstacle occupation.

GOE spectrum. For this purpose we choose a boundary corresponding to the lake beds in Mexico City [Flores *et al.*, 1987]. Spectral statistics for this boundary with Neumann conditions were obtained and the number variance $\sum^2$ is displayed in Figure 6. We find a satisfactory agreement with GOE at short range, while at long range the numerical data show excessive stiffness. This effect is well known [Seligman *et al.*, 1984] and may be explained by the fact that the shortest periodic orbit of a billiard of this shape is fairly long [Berry, 1985]. Note that when randomly distributed obstacles are present this is not the case, as we saw in Figure 5. The reason lies in the appearance of arbitrarily short periodic orbits bouncing between the obstacles. The short-range behavior of spectral fluctuation as well as $\langle \alpha \rangle$ are not significantly altered by this effect.

Taking this boundary with 10% and 30% coverage by randomly distributed obstacles we obtain essentially the same results found for the rectangle. In Figures 7(a) and 7(b), we show typical wave functions, with and without obstacles for $\nu = 0.5$ Hz, the frequency for which the strongest amplification is recorded in Mexico City earthquakes. The value $\langle \alpha \rangle = 0.38$ around this frequency indicates delocalization, even with 10% of the area covered by obstacles. We conclude that for Neumann boundary conditions and for low frequencies dynamical localization plays no significant role.

Using the appropriate scale, the obstacles we have used a linear extension of the order of 150 m. If the SP phase is relevant in the Mexico City seismic response, we further conclude that existing and reasonable amounts of future large underground constructions would induce rather small effects on damage distribution.

Acknowledgments. We would like to thank R. A. Méndez for helpful discussions. This work was supported by DGAPA-UNAM and DGSCA-UNAM and by project PADEP-3314.

REFERENCES

Aki, K., Local site effects on strong ground motion. Proceedings of Earthquake Engineering and Soil Dynamics II, G. T. Div. ASCE, Park City, Utah, 103–155, 1988.

Méndez, R. A., *Cálculo de Modos Resonantes en Cavidades Arbitrarias Bidimensionales*, B.Sc. Thesis, Facultad de Ciencias, UNAM, 1992.

Berry M. V., in *Chaotic Behaviour of Deterministic Systems*, Iooss, Helleman and Stora, Eds., North-Holland, 171–271, 1983.

Berry, M. V., Semiclassical theory of spectral rigidity, *Proc. R. Soc. Lond.* A, **400**, 229–251, 1985.

Brody, T. A., J. Flores, J. Bruce French, P. A. Mello, A. Pandey, and S. S. M. Wong, Random-matrix physics: Spectrum and strength fluctuations, *Rev. Mod. Phys.*, **53**, 385–479, 1981.

Dalichaouch, R., J. P. Armstrong, S. Schultz, P. M. Platzman, and S. L. McCall, Microwave localization by two-dimensional random scattering, *Nature*, **354**, 53–55, 1991.

Flores, J., O. Novaro, and T. H. Seligman, Possible resonance effect in the distribution of earthquake damage in Mexico City, *Nature*, **326**, 783–785, 1987.

Lanczos, C., *The Variational Principles of Mechanics*, 2nd ed., University of Toronto Press, 1952.

Leyvraz, F., J. Quezada, and T. H. Seligman, Novel signature of chaos in quantum-mechanical states, *Phys. Rev. Lett.*, **67**, 2921–2925, 1991.

Mateos, J. L., J. Flores, O. Novaro, T. H. Seligman, and J. M. Alvarez-Tostado, Resonant response models for the Valley of Mexico. II. The trapping of horizontal P-waves, *Geoph. J. Int.*, **113**, 449–462, 1993a.

Mateos, J. L., J. Flores, O. Novaro, J. M. Alvarez-Tostado, and T. H. Seligman, Generation of inhomogeneous P-waves in a layered medium, *Tectonophysics*, **218**, 247–256, 1993b.

Seligman, T. H., J. J. M. Verbaarschot, and M. R. Zirnbauer, Quantum spectra and transition from regular to chaotic classical motion, *Phys. Rev. Lett.*, **53**, 215–217, 1984.

Ulla, N., and C. E. Porter, in *Statistical Theories of Spectra: Fluctuations*, C. E. Porter, Ed., Academic Press, 529–531, 1965.

Ziman, J. M., *Models of Disorder*, Cambridge University Press, 1979.

Domain Decomposition Methods for Model Parallelization

By Ismael Herrera,[1,2] Abel Camacho[1] and Joaquín Hernández[3]

[1]Instituto de Geofísica, Universidad Nacional Autónoma de México, Apartado Postal 22-585, 14000 México, D.F., México

[2]Also at Instituto de Matemáticas, UNAM

[3]Dirección General de Servicios de Cómputo Académico, Universidad Nacional Autónoma de México

This paper is devoted to presenting a brief overview of domain decomposition methods. Overlapping and nonoverlapping procedures are discussed. For this latter kind of method, Poincaré–Steklov operators are presented and a maximum principle is introduced. Conjugated gradient and preconditioned conjugated gradient are considered. Also Schwarz's alternating method is briefly discussed.

1. Introduction

Domain compositon methods for the numerical solution of partial differential equations have received much attention in recent years [Chan *et al.*, 1989a, 1989b; Glowinski *et al.*, 1988, 1990; Keyes *et al.*, 1991; Quarteroni *et al.*, 1992]. At present, this is mainly due to the fact that they constitute a very effective manner of parallelizing numerical models of continuous systems. Parallel computing is already a very important resource in supercomputing and it is expected to be even more important in the future.

There are additional reasons for the interest in domain decomposition methods, such as the following: domains of irregular shape can be decomposed into subdomains of regular shape and regions of relative nonuniformity of the differential operator or roughness of the solutions can be isolated into different subdomains. This paper is devoted to presenting a brief overview of domain decomposition methods.

2. General Discussion

One approach to domain decomposition deals with the systems that are obtained after the differential equations have been discretized, but it is also possible to formulate domain decomposition procedures treating the differential equations before discretization. In this paper this latter approach will be applied because it is more elegant and has the advantage of permitting use of the known properties of partial differential equations in a more direct manner. In addition, it is always possible afterward to give discretized versions of the results so obtained.

In the exposition that follows the different methods will be derived from a unified perspective, which is based on the experience and clarity that have been gained through the considerable amount of work that has been done in recent years. This manner of presenting matters has clear expository advantages, although it does not correspond to the way in which, historically, the methods were developed.

In domain decomposition methods, the region in which the problem is formulated is split into several – usually many – subregions. Given a differential equation, consider its set of solutions at each of the subregions. Then, the main objective of domain decomposition methods is to select a solution at each subregion in such a way that some matching conditions are satisfied. Generally, the methods may be classified into two broad categories, depending on whether such subregions do not overlap – nonoverlapping methods – or do overlap – overlapping methods. The main difference between these two kinds of procedures is the matching conditions. For elliptic equations of second order, for example, the nonoverlapping procedures require that the solution, together with its normal derivative across the common boundaries of the nonoverlapping subregions, be continuous. On the other hand, for the same kind of equations, when the procedure is overlapping, the smoothness conditions are relaxed, since only the solution itself is required to be continuous.

For elliptic differential equations of second order [Keyes and Gropp, 1987], which are the only ones to be considered in some detail in what follows, independently of the kind of domain decomposition that is applied – overlapping or nonoverlapping – knowing the restriction of the solution to the boundaries of the subregions determines uniquely the solution at the interior of the subregions, and the process of extending such restriction from the boundaries to the interior involves solving local problems only. Therefore, domain decomposition procedures frequently aim to obtain that restriction and the problem of obtaining the solution at the boundaries of the subregions may be referred as the "domain decomposition problem."

Time-dependent problems of parabolic type, when time discretization is applied, give rise to elliptic problems at each time step and the discussion presented in this paper applies to them in this manner. Some special methods for this kind of equation profit from the local behavior of the responses [Israeli and Vozovoi, 1993], but even if the methodology is not specialized, when iterative procedures are applied, the locality of the responses for this kind of equation produces rapid convergence.

3. Nonoverlapping Procedures

Consider the most general elliptic equation of second order, in any number of dimensions, written in conservative form

$$\mathcal{L}u \equiv -\nabla \cdot (\underline{\mathbf{a}} \cdot \nabla u) + \nabla \cdot \underline{\mathbf{b}}u + cu = f_\Omega \tag{3.1}$$

To apply domain decomposition methods, the region Ω in which the problem is formulated is partitioned into a collection of disjoint subregions $\{\Omega_1, \dots, \Omega_N\}$ and solutions are constructed separately in each. However, when putting such local solutions together, adequate "matching conditions" must be satisfied. They usually derive from physical requirements of the models. For Eq. (3.1), one is usually led to

$$[u] = 0 \qquad \text{and} \qquad \left[\frac{\partial u}{\partial n}\right] = 0 \quad \text{on } \Sigma \tag{3.2}$$

where Σ is the union of the intersections of the boundaries of the subregions (see Figure 1), and the square brackets stand for the jump – across Σ – of the function contained inside.

Observe, in particular, that when the coefficients are continuous, Eq. (3.2) implies that "diffusive flux" $(\underline{\mathbf{a}} \cdot \nabla u) \cdot \underline{\mathbf{n}}$ and "total flux" $(\underline{\mathbf{a}} \cdot \nabla u - \underline{\mathbf{b}}u) \cdot \underline{\mathbf{n}}$ are continuous.

Consider first the one-dimensional case, for which the region Ω will be the interval $(0, l)$ and the subregions Ω_i will be subintervals (x_{i-1}, x_i), where $i = 1, \dots, N$. At each subinterval Ω_i, let $\overset{i}{u}$ be a function defined on Ω_i and satisfying Eq. (3.1) there. The boundary values which are relevant in Ω_i, constitute a 4-D vector

$$\underline{\overset{i}{S}} = \left(\overset{i}{U}_{i-1}, \overset{i}{V}_{i-1}, \overset{i}{U}_i, \overset{i}{V}_i\right) \tag{3.3}$$

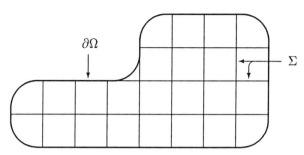

FIGURE 1. The partition of Ω.

where $\overset{i}{U} = \overset{i}{u}(x_j)$, $\overset{i}{V}_j = d\,\overset{i}{u}/dx(x_j)$. Observe that the four components of such a vector are not independent; indeed, there is a 2×4 matrix $\underline{\underline{\overset{i}{R}}}$ and a 2-D vector $\underline{\overset{i}{F}}$ such that

$$\underline{\underline{\overset{i}{R}}} \cdot \underline{\overset{i}{S}} = \underline{\overset{i}{F}}; \quad i = 1, \ldots, N \tag{3.4}$$

This is a system of $2N$ equations in $4N$ unknowns. Introducing the notation $\underline{\overset{i}{S}}_- = (\overset{i}{U}_{i-1}, \overset{i}{V}_{i-1})$ and $\underline{\overset{i}{S}}_+ = (\overset{i}{U}_i, \overset{i}{V}_i)$ allows decomposition of the matrix $\underline{\underline{\overset{i}{R}}}$ into two 2×2 submatrices $\underline{\underline{\overset{i}{R}}}_-$ and $\underline{\underline{\overset{i}{R}}}_+$, so that Eq. (3.4) becomes

$$\underline{\underline{\overset{i}{R}}} \cdot \underline{\overset{i}{S}} = \underline{\underline{\overset{i}{R}}}_- \cdot \underline{\overset{i}{S}}_- + \underline{\underline{\overset{i}{R}}}_+ \cdot \underline{\overset{i}{S}}_+ = \underline{\overset{i}{F}}; \qquad i = 1, \ldots, N \tag{3.5}$$

while Eq. (3.2) is

$$\underline{\overset{i}{S}}_+ = \underline{\overset{i+1}{S}}_- \tag{3.6}$$

Generally, Eqs. (3.5) and (3.6), together with two boundary conditions, constitute a determined system for the $4N$ components of the vectors $\{\underline{\overset{i}{S}}, \ldots, \underline{\overset{N}{S}}\}$. This is the basic system of equations of nonoverlapping methods.

Different methods with nonoverlapping subdomains derive from different approaches to solving this system of equations. A first option is to set, making use of Eq. (3.6),

$$\underline{\overset{i}{S}} = \left(\overset{i}{U}_i, \overset{i}{V}_i\right) = \underline{\overset{i}{S}}_+ = \underline{\overset{i+1}{S}}_-; \qquad i = 1, \ldots N - 1 \tag{3.7}$$

Substituting into Eq. (3.5) one obtains

$$\underline{\underline{\overset{i}{R}}}_- \cdot \underline{\overset{i-1}{S}} + \underline{\underline{\overset{i}{R}}}_+ \cdot \underline{\overset{i}{S}} = \underline{\overset{i}{F}}; \qquad i = 1, \ldots, N \tag{3.8}$$

Equations (3.8), together with two boundary conditions, usually determine the value of the function and its derivative at the $N + 1$ nodes of the partition. The system (3.8) is a 2×2 block bidiagonal system and is the basis for application of direct methods of solution.

If a direct approach is used, one solves for the function and its derivative at each of the nodes. However, to develop domain decomposition methods, most frequently a trial and error, or search, approach is applied. To this end, observe that the Dirichlet problem at each one of the subintervals is well posed and the values of the derivative at the end points of the subintervals are determined in this manner; the essential process may be described as follows: Choose a collection of values $\{U^1, \ldots, U^{N-1}\}$, set $\overset{i}{U_i} = \overset{i+1}{U}_i = \overset{i}{U}$, solve for $\overset{i}{V}_{i-1}$ *and* $\overset{i}{V}_i$ each one of the equations of system (3.5), and compute

$$J_i = \overset{i+1}{V}_i - \overset{i}{V}_i; \qquad i = 1, \ldots, N - 1 \tag{3.9}$$

Repeat this process, choosing new collections of values $\{U^1, \ldots, U^{N-1}\}$ until $J_i = 0$, $\forall i = 1, \ldots, N - 1$.

Extension of the previous discussion to more dimensions is relatively straightforward. If one defines again the function $J(x, y)$ by $J \equiv [\partial u/\partial n]$, on Σ, and assumes the Dirichlet problem is well posed in each of the subregions, then J is a functional of the values of the function $u(x, y)$ on Σ. When a direct approach is used, the resulting matrices are generally sparse because the value of J at a given point has a localized domain of dependence. Thus, for example, consider two neighboring subregions, Ω_i and Ω_j, in Figure 2, then at any point of the common boundary $\Sigma_{i,j} = \partial\Omega_i \cap \partial\Omega_j$, the normal derivative on one side is a functional of the values of u on $\partial\Omega_i$, while on the other side it is a functional of the values on $\partial\Omega_j$.

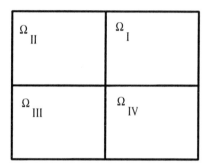

FIGURE 2. The elemental subregions.

Hence, the domain of dependence of J at that point is $\partial\Omega_i \cup \partial\Omega_j$. This would lead to a block heptadiagonal structure of the matrix when a direct aproach is used. However, the number of significant diagonals may be increased by the continuity conditions at corners (Figure 1).

When a search approach is used, as when iterative procedures are applied, one searches for a function $U(x, y)$, defined on Σ, for which $J \equiv 0$. The search approach is the basis of all iterative methods. The different strategies that are followed by the iterative procedures are intended to make the search more efficient. Maybe the most elemental iterative procedure is the Jacobi method [Allen *et al.*, 1988] and more sophisticated methods frequently were derived with the intention of improving efficiency. One which has been very successful is the conjugate gradient method [Hestens and Stiefel, 1952], which in its original form can be applied to positive definite matrices only.

In the case of elliptic differential operators of second order, it is possible to associate positive quadratic forms to nonoverlapping domain decompositions, when the differential operator is symmetric and positive: $\mathbf{b} = 0$, $c \geq 0$, in Eq. (3.1). Indeed, this becomes possible by introducing Poincaré–Steklov operators [Agoshkov, 1987]. A method for doing this, which is more direct and simple and was introduced in Herrera *et al.* [1994], is explained next in a revised form.

For simplicity Dirichlet boundary conditions will be considered exclusively, but the procedure has more general validity. Let D_H be the subspace of functions of D, which are continuous, vanish on the boundary, and satisfy the homogeneous differential equation in Ω. Then, given $\hat{u}_H$ and $\hat{v}_H \in D_H$, the following identity holds:

$$-\int_\Sigma \hat{v}_H[\underline{\underline{a}} \cdot \nabla \hat{u}_H] \cdot \underline{n} \, d\underline{x} = \int_\Omega (\nabla \hat{u}_H \cdot \underline{\underline{a}} \cdot \nabla \hat{v}_H + c\hat{u}_H \hat{v}_H) d\underline{x} \qquad (3.10a)$$

which exhibits the symmetry of the expressions involved. In addition

$$-\int_\Sigma \hat{u}_H[\underline{a} \cdot \nabla \hat{u}_H] \cdot \underline{n} \, d\underline{x} = \int_\Omega (\nabla \hat{u}_H \cdot \underline{\underline{a}} \cdot \nabla \hat{u}_H + c\hat{u}_H^2) d\underline{x} \geq 0 \qquad (3.10b)$$

and the equality holds, if and only if $\hat{u}_H \equiv 0$ in Ω. If $u \in D$ is a solution of the boundary value problem then

$$-\int_\Sigma \hat{u}_H[\underline{\underline{a}} \cdot \nabla \hat{u}_H] \cdot \underline{n} \, d\underline{x} = \int_\Omega \hat{u}_H \mathcal{L}u \, d\underline{x} - \int_{\partial\Omega} u(\underline{a} \cdot \nabla \hat{u}_H) \cdot \underline{n} \, d\underline{x}$$

$$= \int_\Omega \hat{u}_H f_\Omega \, d\underline{x} - \int_{\partial\Omega} u_\partial(\underline{\underline{a}} \cdot \nabla \hat{u}_H) \cdot \underline{n} \, d\underline{x} \qquad (3.11)$$

where u_∂ is the boundary value of u. Observe that when $\hat{u}_H \in D_H$ is given, evaluation of the last two integrals is possible because both f_Ω and u_∂ are data of the problem.

Assume for the time being that there is a function $u_H \in D_H$, such that $u_H = u$ on Σ; then for any $\hat{u}_H \in D_H$, one has

$$-\int_\Sigma \hat{u}_H[\underline{a} \cdot \nabla \hat{u}_H] \cdot \underline{n} \, d\underline{x} + 2\int_\Sigma u_H[\underline{\underline{a}} \cdot \nabla \hat{u}_H] \cdot \underline{n} \, d\underline{x} \geq \int_\Sigma u_H[\underline{\underline{a}} \cdot \nabla u_H] \cdot \underline{n} \, d\underline{x} \qquad (3.12)$$

and the minimum of the left-hand side is attained if and only if $\hat{u}_H = u_H = u$ on Σ. It is finally observed that the left-hand member of (3.12) can be expressed by means of any of the two following functionals

$$\Im_1(\hat{u}_H) \equiv \int_\Omega \{\nabla \hat{u}_H \cdot \underline{\underline{a}} \cdot \nabla \hat{u}_H + c\hat{u}_H^2\} d\underline{x} - 2\left\{\int_\Omega \hat{u}_H f_\Omega \, d\underline{x} - \int_{\partial\Omega} u_\partial(\underline{a} \cdot \nabla \hat{u}_H) \cdot \underline{n} \, d\underline{x}\right\} \qquad (3.13a)$$

and

$$\Im_2(\hat{u}_H) \equiv -\int_\Sigma \hat{u}_H[\underline{a} \cdot \nabla \hat{u}_H] \cdot \underline{n} \, d\underline{x} - 2\left\{\int_\Omega \hat{u}_H f_\Omega \, d\underline{x} - \int_{\partial\Omega} u_\partial(\underline{a} \cdot \nabla \hat{u}_H) \cdot \underline{n} \, d\underline{x}\right\} \qquad (3.13b)$$

Our conclusion is that a function $u_H \in D_H$ satisfies the condition $u_H = u$, on Σ – here, $u \in D$ is a solution of the boundary value problem – if and only if the functionals $\Im_1$ and $\Im_2$, which are identical, attain their minimum.

Taking the variation of these functionals two variational formulations follow; the first

$$\int_\Omega (\nabla \hat{u}_H \cdot \underline{\underline{a}} \cdot \nabla v_H + c\hat{u}_H v_H) d\underline{x} = \int_\Omega f_\Omega v_H \, d\underline{x} - \int_{\partial\Omega} u_\partial(\underline{\underline{a}} \cdot \nabla v_H) \cdot \underline{n} \, d\underline{x} \qquad (3.14a)$$

characterizes any function $u \in D$ which takes, on Σ, the values of the solution u of the boundary value problem, while the second one

$$-\int_\Sigma u_H[\underline{a} \cdot \nabla v_H] \cdot \underline{n} \, d\underline{x} = \int_\Omega v_H f_\Omega \, d\underline{x} - \int_{\partial\Omega} u_\partial(\underline{a} \cdot \nabla v_H) \cdot \underline{n} \, d\underline{x} \qquad (3.14b)$$

characterizes directly the values of u on Σ. Of course, the full statement of these variational principles requires that Eqs. (3.14) be satisfied for every $v_H \in D_H$.

4. Overlapping Procedures

Going back to the one-dimensional case of Section 3, the domain decomposition of the interval may be defined as the collection of subintervals $\Omega_i = (x_{i-1}, x_{i+1})$, where

$i = 1, \ldots, N - 1$. These subintervals are not disjoint. Again, the domain decomposition problem may be transformed into finding the values of the solution of the problem at $\{x_1, \ldots, x_{N-1}\}$, which in turn may be formulated as a search problem. Let $\{U_1, \ldots, U_{N-1}\}$ be a collection of values which are being tested as possible solutions of this problem. Consider the Dirichlet problem in Ω_i, with boundary values U_{i-1} and U_{i+1}, at x_{i-1} and x_{i+1}, respectively. The value of the solution of this problem at x_i is linear in U_{i-1}, U_{i+1} and f_Ω. Thus, it can be written as $L^i_- U_{i-1} + L^i_+ U_{i+1} + \mathcal{G}^i(f_\Omega)$, and the condition that characterizes the solution is

$$L^i_- U_{i-1} - U_i + L^i_+ U_{i+1} = -\mathcal{G}^i(f_\Omega) \tag{4.1}$$

This is a tridiagonal system. A first point to be observed is that only the values of the function occur in the system and the derivatives are not involved. This is typical of overlapping procedures.

Let Σ_{ij} be the set of internal boundaries contained in Ω_{ij} and $\Sigma = \bigcup_{ij} \Sigma_{ij}$. Then $\Sigma = (\bigcup_{ij} \partial\Omega_{ij}) - \partial\Omega$. When $U(x, y)$ is a function defined on Σ, the symbols $U_{\Sigma_{ij}}$ and $U_{\partial_{ij}}$ will be used to represent the restrictions of such a function to Σ_{ij} and $\partial\Omega_{ij}$, respectively. On every Ω_{ij}, consider the Dirichlet problem for which the boundary data is $U_{\partial_{ij}}$. Then the restriction of the solution of this problem to Σ_{ij} is linear on $U_{\partial_{ij}}$ and on f_Ω, and can be written as $\mathcal{L}(U_{\partial_{ij}}) + \mathcal{G}(f_\Omega)$. Any function U defined on Σ is a solution of the domain decomposition problem if and only if

$$U_{\Sigma_{ij}} = \mathcal{L}(U_{\partial_{ij}}) + \mathcal{G}(f_\Omega) \qquad \text{on} \quad \Sigma \tag{4.2}$$

and the resulting system of equations is at most nine-diagonal. The size of the blocks depends on the kind of discretization applied.

5. Conjugate Gradient Method

If the search space is a linear space, a basis for it can be constructed and the solution be found by successively trying each of the members of the basis. If the dimension of the space is N, this process would involve N steps: one for each new search direction. The computations required are simplified if an orthogonal basis is available. However, when N is large, construction of such a basis is generally quite expensive, since each new element has to be orthogonalized with respect to each of the previous ones. A fundamental advantage of the conjugate gradient method (CGM, or simply CG) is that it supplies a simple manner of constructing new search directions which are orthogonal to all those that have already been tried, except the very last one. Consider the equation

$$\underline{\underline{A}}\, \underline{U} = \underline{b} \tag{5.1}$$

where $\underline{\underline{A}}$ is positive definite and symmetric with respect to an inner product and write $\underline{U}_s$ for the unique solution of (5.1). The procedure for constructing an orthogonal basis may be described as follows:

(a) Choose $\underline{U}^0$ arbitrarily.

(b) Define the error $\underline{e}^k = \underline{U}_s - \underline{U}^0$.

(c) Choose the first search direction $\underline{p}^1 = \underline{\underline{A}}\,\underline{e}^0 = \underline{b} - \underline{\underline{A}}\,\underline{U}^0$.

(d) In the space spanned by $\{\underline{p}^1, \ldots, \underline{p}^k\}$, where the system $\{\underline{p}^1, \ldots, \underline{p}^k\}$ is orthogonal, choose $\underline{U}^k$ so that $\underline{e}^k \perp \text{span} \{\underline{p}^1, \ldots, \underline{p}^k\}$.

   Note: This requires $\underline{U}^k = \underline{U}^{k-1} + \alpha^k \underline{p}^k$ with $a^k = (\underline{e}^k, \underline{p}^k)/(\underline{p}^k, \underline{p}^k)$.

(e) Incorporate $\underline{\underline{A}}\,\underline{e}^k$ in the search space.

Note: Since $\underline{\underline{A}}\,\underline{e}^k$ is orthogonal to $\{\underline{p}^1,\,\ldots,\,\underline{p}^k\,\}$, the only requirement for the new search direction $\underline{p}^{k+1}$ is that it be orthogonal to $\underline{p}^k$. Thus, $\underline{p}^{k+1} = \underline{\underline{A}}\underline{e}^k - \beta^{k+1}\underline{p}^k$, with $\beta = (\underline{\underline{A}}\underline{e}^k,\,\underline{p}^k)/(\underline{p}^k,\,\underline{p}^k)$.

Final note: The term $\alpha^k = (\underline{e}^k,\,\underline{p}^k)/(\underline{p}^k,\,\underline{p}^k)$ is not computable for an arbitrary inner product since $\underline{e}^k$ is not known. A suitable inner product is $(\underline{e}^k,\,\underline{p}^k) \equiv \underline{e}^k\cdot\underline{\underline{A}}\underline{p}^k = \underline{p}^k\cdot\underline{\underline{A}}\underline{e}^k = \underline{p}^k\cdot\underline{r}^k = \underline{p}^k\cdot(\underline{b}-\underline{\underline{A}}U^k)$, where the dot stands for the standard Euclidean inner product. For an algorithm derived in this manner, suitable for numerical implementation, see, for example, Allen [1988].

6. Preconditioners

Another important property of CGM is that the manner of generating the new search directions is not random; on the contrary, it is related to the solution of the problem since $\underline{\underline{A}}\,\underline{e}^k$ is incorporated at every step. Indeed, the angle between $\underline{e}^k$ and $\underline{\underline{A}}\,\underline{e}^k$ is controlled by properties of the matrix $\underline{\underline{A}}$; more specifically, it has to be small if the condition number is small. In other words, the selection of the search direction is very good when the condition number is small. It can be shown [Golub and Van Loan, 1983, Section 10.2] that

$$\|u - \underline{U}^k\| \leq \|u - \underline{U}^0\| \left[\frac{\rho - 1}{\rho + 1}\right]^{2k} \tag{6.1}$$

where ρ is the square root of the condition number of $\underline{\underline{A}}$. Also, $\|\,\|$ is used to indicate the energy norm associated with $\underline{\underline{A}}$. From this relation it may be concluded that CGM converges rapidly when the condition number is not large. However, if ρ is large, then $(\rho - 1)/(\rho + 1)$ is close to 1 and the performance of CG may be poor. When applying the conjugate gradient method, the domain decomposition procedure may involve many subdomains, and frequently this leads to systems of equations which are poorly conditioned. In such cases means of diminishing the condition number of the system have to be sought. One way of achieving such a reduction is by the use of "preconditioners."

Before proceeding to discuss the *preconditioned* conjugate gradient (PCG) method, some comments are in order, in particular, when a condition number must be considered large. If we are dealing with a system of 1000 equations, then a condition number such that 100 iterations are required for convergence is pretty large, but probably satisfactory, since it reduces by a factor of $1/10$ the number of search directions required to find the solution.

PCG consists in choosing a preconditioner – i.e., a matrix $\underline{\underline{B}}$, positive-definite and symmetric – and writing the equivalent system

$$\underline{\underline{B}}^{-1}\underline{\underline{A}}\underline{U} = \underline{\underline{B}}^{-1}\underline{b} \tag{6.2}$$

This system is symmetric in the inner product $(\underline{U},\,\underline{V}) = \underline{U}\cdot\underline{\underline{B}}\underline{V}$ and CGM will converge faster when applied to (6.2) than when applied to the original system, if ρ', the square root of the condition number of the modified system, is smaller than ρ: the closer to 1, the faster.

Because of the importance of reducing the condition number effectively, a lot of work has been done to develop efficient preconditioners. Noticing that the only matrix with condition number equal to 1 is the identity matrix, preconditioners may be thought of as approximate inverses of the original matrix $\underline{\underline{A}}$. Considerable progress has been made in the understanding of the problems associated with the construction of preconditioners for systems occurring in domain decomposition methods (see, for example, Bramble *et al.* [1986, 1987]).

Actually, the best strategy in the construction of such preconditioners depends on many factors related not only to the nature of the system considered but also to the frequency with which the same system will have to be solved (several accounts of the matters involved may be found in Glowinski *et al.* [1988, 1990], Chan *et al.* [1989a, 1989b] and Keyes *et al.* [1991]).

Acknowledgments. This work was supported by a grant from the Universidad Nacional Autónoma de México and Cray Research, Inc., and by a CONACYT-NSF 1994 grant. We would also like to thank the Dirección General de Servicios de Cómputo Académico for its support and the free use of UNAM's CRAY Y-MP. Ismael Herrera holds a chair of excellence (Cátedra Patrimonial de Excelencia: Nivel I) from CONACYT.

REFERENCES

Agoshkov, V. I., Poincaré-Steklov's operators and domain decomposition methods in finite dimensional spaces. *Domain Decomposition Methods for Partial Differential Equations*, SIAM, Ecole Nationale des Ponts et Chaussees, Paris, France, 73–112, January, 1987.

Allen, M. B., I. Herrera, and G. F. Pinder, *Numerical Modeling in Science and Engineering*, Wiley-Interscience, 1988.

Bramble, J. H., J. E. Pasciak, and A. H. Schatz, The construction of preconditioners for elliptic problems by substructuring, I, *Mathematics of Computation*, **175**(47), 103–134, 1986.

Bramble, J. H., J. E. Pasciak, and A. H. Schatz, The construction of preconditioners for elliptic problems by substructuring, II, *Mathematics of Computation*, **179**(49), 1–16, 1987.

Chan, T. F., R. Glowinski, J. Perriaux, and O. B. Widlund, Eds., *Domain Decomposition Methods for Partial Differential Equations*, Vol. II, SIAM, 1989a.

Chan, T. F., R. Glowinski, J. Perriaux, and O. B. Widlund, Eds., *Domain Decomposition Methods for Partial Differential Equations*, Vol. III, SIAM, 1989b.

Glowinski R., H. G. Golub, G. A. Meurant, and J. Periaux, Eds., *Domain Decomposition Methods for Partial Differential Equations*, Vol. I, SIAM, 1988.

Glowinski, R., Y. A. Kuznetsov, G. Meurant, J. Periaux, and O. F. Widlund, Eds., *Domain Decomposition Methods for Partial Differential Equations*, Vol. IV, SIAM, 1990.

Golub, G. H., and C. F. Van Loan, *Matrix Computations*, Johns Hopkins University Press, 1983.

Herrera, I., J. Guarnaccia, and G. F. Pinder, Domain decomposition method for collocation procedures. *Computational Methods in Water Resources X*, Vol. 1, A. Peters., et al., Eds., Kluwer Academic, 273–280, 1994.

Hestenes, M. R., and E. Stiefel, Method of conjugate gradients for solving linear systems, *J. Res. Nat. Bur. Standards*, **49**, 409–435, 1952.

Israeli, M., and L. Vozovoi, Domain decomposition methods for solving parabolic PDEs on multiprocessors, *Applied Numerical Mathematics*, **12**, 193–212, North–Holland, 1993.

Keyes, D. E., and W. D. Gropp, A comparison of domain decomposition techniques for elliptic partial differential equations and their parallel implementation, *SIAM J. Sci. Stat. Comput.*, **8**(2), 166–201, 1987.

Keyes, D. E., T. F. Chan, G. Meurant, J. S. Scroggs, and R. G. Voigt, Eds., *Domain Decomposition Methods for Partial Differential Equations*, Vol. V, SIAM, 1991.

Quarteroni, A., J. Periaux, Y. A. Kuznetsov, and O. F. Widlund, Eds., *Domain Decomposition Methods in Science and Engineering*, Contemporary Mathematics **157**, The American Mathematical Society, 1992.

Parallelization Using TH–Collocation

By Ismael Herrera,[1,3] Joaquín Hernández,[2] Abel Camacho[1] and
Jaime Garfias[4]

[1]Instituto de Geofísica, Universidad Nacional Autónoma de México, Apartado Postal 22-585,
14000 México, D.F., México

[2]Dirección General de Servicios de Cómputo Académico, Universidad Nacional Autónoma de
México, Apartado Postal 20-059, 01000 México, D.F., México

[3]Also at Instituto de Matemáticas, UNAM

[4]Universidad Autónoma del Estado de México

In recent years, the innovation of massively parallel processor systems has created the need for
new algorithms for these architectures. The TH–collocation method is very suitable for parallelization. In this paper, the implementation of the TH–collocation method on a massively
parallel system is carried out by using the CRAY T3D emulator and a CRAY T3D supercomputer. A distributed memory programming model is used, as is explained here; furthermore,
the analysis points to particular procedures that produce optimal accuracy. The techniques for
the optimization of this program, using the CRAY utilities, are discussed. Example calculations
illustrate the computational procedure and verify the theoretical convergence rates.

1. Introduction

Field-scale simulations of fluids in porous media often involve problems that are so
large that solution on the full computational domain is either impossible or extremely
inefficient. Parallelization techniques [Bramble *et al.*, 1986, 1988; Ewing, 1990] have
been developed to allow these large processes to be split into pieces that can be solved
independently and then put back together to give an approximation of the total solution.
The lack of dependence of the solution processes on the separate parts of the problem
gives extreme flexibility to the methods and allows efficient use of parallel architecture
supercomputers. The methods can simply separate subdomain solutions to divide the
computational effort or can allow the use of different discretizations or even different
model equations on the separate domains. Thus, a simplified description of the physics
can be used in regions where the simplification is valid, and more rigorous models can
be used locally [Buzbee and Dorr, 1974].

Many applications of fluid flow in porous media involve both large-scale processes and
highly localized phenomena that are often critical to the physical behavior of the flow.
For large-scale problems, it is frequently impossible to use a uniform or quasi-uniform
grid that is sufficiently fine to resolve the local phenomena. Since these local processes are
often dynamic, efficient numerical simulations require the ability to perform dynamic self-
adaptive local grid refinement. Normal introduction of local grid refinement techniques
destroys the vectorization capabilities of supercomputers and hence their efficiency. Parallel techniques possess enormous potential for efficient local accuracy improvements in
many large-scale problems.

In this paper we will discuss the collocation discretization and the use of a parallel
system using a CRAY T3D supercomputer. All of these computations illustrate the
enormous potential for new advances in the use of supercomputers in simulation by using
local refinement techniques. Finally, example calculations are presented to illustrate the
methodology.

2. Preliminary Concepts

In this section, Herrera's algebraic theory of boundary value problems [Herrera 1977, 1984, 1985a, 1985b; Herrera *et al.*, 1982, 1985] is briefly explained.

Consider a region Ω and the linear spaces D_1 and D_2 of trial and test functions defined in Ω, respectively. Assume further that functions belonging to D_1 and D_2 may have jump discontinuities across some internal boundaries whose union will be denoted by Σ. For example, in applications of the theory to finite element methods, the set Σ will be the union of all the interelement boundaries. In this setting, the general boundary value problem to be considered is one with prescribed jumps across Σ. The differential equation is

$$\mathcal{L}u = f_\Omega \quad \text{in} \quad \Omega \tag{2.1}$$

where Ω may be a purely spatial region or, more generally, a space–time region. Certain boundary and jump conditions are specified on the boundary $\partial\Omega$ of Ω and on Σ, respectively. When Ω is a space–time region, such conditions generally include initial conditions. In the literature on mathematical modeling of macroscopic physical systems, there are a variety of examples of initial-boundary value problems with prescribed jumps. To mention just one class, problems of elastic wave diffraction can be formulated as such [Herrera *et al.*, 1985; Herrera, 1986]. The jump conditions that the solution sought must satisfy across Σ, in order to define a well-posed problem, depend on the specific applications and on the differential operator considered. For example, for elliptic problems of second order, continuity of the solution sought and its normal derivative is usually required, but the problem in which the solution and its normal derivative jump across Σ in a prescribed manner is also well posed [Herrera, 1986].

The definition of a formal disjoint requires that a differential operator $\mathcal{L}$ and its formal adjoint $\mathcal{L}^*$ satisfy the condition that $w\mathcal{L}u - u\mathcal{L}^*w$ be a divergence, i.e.,

$$w\mathcal{L}u - u\mathcal{L}^*w = \nabla \cdot \{\underline{\mathcal{D}}(u, w)\} \tag{2.2}$$

for suitable vector-valued bilinear function $\underline{\mathcal{D}}(u, w)$. Integration of Eq. (2.2) over Ω and multiplication of the generalized divergence theorem [Allen *et al.*, 1988] yield

$$\int_\Omega \{w\mathcal{L}u - u\mathcal{L}^*w\}\, dx = \int_{\partial\Omega} \mathcal{R}_\partial(w, u)\, dx + \int_\Sigma \mathcal{R}_\Sigma(u, w)\, dx \tag{2.3}$$

where

$$\mathcal{R}_\partial(u, w) = \underline{\mathcal{D}}(u, w) \cdot \underline{\mathbf{n}} \quad \text{and} \quad \mathcal{R}_\Sigma(u, w) = -[\underline{\mathcal{D}}(u, w)] \cdot \underline{\mathbf{n}} \tag{2.4}$$

Here the square brackets stand for the "jumps" across Σ of the function contained inside, i.e., limit on the positive side minus limit on the negative side. Here, as in what follows, the positive side of Σ is chosen arbitrarily and then the unit normal vector $\mathbf{n}$ pointing toward the positive side of Σ is taken. Observe that generally $\mathcal{L}u$ will not be defined on Σ, since u and its derivatives may be discontinuous. Thus, in this paper, it is understood that integrals over Ω are carried out excluding Σ. Consequently, differential operators will always be understood in an elementary sense and not in a distributed sense.

In the general theory of partial differential equations, Green's formulas are used extensively. For the construction of such formulas it is standard to introduce a decomposition of the bilinear function $\mathcal{R}_\partial$ (see, for example, Lions and Magenes [1972]). Indicating transposes of bilinear forms by means of an asterisk, the general form of such a decomposition is

$$\mathcal{R}_\partial(u, w) = \underline{\mathcal{D}}(u, w) \cdot \underline{\mathbf{n}} = \mathcal{B}(u, w) - \mathcal{C}^*(u, w) \tag{2.5}$$

where $\mathcal{B}(u, w)$ and $\mathcal{C}(w, u) = \mathcal{C}^*(u, w)$ are two bilinear functions. When considering initial-boundary value problems, the definitions of these bilinear forms depend on the type of boundary and initial conditions prescribed. A basic property required of $\mathcal{B}(u, w)$ is that for any u that satisfies the prescribed boundary and initial conditions, $\mathcal{B}(u, w)$ is a well-defined linear function of w, independent of the particular choice of u. This linear function will be denoted g_∂ (thus its value for any given function w will be $g_\partial(w)$), and the boundary conditions can be specified by requiring that $\mathcal{B}(u, w) = g_\partial(w)$ for every $w \in D_2$ (or, more briefly, $\mathcal{B}(u, \cdot) = g_\partial$). For example, for the Dirichlet problem of the Laplace equation, it will be seen later that $\mathcal{B}(u, w)$ can be taken to be $u(\partial u/\partial n)$ on $\partial\Omega$. Thus, if u is the prescribed value of u on $\partial\Omega$ one has $\mathcal{B}(u, w) = u_\partial(\partial w/\partial n)$ for any function u that satisfies the boundary conditions. Thus $g(w) = u_\partial(\partial w/\partial u)$ in this case.

The linear function $\mathcal{C}^*(u, \cdot)$, on the other hand, cannot be evaluated in terms of the prescribed boundary values, but it also depends exclusively on certain boundary values of u (the "complementary boundary values"). Generally, such boundary values can be evaluated only after the initial-boundary value problem has been solved. Taking the example of the Dirichlet problem for the Laplace equation, as before, $\mathcal{C}^*(u, w) = w(\partial u/\partial n)$ and the complementary boundary values correspond to the normal derivative on $\partial\Omega$.

In a similar fashion, convenient formulations of boundary value problems with prescribed jumps require constructing Green's formulas in discontinuous fields. This can be done by introducing a general decomposition of the bilinear function $\mathcal{R}_\Sigma(u, w)$, whose definition is pointwise. The general theory includes the treatment of differential operators with discontinuous coefficients [Herrera *et al.*, 1985]. However, in this paper, only continuous coefficients will be considered. In this case, such a decomposition is easy to obtain, and it stems from the algebraic identity

$$[\underline{\mathcal{D}}(u, w)] = \underline{\mathcal{D}}([u], \dot{w}) + \underline{\mathcal{D}}(\dot{u}, [w]) \tag{2.6}$$

where

$$[u] = u_+ - u_-, \qquad \dot{u} = \frac{u_+ + u_-}{2} \tag{2.7}$$

The desired decomposition is obtained by combining the second of Eqs. (2.4) with (2.6):

$$\mathcal{R}_\Sigma(u, w) = \mathcal{J}(u, w) - \mathcal{K}^*(u, w) \tag{2.8}$$

with

$$\mathcal{J}(u, w) = -\underline{\mathcal{D}}([u], \dot{w}) \cdot \mathbf{n} \tag{2.9a}$$

$$\mathcal{K}^*(u, w) = \mathcal{K}(w, u) = \underline{\mathcal{D}}(\dot{u}, [w]) \cdot \mathbf{n} \tag{2.9b}$$

An important property of the bilinear function $\mathcal{J}(u, w)$ is that when the jump of u is specified, it defines a unique linear function of w, which is independent of the particular choice of u. When considering initial-boundary value problems with prescribed jumps, the linear function defined by the prescribed jump in this manner will be denoted by J_Σ; thus its value for any given function w will be $J_\Sigma(w)$ and the jump conditions at any point of Σ can be specified by means of the equation $\mathcal{J}(u, \cdot) = J_\Sigma$. In problems with prescribed jumps, the linear functional $\mathcal{K}^*(u, \cdot)$ plays a role similar to that of the complementary boundary values $\mathcal{C}^*(u, \cdot)$. It can be evaluated only after the initial-boundary value problem has been solved and certain information about the average of the solution and its derivatives on Σ is known. Such information will be called the "generalized averages."

Introducing the notation

$$\langle Pu, w \rangle = \int_\Omega w \mathcal{L} u \, dx, \qquad \langle Q^* u, w \rangle = \int_\Omega w \mathcal{L}^* u \, dx \qquad (2.10a)$$

$$\langle Bu, w \rangle = \int_{\partial\Omega} \mathcal{B}(u, w) \, dx, \qquad \langle C^* u, w \rangle = \int_{\partial\Omega} \mathcal{C}(w, u) \, dx \qquad (2.10b)$$

$$\langle Ju, w \rangle = \int_\Sigma \mathcal{J}(u, w) \, dx, \qquad \langle K^* u, w \rangle = \int_{\partial\Omega} \mathcal{K}(w, u) \, dx \qquad (2.10c)$$

Equation (2.3) can be written as

$$\langle Pu, w \rangle - \langle Q^* u, w \rangle = \langle Bu, w \rangle - \langle C^* u, w \rangle + \langle Ju, w \rangle - \langle K^* u, w \rangle \qquad (2.11)$$

This is an identity between bilinear forms and can be written more briefly, after rearranging, as

$$P - B - J = Q^* - C^* - K^* \qquad (2.12)$$

This is the Green–Herrera formula for operators in discontinuous fields [Ewing, 1990; Herrera, 1985a].

The initial-boundary value problem with prescribed jumps can be formulated pointwise by means of Eq. (2.1) together with

$$\mathcal{B}(u, \cdot) = g_\partial \qquad \text{and} \qquad \mathcal{J}(u, \cdot) = j_\Sigma \qquad (2.13)$$

In order to associate a variational formulation with this problem, define the linear functionals $f, g, j \in D_2^*$ by means of

$$\langle f, w \rangle = \int_\Omega w f_\Omega \, dx, \qquad \langle g, w \rangle = \int_{\partial\Omega} g_\Omega(w) \, dx, \qquad \langle j, w \rangle = \int_\Sigma j_\Sigma(w) \, dx \qquad (2.14)$$

Then the variational formulation of the initial-boundary value problem with prescribed jumps is

$$Pu = f, \qquad Bu = g, \qquad Ju = j \qquad (2.15)$$

The bilinear functional J just constructed, as well as B, are boundary operators for P, which are fully disjoint. (For the definitions of the concepts that appear in italics here, the reader is referred to Herrera's original papers [Herrera, 1977, 1984, 1985a, 1985b, 1986; Herrera *et al.*, 1982, 1985]). When this is the case, the system of equations (2.15) is equivalent to the single variational equation

$$\langle (P - B - J)u, w \rangle = \langle f - g - j, w \rangle \qquad \forall w \in D_2 \qquad (2.16)$$

This is said to be the "variational formulation in terms of the data of the problem," because Pu, Bu and Ju are prescribed. Making use of formula (2.12), the variational formulation (2.16) is transformed into

$$\langle (Q^* - C^* - K^*)u, w \rangle = \langle f - g - j, w \rangle \qquad \forall w \in D_2 \qquad (2.17)$$

This is said to be the "variational formulation in terms of the sought information" because $Q^* u, C^* u$ and K^* are not prescribed. The variational formulations (2.16) and (2.17) are equivalent by virtue of the identity (2.12). The linear functionals $Q^* u, C^* u$ and $K^* u$ supply information about the sought solution at points in the interior of the region Ω, the complementary boundary values at $\partial\Omega$, and the generalized averages of the solution at Σ, respectively, as can be verified by inspection of Eqs. (2.10), and are illustrated in the examples that follow.

Localized adjoint methods are based on the following observations: When the method of weighted residuals is applied, an approximate solution $\hat{u} \in D_1$ satisfies

$$\langle (P - B - J)\hat{u}, w^\alpha \rangle = \langle f - g - j, w^\alpha \rangle, \qquad \alpha = 1, \ldots, N \qquad (2.18)$$

where $\{w^1, \ldots, w^N\} \subset D_2$ is a given system of weighting functions. However, these equations, when they are expressed in terms of the sought information, become

$$\langle (Q^* - C^* - K^*)\hat{u}, w^\alpha \rangle = \langle f - g - j, w^\alpha \rangle, \qquad \alpha = 1, \ldots, N \qquad (2.19)$$

Since the exact solution satisfies (2.17), it must be that

$$\langle (Q^* - C^* - K^*)\hat{u}, w^\alpha \rangle = \langle (Q^* - C^* - K^*)u, w^\alpha \rangle = \langle f - g - j, w^\alpha \rangle, \qquad \alpha = 1, \ldots, N \quad (2.20)$$

either in this form or in the form

$$\langle (Q^* - C^* - K^*)(\hat{u} - u), w^\alpha \rangle = 0, \qquad \alpha = 1, \ldots, N \qquad (2.21)$$

Equations (2.20) can be used to analyze the information about the exact solution that is contained in an approximate one. In localized adjoint methods, these observations have been used as a framework for selecting more convenient test functions.

In Section 3, attention will be restricted to functions which are continuous across Σ, with a possibly discontinuous first derivative. Then, for the most general elliptic operator of second order, to be considered there, we have

$$\mathcal{B}(u, w) = \left(\mathbf{a}_n \frac{\partial w}{\partial n} + b \cdot n \right) u, \qquad \mathcal{C}(w, u) = \mathbf{a}_n w \frac{\partial u}{\partial n} \qquad (2.22)$$

while

$$\mathcal{J}(u, w) = \mathbf{a}_n w \left[\frac{\partial u}{\partial n} \right], \qquad \mathcal{K}(w, u) = \mathbf{a}_n u \left[\frac{\partial w}{\partial n} \right] \qquad (2.23)$$

with $\mathbf{a}_n = \underline{\mathbf{n}} \cdot \underline{\underline{\mathbf{a}}} \cdot \underline{\mathbf{n}}$.

3. TH–Collocation

To illustrate Trefftz–Herrera domain decompositions, let us discuss the general elliptic differential equation of second order defined by

$$\mathcal{L}u \equiv -\nabla \cdot (\underline{\underline{\mathbf{a}}} \cdot \nabla u) + \nabla \cdot (\underline{\mathbf{b}} \, u) + c \, u \qquad (3.1a)$$

whose formal adjoint is

$$\mathcal{L}^* w \equiv -\nabla \cdot (\underline{\underline{\mathbf{a}}} \cdot \nabla w) - \underline{\mathbf{b}} \cdot \nabla w + c \, w \qquad (3.1b)$$

in two dimensions. Referring to Figure 1 of Herrera *et al.* [this volume], $C^0(\Omega)$ weighting functions which are bicubic – i.e., cubic in x and in y, separately – on the elemental rectangles, will be used. When standard collocation is applied, it is usual to represent such polynomials by means of Hermites; however, that will not be done here since it is not convenient for our purposes.

Local coordinates will be introduced; more specifically, linear mappings will be used for transforming some subregions of Ω into the unit square $\Omega_I = [0, 1] \times [0, 1]$. Then the formal adjoint, Eq. (4.1b), is transformed into

$$\mathcal{L}^* w \equiv -\nabla (\underline{\underline{\mathbf{A}}} \cdot \nabla w) - \underline{\mathbf{B}} \cdot \nabla w + \mathbf{C}w \qquad (3.2)$$

where $\underline{\underline{\mathbf{A}}}$, $\underline{\mathbf{B}}$ and $\mathbf{C}$ are related to $\underline{\underline{\mathbf{a}}}$, $\underline{\mathbf{b}}$ and c by means of the mapping. Bicubic polynomials defined on Ω_1 constitute a linear space, denoted by Π, of dimension $4 \times 4 = 16$. Define ξ and η by $\xi = x - 1$ and $\eta = y - 1$; then the subspace whose members contain the factor $\xi\eta$, denoted by $\Pi_{\xi\eta}$, has dimension 9, and the subspace whose members vanish on the whole boundary $\partial\Omega_I$ of the unit square, denoted by Π_0, has dimension 4. Observe that $\Pi_0 \subset \Pi_{\xi\eta}$ and that the dimension of algebraic complements of Π_0 with respect to $\Pi_{\xi\eta}$, is 5. In Table 1, the functions $N^\nu(x, y)$, $\nu = 1, \ldots, 4$ and $B^\nu(x, y)$, $\nu = 0, \ldots, 4$

have been defined. The system $\{N^\nu\}$ constitutes a basis of Π_0, while $\{B^\nu\}$ generates an algebraic complement of Π_0 with respect to $\Pi_{\xi\eta}$.

For constructing the weighting functions, Gaussian collocation is applied to the polynomial representation

$$W_I^\nu(X_1, X_2) = B^\nu(X_1, X_2) + \sum_{j=1}^{4} C_{Ij}^\nu N^j(X_1, X_2) \tag{3.3}$$

and the four coefficients C_j^ν are determined by the condition $\mathcal{L}^* W^\nu = 0$, which is imposed at four Gaussian points of the unit square Ω_I. In principle, five approximate solutions of the adjoint differential equation can be constructed in this manner, corresponding to $\nu = 0, \ldots, 4$. However, only the solutions associated with $\nu = 0, 1, 2$ will be used in the sequel, because otherwise the resulting system of test functions would not be linearly independent since some repetition would occur when developing the test functions associated with neighboring nodes. This yields three approximate solutions, and similar sets of solutions can be constructed in the remaining subregions Ω_{II}, Ω_{III} and Ω_{IV} (Figure 2 of Herrera *et al.* [this volume]) of the square $[-1, 1] \times [-1, 1]$.

When putting together such solutions in a continuous manner, five approximate solutions, which will be denoted by $W^\nu(X_1, X_2)$, $\nu = 0, \ldots, 4$, are obtained. They will be characterized as shown in Figure 1.

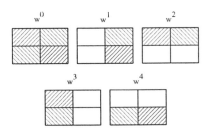

FIGURE 1. The supports of the weighting functions.

$W^0(0, 0) = 1$ and its support is the whole square $[-1, 1] \times [-1, 1]$; $W^1(X_1, X_2)$ vanishes identically on $\Omega_{II} \cup \Omega_{III}$; thus, its support is $\Omega_I \cup \Omega_{IV}$; $W^2(X_1, X_2)$ vanishes identically on $\Omega_{III} \cup \Omega_{IV}$; thus, its support is $\Omega_I \cup \Omega_{II}$; $W^3(X_1, X_2)$ vanishes identically on $\Omega_I \cup \Omega_{IV}$; thus, its support is $\Omega_{II} \cup \Omega_{III}$; $W^4(X_1, X_2)$ vanishes identically on $\Omega_I \cup \Omega_{II}$, thus, its support is $\Omega_{III} \cup \Omega_{IV}$.

Let $\partial_\nu \Omega$ ($\nu = 1, \ldots, 4$) be the left, lower, right and upper boundaries of Ω, respectively, and observe that when a node belongs to $\partial_\nu \Omega$, only W^ν vanishes on the boundary. In conclusion, with every interior node (x_i, y_j) of the partition there are associated five test functions, satisfying approximately

$$(Q - C)w = 0 \tag{3.4}$$

which will be denoted by w_{ij}^ν ($\nu = 0, \ldots, 4$); at boundary nodes there is only one; and at corner nodes, none. If the number of elements in the x and y directions is I and J, respectively, then this yields a system of $(I + 1)(J + 1) - 4$ linearly independent test functions.

4. The Trial Functions

The test functions that have been developed have the property of concentrating all the information on the internal boundaries Σ, and, correspondingly, the trial functions

TABLE 1. Definition of $N^\nu(x, y)$ and $B^\nu(x, y)$.

ν	0	1	2	3	4
$N^\nu(x, y)$		$\xi\eta xy$	$\xi\eta x^2 y$	$\xi\eta xy^2$	$\xi\eta x^2 y^2$
$B^\nu(x, y)$	$\xi\eta$	$\xi\eta x^2$	$\xi\eta y^2$	$\xi\eta x$	$\xi\eta y$

will supply information on Σ, exclusively. In particular, a function $\hat{u}_H \in \hat{D}_{1H}$ will be constructed by collocation which satisfies, approximately, the first of Eqs. (2.3). The following notation is adopted: When p and q are functions defined in Ω, the relation $p \cong q$ means that $p = q$ at each of the Gaussian points of the elemental rectangles. The basic trial functions will be denoted by Φ_{ij}^ν, where the ranges of i, j and ν are the same as for the test functions; they satisfy $(P - B - J)u = 0$. More precisely, they satisfy $\mathcal{L}\Phi_{ij}^\nu \cong 0$, are continuous, and vanish on the external boundary $\partial\Omega$. The construction of such functions is the same as that of weighting functions, except that $\mathcal{L}^*$ is replaced by $\mathcal{L}$. The approximate solution has the expression

$$\hat{u}_H = \sum U_{ij}^\nu \Phi_{ij}^\nu \qquad (4.1)$$

where the coefficients U_{ij}^ν are determined by the system of Eqs. (5.3). This is

$$M_{klij}^{\mu\nu} U_{ij}^\nu = F_{kl}^\mu \qquad (4.2)$$

where summation convention is understood – repeated indices are summed over their ranges – and

$$M_{klij}^{\mu\nu} = -\int_\Sigma \kappa^0(w_{kl}^\mu, \Phi_{ij}^\nu)dx = -\int_\Sigma a_n \Phi_{ij}^\nu \left[\frac{\partial w}{\partial n}\right] dx \qquad (4.3a)$$

while

$$F_{kl}^\nu = \langle f - g, w_{kl}^\mu \rangle = \int_\Omega w_{kl}^\mu f_\Omega \, dx - \int_{\partial\Omega} u_\partial(\underline{\underline{a}} \cdot \nabla w_{kl}^\mu + \underline{b}\, w_{kl}^\mu) \cdot \underline{n} \, dx \qquad (4.3b)$$

It is interesting to observe that

$$-\int_\Sigma \kappa^0(w_{kl}^\mu, \Phi_{i,j}^\nu)dx = \int_\Omega \underline{\underline{a}} \cdot \nabla w_{kl}^\mu \cdot \Phi_{i,j}^\nu - (\underline{b} \cdot \nabla w_{kl}^\mu + c\, w_{kl}^\mu)\Phi_{ij}^\nu \, dx \qquad (4.4)$$

which allows replacing the surface integral over Σ by a volume integral, which needs to be carried out over the intersection of the supports of w_{kl}^μ and Φ_{ij}^ν only. If Gaussian integration is applied, the fact that collocation points and the pivotal points for the quadrature coincide makes the process very economical. As has already been mentioned, the function $\hat{u}_H$ gives an approximate solution on Σ. If we wish to extend this information to the interior of the subregions, it is necessary to construct $\hat{u}_p$ such that $\mathcal{L}u_p \cong f_\Omega$ and vanishes on Σ. This involves local problems only and can also be done by collocation and constitutes postprocessing of the solution. Then $\hat{u}_p + \hat{u}_H$ is an approximation to u everywhere. Finally, it must be mentioned that when $\mathcal{L}$ is positive definite, so is the system of equations (5.2).

5. Results

A wide variety of computations have been performed on supercomputers using the techniques described in the previous sections. Many of these computational experiments were of such size and complexity as to test the computational limits of the computers used.

Of course, the implementation of any algorithm in a parallel processing environment is extremely sensitive to the data structure and its use. The memory characteristics of the computer govern the algorithms and their efficiencies. The relative success of the compilers for shared memory machines is indicative of the greater difficulty in developing efficient parallel algorithms on distributed memory machines for the complex problems arising in reservoir simulation applications.

The CRAY supercomputer offers multitasking for implementation of parallel processing, thus allowing the user to provide separate routines for tasks to be run in parallel. In addition, CRAY also allows "microtasking" for parallelism with each separate routine. However, the data structure determines the efficiency of each of these types of parallel algorithms.

To demonstrate the applicability of the computational algorithm, and to verify the theoretical results of the sections, several numerical examples are solved. A completely general computing program was developed and implemented on a CRAY T3D supercomputer.

The procedure is very easy to program and turns out to be quite versatile. Optimizing a program begins by finding where the main workload is located, that is to say, locating which parts of the code are the most time-consuming. This was done by using the CRAY utilities Flowview and Perfview. Compiler options and directives were used to refine the optimization. A very useful one was the ATexpert tool; with this utility one can visualize the parts of the program that run in parallel and estimate its speedup. Figure 2, based on an actual ATexpert window, illustrates this.

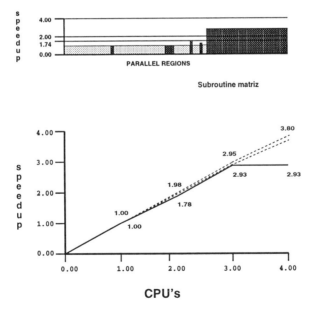

FIGURE 2. Serial and parallel portions (upper panel) and overall speedup (lower panel), as determined by ATexpert.

Numerical results computed using this methodology are shown in the convergence plots of Figure 3. The figure shows the solution error as a function of grid spacing for cases using different choices of interpolation knots. All convergence rates correspond to the theoretical predictions of the previous section.

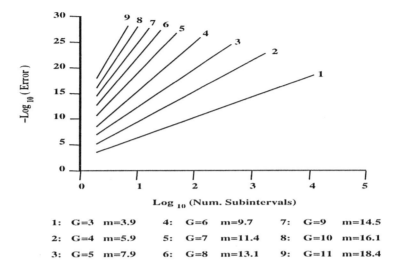

1:	G=3	m=3.9	4:	G=6	m=9.7	7:	G=9	m=14.5
2:	G=4	m=5.9	5:	G=7	m=11.4	8:	G=10	m=16.1
3:	G=5	m=7.9	6:	G=8	m=13.1	9:	G=11	m=18.4

FIGURE 3. Solution errors for $u'' + (2x^2 + 4x)/[(x+1)(x^2 + 2x + 2)]u' + [1 - 2/(x^2 + 2x + 2)]u = 1/(x^2 + 2x + 2)$ with Dirichlet's conditions $u(0) = 0.5$, $u(1) = 0.2$. The solution is $u(x) = 1/(x^2 + 2x + 2)$; the experimental slope is m.

The computer program was tested by applying it to three equations, but in this paper only the results for the first equation are presented; the boundary conditions were Dirichlet conditions. For every one of the examples, one can choose arbitrarily the number E of subintervals and the number n of collocations points. When n is fixed, this defines a straight line of slope $2n - 1$.

Figure 3 displays the negative value of the error decimal logarithm, at the nodes, versus the decimal logarithm of the number of subintervals considered in a certain position. As expected, when the weight function degree is imposed, straight lines are obtained as the number of subintervals varies. The experimental slopes of these straight lines also appear for G = 3, 4, 5, 6, 7, 8, 9, 10 and 11.

6. Conclusion

The algebraic theory for numerical methods, as developed by Herrera [Ewing, 1990; Herrera, 1977, 1984, 1985a; Herrera *et al.*, 1982], provides a broad theoretical framework for the development and analysis of numerical approximations. Analysis of the method provides error estimates. Furthermore, the analysis points to particular procedures that produce optimal accuracy. Examples developed in this paper illustrate the computational procedure and verify the theoretical convergence rates. Actually, the method presented here has considerable generality and its applicability is not restricted, by any means, to cases where the discretization procedure used is collocation.

From a more general perspective, the results of this paper illustrate some of the advantages of an approach for developing algorithms to treat numerically ordinary differential equations that the authors are advocating, whose basic ingredients consist of (a) identifying the information about the sought solution contained in the approximate one and (b) using this insight to choose the interpolation procedure.

Finally, we want to emphasize that parallelization using optimal collocation forms a general and powerful framework for investigating and comparing a wide variety of numerical methods for problems for ordinary differential equations. The framework motivates different choices of test functions to approximate different properties of the unknowns, such as fluxes. The general theory is expanding to provide more insight. These techniques appear to have enormous flexibility and potential for treating many applications of fluid flow in porous media.

Acknowledgments. This work was supported by a grant from Universidad Nacional Autónoma de México and Cray Research, Inc. We would also like to thank Dirección General de Servicios de Cómputo Académico for its support and the use of the CRAY Y-MP under its charge and to Consejo Nacional de Ciencia y Tecnología for its CONACYT-NSF 1994 grant. Ismael Herrera holds a chair of excellence (Cátedra Patrimonial de Excelencia Nivel I) from CONACYT.

REFERENCES

Allen, M. B., I. Herrera, and G. F. Pinder, *Numerical Modeling in Science and Engineering*, Wiley-Interscience, 1988.

Bramble, J. H., R. E. Ewing, J. E. Pasciak, and A. H. Scatz, A preconditioning technique for the efficient solution of problem with local grid refinement, *Comp. Meth. Appl. Mech. Eng.*, **67**, 149–159, 1988.

Bramble, J. H., J. E. Pasciak, and A. H. Scatz, An iterative method for elliptic problem on regions partitioned into substructures, *Math. Comp.*, **46**, 361–370, 1986.

Buzbee, B. L., and F. W. Dorr, The direct solution of biharmonic equation on rectangular regions and the Poissons equation on irregular regions, *SIAM J. Numer. Anal.*, **11**, 753–763, 1974.

Ewing, R. E., A survey of domain decomposition techniques and their implementation, *Adv. Water Resources*, **13**, No. 3, 1990.

Herrera, I., Soil–structure interaction as a diffraction problem, Proc. Sixth World Conference on Earthquake Engineering, New Delhi, India, 19–24, 1977.

Herrera, I., *Boundary Methods: An Algebraic Theory*, Pitman, 1984.

Herrera, I., Unified formulations of numerical methods. Part I. Green's formulas for operators in discontinuous fields, *Numer. Meth. Partial Diff. Equ.*, **1**, 25, 1985a.

Herrera, I., Unified formulations of numerical methods. Part II. Finite elements, boundary methods and its coupling, *Numer. Meth. Partial Diff. Equ.*, **1**, 241, 1985b.

Herrera, I., Some unifying concepts in applied mathematics, *New Directions in Pure, Applied and Computational Mathematics*, R. E. Ewing, K. I. Gross, and C. F. Martin, Eds., Springer-Verlag, 79–88, 1986.

Herrera, I., Trefftz method, domain decomposition and TH–collocation. *Computational Methods in Water Resources XI, Vol. 1, Computational Methods in Subsurface Flow and Transport Problems*, A. A. Aldama, J. Aparicio, C. A. Brebbia, W. G. Gray, I. Herrera, and G. F. Pinder, Eds., Computational Mechanics Publications, 515–523, 1996.

Herrera, I., L. Chargoy, and G. Alduncin, Unified formulation of numerical methods. Part III. Finite differences and ordinary differential equations, *Numer. Meth. Partial Diff. Equ.*, **1**, 241r, 1985.

Herrera, I., F. J. Sanchez Sesma, and J. Aviles, A boundary method for elastic wave diffraction: Application to scattering of SH-waves by surface irregularities, *Bull. Seismol. Soc. Am.*, **72**, 473, 1982.

Lions, J. L., and E. Magenes, *Non-Homogeneous Boundary Value Problems and Applications*, Springer-Verlag, 1972.

A Method for Simultaneous Estimation of Multiphase Relative Permeability and Capillary Pressure Functions

By G. Mejía,[1] A. Watson,[2] K. Mohanty[3] and J. Nordtvedt[2]

[1]Centro de Calidad Ambiental, ITESM, Suc. de Correos J, 64849 Monterrey, NL, México

[2]Department of Chemical Engineering, Texas A&M University, College Station, TX 77843, USA

[3]Department of Chemical Engineering, University of Houston, Houston, TX 77204-4792, USA

Simulation of unsteady-state multiphase flow through porous media is a powerful tool to describe reservoir behavior and migration of nonaqueous phase liquids in the subsurface. Good predictions are obtained if accurate estimates of relative permeabilities and capillary pressure functions (flow functions) are provided to numerical simulators. Although several methods exist to obtain estimates of two- and three-phase relative permeability functions, they have several assumptions that make their use limited and their results subject to controversy. In this paper, we present a method for simultaneous estimation of multiphase flow functions. The method is an extension of a regression-based method developed for estimating two-phase flow functions using data from dynamic displacement experiments. In our method, we present and discuss a new solution of the multiphase flow model, using B-splines in representing the flow functions. Our method has few of the assumptions made by other methods. The mathematical model considers three dimensions and heterogeneities in porosity, permeability, and initial and residual saturations of the media. Flow functions are assumed to be dependent on two saturations in the three-phase case. As a new feature, the method can utilize in situ saturation data in the estimation procedure. The computer code is written in Fortran and was designed to take full advantage of vectorization in a CRAY Y-MP supercomputer. Results of this method have been obtained for data analyzed at Texas A&M University and ARCO in Plano, Texas. The results obtained show that the parameter estimation procedure is a reliable technique for obtaining accurate estimates of two- and three-phase flow functions.

1. Introduction

Numerical simulators are used to predict migration of pollutants in the ground or to study behavior of oil reservoirs. These simulators are computer codes which solve a mathematical model of the flow system; such a model consists of the energy, mass, and momentum equations with adequate initial and boundary conditions. Although a model which considers the exact geometry of porous media is desirable, this approach would require a great amount of data describing the porous structure as well as disk memory and computational effort to solve the model. Since this approach is impractical, a more viable approximation is to develop a continuous representation, where properties and state variables correspond to local volume elements, large compared with pore sizes but small compared with the porous system. Furthermore, most oil recovery processes – water flooding, gas drainage, water-alternating-gas CO_2 flooding [Collins, 1976] and recovery of nonaqueous phase liquids (NAPLs) from aquifers by pump-and-treat and vacuum extraction – involve multiphase flow [Luckner and Schestakow, 1991; Abriola, 1984], where usually flow through the medium is three-phase: a gas, an aqueous, and a nonaqueous phase. In these cases, relative permeabilities and capillary pressure functions (flow functions) are essential inputs to numerical simulators to achieve reliable predictions for any recovery method.

Relative permeabilities and capillary pressure are functions of saturations, i.e., the fractions of each phase present in the void volume of the porous media [Luckner and Schestakow, 1991; Donaldson and Dean, 1966]. Despite the great importance of describing three-phase flow, a model which considers relative permeabilities and capillary pressure as functions of two saturations does not exist. Several unsteady-state experiments [Saraf *et al.*, 1982; Sarem, 1966; Donaldson and Dean, 1966] show that isoperms may be concave or convex, i.e., functions of two saturations. Moreover, existing models to predict migration of contaminants in the subsurface neglect resistence of the gaseous phase. This may lead to serious errors in predicting two- or three-phase flow through porous media [Kool *et al.*, 1987; Parker *et al.*, 1988].

Unsteady-state displacement experiments are used to estimate relative permeability and capillary pressure functions of rock samples. In the two-phase case, drainage and imbibition displacement experiments are performed to estimate the flow functions. In the drainage case, the sample is flooded with the wetting phase, and then it is displaced by injecting the nonwetting phase. The imbibition case is the opposite (the wetting phase displaces the nonwetting phase). The latter type of displacement can happen without application of a pressure gradient, i.e., just by capillary forces (spontaneous imbibition). Production and pressure drop data are measured during the transient time of the experiment. This procedure allows a great amount of data to be obtained in one experiment. In situ data may be collected using computed tomographic (CT) scanning at any time during the displacement experiment [Mohanty and Miller, 1991]. In the three-phase case, we have a wetting, an intermediate-wetting, and a nonwetting phase. Oak *et al.* [1990] reviewed several procedures followed to perform the displacement experiment. In their work, they designed the experiments as a combination of drainage and imbibition in accordance with the displacement of the phases.

When multiphase unsteady-state flow is considered, the mathematical model is a set of coupled nonlinear partial differential equations. If heterogeneities are considered and a two- or three-dimensional representation is used, the model requires that a large number of computer calculations be made. Recent development of modern supercomputers has made it possible to solve these complex models in a reasonable amount of time. In this paper we present the solution of the multiphase flow equations, assuming, for the first time for the three-phase case, that relative permeabilities and capillary pressures are functions of two saturations. The computer code was designed to take advantage of vectorization in a CRAY Y-MP supercomputer.

2. Mathematical Model

Darcy's law was originally developed for a single-phase flow through a porous material [Collins, 1976]. This law gives the relationship between the flow rate and pressure gradient, and, in the case of multiphase flow, this relationship for component i is given by

$$\mathbf{u_i} = -\frac{\mathbf{K}k_{ri}}{\mu_i}\left(\nabla P_i + \rho_i \frac{\mathbf{g}}{g_c}\right) \tag{1}$$

where $\mathbf{g}$ is the gravitational acceleration vector, $\mathbf{u}$ is the velocity vector, $\mathbf{K}$ is the absolute permeability tensor, μ is the viscosity, k_r is the relative permeability, ρ is the density, P is the pressure, and g_c is a unit conversion constant. This equation and the continuity equation are used to model unsteady-state multiphase flow through porous media

[Collins, 1976]. The mass balance equation for component i is

$$- \nabla \cdot \rho_i \mathbf{u_i} = \frac{\partial}{\partial t}(\rho_i \phi S_i) + \tilde{q}_i \tag{2}$$

where ϕ is the porosity of the medium, S is the saturation, t is time, and $\tilde{q}_i$ represents a sink of phase i in the differential volume considered.

Using Eqs. (1) and (2), Aziz and Settari [1979] developed the following equations to describe immiscible three-phase flow through porous media:

$$\nabla \cdot \left(\frac{K k_{rw}}{\hat{V}_w \mu_w} \left(\nabla P_w - \rho_w \frac{g}{g_c} \nabla z \right) \right) = \frac{\partial(\phi S_w / \hat{V}_w)}{\partial t} + q_w \tag{3}$$

$$\nabla \cdot \left(\frac{K k_{ro}}{\hat{V}_o \mu_o} \left(\nabla P_o - \rho_o \frac{g}{g_c} \nabla z \right) \right) = \frac{\partial(\phi S_o / \hat{V}_o)}{\partial t} + q_o \tag{4}$$

$$\nabla \cdot \left(\frac{K k_{rg}}{\hat{V}_g \mu_g} \left(\nabla P_g - \rho_g \frac{g}{g_c} \nabla z \right) \right) = \frac{\partial(\phi S_g / \hat{V}_g)}{\partial t} + q_g \tag{5}$$

$$P_{cow} = P_o - P_w \tag{6}$$

$$P_{cgo} = P_g - P_o \tag{7}$$

$$S_o + S_w + S_g = 1 \tag{8}$$

The subscripts o, w, and g refer to oil, water, and gas; $\hat{V}$ is the formation volume factor; z is the direction parallel to the gravity force; and P_c is the capillary pressure. These equations may be simplified to represent two-phase flow through porous media. This model is flexible for representing the displacement experiment since different values of permeability, porosity, and initial water saturation may be specified for different sections in the core. The model also considers capillary pressure and gravity effects. Systems in one, two, and three dimensions can be modeled. As indicated by Eq. (8), only two fluid saturations are independent. Several finite difference approaches have been proposed to solve the set of coupled, nonlinear partial differential equations. We solved the system using a variation of the simultaneous solution method [Aziz and Settari, 1979].

In the simultaneous solution method, Eqs. (3) to (5) are solved for the three pressures (P_w, P_o, and P_g), then Eqs. (6) to (8) are used to find the three saturation values. In this paper, we solved Eqs. (3) to (5) for two saturations (S_w and S_o) and one pressure (P_o), using a fully implicit finite difference approach. The remaining two pressures were calculated with Eqs. (6) and (7) and the gas saturation (S_g) was obtained by Eq. (8). In the following paragraphs we present the steps followed to solve the set of coupled partial differential equations (PDEs).

Introducing Eq. (6) into Eq. (3) and Eqs. (7) and (8) into Eq. (5), Eqs. (3) to (5) were solved for S_w, S_o, and P_o. In the solution we considered that some gas might be dissolved in the oil phase. For this purpose, Eq. (4) was multiplied by the solution gas/oil ratio [Aziz and Settari, 1979] $R_s = f(P_o)$ and the resulting equation was added to Eq. (5). R_s gives the mass transfer between the oil and gas phases as a function of the oil pressure. With these modifications in Eqs. (3) to (5) we obtained the following set of coupled PDEs:

$$\nabla \cdot \left[\lambda_w \left(\nabla(P_o - P_{cow}(S_w, S_o)) - \gamma_w \nabla z \right) \right] = \frac{\partial}{\partial t}\left(\frac{\phi S_w}{\hat{V}_w} \right) + q_w \tag{9}$$

$$\nabla \cdot \left[\lambda_o \left(\nabla P_o - \gamma_o \nabla z \right) \right] = \frac{\partial}{\partial t}\left(\frac{\phi S_o}{\hat{V}_o} \right) + q_o \tag{10}$$

$$\nabla \cdot \left[R_s \lambda_o \left(\nabla P_o - \gamma_o \nabla z \right) + \lambda_g \left(\nabla (P_o + P_{cgo}(S_w, S_o)) - \gamma_g \nabla z \right) \right]$$

$$= \frac{\partial}{\partial t} \left[\phi \left(R_s \frac{S_o}{\hat{V}_o} + \frac{(1 - S_w - S_o)}{\hat{V}_g} \right) \right] + R_s q_o + q_g \tag{11}$$

where

$$\lambda_i = \frac{\mathbf{K} k_{ri}}{\mu_i \hat{V}_i} \tag{12}$$

and

$$\gamma_i = \rho_i \frac{g}{g_c} \tag{13}$$

Equations (9) to (11) can be simplified as appropiate to represent two-phase flow.

In Eq. (12), λ_i is known as the mobility of phase i. This term was computed by using upstream weighting. In the same equation, we assumed that the permeability $\mathbf{K}$ is a diagonal tensor given by

$$\mathbf{K} = \begin{pmatrix} k_x & 0 & 0 \\ 0 & k_y & 0 \\ 0 & 0 & k_z \end{pmatrix} \tag{14}$$

where the eigenvalue k_j of $\mathbf{K}$ is the absolute permeability in direction j. With this representation of $\mathbf{K}$ it is assumed that the reference axes of the core sample are collinear with the eigenvectors of $\mathbf{K}$. This is a good assumption for practical problems and for core samples under study. The medium is called isotropic if $k_x = k_y = k_z$. In the finite difference expansion of Eqs. (9) to (11) we assumed that the medium was anisotropic (in general $k_x \neq k_y \neq k_z$) and that $\mathbf{K}$ could be different for each grid block.

We discretized Eqs. (9) to (11) by introducing the dimensionless variables $x_1 = x/L_1$, $x_2 = y/L_2$, and $x_3 = z/L_3$, where x, y, and z are the main reference coordinates and L_1, L_2, and L_3 are characteristic dimensions of the core. In Eqs. (9) to (11), spatial derivatives of the capillary pressure functions must be evaluated. Since the capillary pressures are functions of two saturations and the saturations are functions of position along the core, the spatial derivatives of P_{ci} with respect to x_j were computed by the chain rule

$$\left(\frac{\partial P_{ci}}{\partial x_j} \right) = \left(\frac{\partial P_{ci}}{\partial S_w} \right) \left(\frac{\partial S_w}{\partial x_j} \right) + \left(\frac{\partial P_{ci}}{\partial S_o} \right) \left(\frac{\partial S_o}{\partial x_j} \right) \tag{15}$$

In this equation, we calculated $(\partial P_{ci}/\partial S_w)$ and $(\partial P_{ci}/\partial S_o)$ from their B-spline representation. The spatial derivatives $(\partial S_w/\partial x_j)$ and $(\partial S_o/\partial x_j)$ were calculated using an implicit finite difference approach.

Introducing Eqs. (12) to (15) into Eqs. (9) to (11), we discretized the resulting three equations, using a fully implicit finite difference approach, for an arbitrary grid in the three dimensionless directions. For example, the discretized left-hand side of Eq. (9) in direction x_1 is

$$\left(\frac{1}{L_1} \right)^2 \frac{\partial}{\partial x_1} \left[\frac{k_x k_{rw}}{\mu_w \hat{V}_w} \left(\left(\frac{\partial P_o}{\partial x_1} \right) \right. \right.$$

$$\left. \left. - \left(\frac{\partial P_{cow}}{\partial S_w} \right) \left(\frac{\partial S_w}{\partial x_1} \right) - \left(\frac{\partial P_{cow}}{\partial S_o} \right) \left(\frac{\partial S_o}{\partial x_1} \right) - \rho_w \frac{g}{g_c} \left(\frac{\partial z}{\partial x_1} \right) \right) \right]$$

$$\approx \left(\frac{1}{L_1} \right)^2 \left(\frac{1}{x_{1,i+\frac{1}{2}} - x_{i-\frac{1}{2}}} \right) \left[\left(\frac{k_x k_{rw}}{\mu_w \hat{V}_w} \right)_{i+\frac{1}{2}} \left(\left(\frac{P_{o,i+1} - P_{o,i}}{x_{1,i+1} - x_{1,i}} \right) \right. \right.$$

$$-\left(\frac{\partial P_{cow}}{\partial S_w}\right)_{i+\frac{1}{2}}\left(\frac{S_{w,i+1}-S_{w,i}}{x_{1,i+1}-x_{1,i}}\right) - \left(\frac{\partial P_{cow}}{\partial S_o}\right)_{i+\frac{1}{2}}\left(\frac{S_{o,i+1}-S_{o,i}}{x_{1,i+1}-x_{1,i}}\right)$$

$$-\frac{g}{g_c}\rho_{w,i+\frac{1}{2}}\left(\frac{z_{i+1}-z_i}{x_{1,i+1}-x_{1,i}}\right)\right) - \left(\frac{k_x k_{rw}}{\mu_w \hat{V}_w}\right)_{i-\frac{1}{2}}\left(\left(\frac{P_o-P_{o,i-1}}{x_{1,i}-x_{1,i-1}}\right)\right.$$

$$-\left(\frac{\partial P_{cow}}{\partial S_w}\right)_{i-\frac{1}{2}}\left(\frac{S_{w,i}-S_{w,i-1}}{x_{1,i}-x_{1,i-1}}\right) - \left(\frac{\partial P_{cow}}{\partial S_o}\right)_{i-\frac{1}{2}}\left(\frac{S_{o,i}-S_{o,i-1}}{x_{1,i}-x_{1,i-1}}\right)$$

$$-\frac{g}{g_c}\rho_{w,i-\frac{1}{2}}\left(\frac{z_i-z_{i-1}}{x_{1,i}-x_{1,i-1}}\right)\right) \Bigg] \tag{16}$$

We obtained expressions similar to Eq. (17) for directions x_2 and x_3. The left-hand sides in Eqs. (10) and (11) were discretized in a similar way.

We discretized in time the right-hand sides of Eqs. (9) to (11) assuming that the medium was incompressible (ϕ is constant). For example, the right-hand side of Eq. (9) was expanded as follows:

$$\frac{\partial}{\partial t}\left(\frac{\phi}{\hat{V}_w}S_w\right) + q_w$$

$$\approx \phi\left[\frac{1}{\hat{V}_w}\left(\frac{S_{w,i}^{n+1}-S_{w,i}^n}{t^{n+1}-t^n}\right) + S_{w,i}^n\left(\frac{(1/\hat{V}_w)_i^{n+1}-(1/\hat{V}_w)_i^n}{t^{n+1}-t^n}\right)\right] + q_w \tag{17}$$

Collecting similar terms from the finite difference approach, we obtained a set of linear equations to be solved simultaneously for S_w, S_o, and P_o. In matrix form, the set of linear equations can be written as follows:

$$\mathbf{Ax} = \mathbf{b} \tag{18}$$

The values of the coefficients a_{ij} in Eq. (18) depend on the relative permeability and capillary pressure functions; therefore, these coefficients are functions of S_w, S_o, and the partial derivatives of P_o. Using the implicit approach, the a_{ij}'s are updated in each iteration when the linear equations are solved. Hence, the problem solved in Eq. (18) is

$$\mathbf{A(x)}_n^{i-1}\mathbf{x}_n^i = \mathbf{b} \tag{19}$$

where n refers to the time step t^n and i to the number of update in the coefficients. Then the matrix $\mathbf{A}$ is evaluated with the values of S_w, S_o, and P_o at iteration $i-1$. With this matrix, new values of S_w, S_o, and P_o are calculated. The update is performed until the values of S_w, S_o, and P_o converge.

To solve Eq. (19), initial and boundary conditions were specified [Aziz and Settari, 1979]. The saturation values along the core are the initial conditions:

$$S_w(x, t=0) = S_{wi}(x) \tag{20}$$

$$S_o(x, t=0) = S_{oi}(x) \tag{21}$$

At the inlet face and in the envelope of the core, reflection boundary conditions apply:

$$\left(\frac{\partial P_o}{\partial x}\right) - \gamma\left(\frac{\partial z}{\partial x}\right) = 0 \tag{22}$$

$$\left(\frac{\partial S_w}{\partial x}\right) = \left(\frac{\partial S_o}{\partial x}\right) = 0 \tag{23}$$

At the outlet face, before the capillary pressures are zero, reflection boundary conditions apply (Eqs. (22) and (23)). If P_{cgo} or P_{cow} becomes zero, then $P_o = P_{ref}$, or $P_w = P_{ref}$ respectively, where P_{ref} is the reference pressure at the outlet face of the core sample.

A representation for the relative permeability and capillary pressure curves as functions of saturation must be assumed to solve the model. The coefficients in these representations are estimated by the parameter estimation method presented in this paper. Since, in general, the relative permeabilities and capillary pressure are functions of two saturations, we used bivariate normalized splines (B-splines) to represent the flow functions [de Boor, 1978]. For example, the following expression is a representation for a relative permeability function:

$$k_{ri}(S_w, S_g) = \sum_{j=1}^{N_i} a_j^i B_j^m(S_w, S_g) \qquad i = o, w, g \qquad (24)$$

where a_j^i is the coefficient of the B-spline B_j in the representation of function i. Watson *et al.* [1988] showed that B-splines can accurately represent two-phase flow functions varying N_i, m, or B_j in Eq. (24). For the three-phase case B_j and m can be selected and N_i can be chosen as large as necessary to represent the relative permeability functions accurately [de Boor, 1978].

3. Estimation Procedure

In the parameter estimation procedure [Watson *et al.*, 1988] the coefficients in the B-spline representation of the flow functions are estimated in terms of a minimization of an objective function. This objective function is the sum of squared differences between measured data and predicted data with the model of the experiment:

$$J(\vec{\beta}) = [\vec{y} - \vec{f}(\vec{\beta})]^T W [\vec{y} - \vec{f}(\vec{\beta})] \qquad (25)$$

In this equation, $\vec{y}$ is the vector of measured data in the experiment. The vector $\vec{f}(\vec{\beta})$ corresponds to the values predicted with the mathematical model of the experiment. These predicted values are computed with the current estimates of the unknown vector of parameter values $\vec{\beta}$. W is a weighting matrix chosen according to statistical criteria. Usually, it is a diagonal matrix with entries equal to the inverse of the variance of the data [Weisberg, 1985].

All the coefficients of the relative permeabilities and capillary pressure functions to be estimated are included in the vector of parameters $\vec{\beta}$. Representing the flow functions by using B-splines, the vector $\vec{\beta}$ in Eq. (25) is

$$\vec{\beta} = [a_1^o, a_2^o, \ldots, a_{N_o}^o, a_1^w, \ldots, a_{N_w}^w, a_1^g, \ldots, a_{N_g}^g, c_1^1, \ldots, c_{N_1}^1, c_1^2, \ldots, c_{N_2}^2]^T \qquad (26)$$

where o, w and g represent oil, water and gas, respectively, and c_k^j are the coefficients for the two capillary pressure representations ($j = 1, 2$). In the parameter estimation procedure it may be convenient that the estimates of the functions have some characteristic shape. For example, it is desirable that the flow functions be monotonic with saturation. This can be implemented by using linear inequality constraints with the minimization of Eq. (25) [Richmond, 1988; Richmond and Watson, 1990]

$$G\vec{\beta} \geq \vec{g} \qquad (27)$$

The form of the matrix G and the vector $\vec{g}$ is defined knowing that the function is monotonic if the sequence of coefficients in the B-spline representation is monotonic [de Boor, 1978].

We used a Levenberg–Marquardt algorithm with trust region subject to linear inequality constraints to minimize Eq. (25) [Watson *et al.*, 1988]. Because of the relatively large amounts of computer time that can be required to solve the estimation procedure, it is desirable that the algorithm to be used in the minimization of Eq. (25) be as efficient as possible. The CRAY supercomputer was used to solve the estimation procedure in a reasonable time.

4. Discussion

We have used the parameter estimation procedure in several simulated and experimental drainage and imbibition two-phase dynamic displacement experiments [Richmond, 1988; Richmond and Watson, 1990] and centrifuge experiments [Nordtvedt *et al.*, 1993]. Good estimates of the flow functions were obtained by production and pressure drop data. However, in some cases local minima were reached. Simulated cases were created by assuming true flow functions, simulating an experiment and randomizing collected data. The randomized data are used to estimate the true functions given an initial guess. With new CT scanning techniques, in situ saturation data can be measured from displacement experiments. For the same simulated case where local minima were reached, these were not found and better estimates of the flow functions were obtained when in situ saturation data were included in the estimation procedure [Mejia-Velazquez, 1992]. Experimental production, pressure drop and in situ saturation data were used at ARCO to estimate flow functions for two samples. In one sample we modeled the displacement experiment in one and two dimensions, at low and high flow rates. We analyzed the other sample with a one-dimensional representation. Information about heterogeneities of the samples was included in the estimation procedure. In all cases we obtained a better prediction of the flow experiments than results that consider the sample as homogeneous and one-dimensional and neglect capillary effects in the estimation procedure [Mejia *et al.*, 1995]. Results for a three-phase simulated case were obtained. The method showed good estimates of the flow functions [Mejia-Velazquez, 1992], indicating its great potential to describe multiphase displacement experiments.

Central processing unit (CPU) time to estimate the flow functions in a CRAY Y-MP ranged from 2 to 3 minutes for a two-phase flow, homogeneous, one-dimensional, 10 grid simulated experiment, to several hours for an experimental two-phase flow, heterogeneous, two-dimensional, 9×24 grid system. In the three-phase simulated experiment, for a one-dimensional, homogeneous, 10 grid system, CPU time ranged from 2000 to 3000 seconds, depending on the initial guess. Although the method has proved successful in estimating two-phase flow functions, more three-phase cases need to be analyzed to study its performance. The model may be modified also for field optimization and assessment of different subsurface contamination scenarios.

5. Conclusions

• We have presented and discussed a new method for estimating simultaneously relative permeability and capillary pressure flow functions. The method is an extension of a regression-based procedure and has few of the assumptions made by other methods. It seems to be a reliable technique for obtaining good estimates of the flow functions.

• The mathematical model considers heterogeneities in porosity, residual and initial saturations, as well as anisotropy in permeability. Accounting for all known heterogeneities reduces modeling errors. For the first time in the three-phase case, in the mathematical model the flow functions are assumed to be dependent on two saturations.

• As a new feature, the model can use in situ saturation in addition to production and pressure drop data in the estimation procedure. Several works show that using in situ saturation data in the estimation produces better estimates of the flow functions and better prediction of displacement of phases; also local minimum values of the objective function may be avoided.

- The computer code is designed to take advantage of vectorization capabilities in a CRAY Y-MP. The use of a supercomputer allows one to obtain results in a reasonable amount of time since many computations are necessary to reach an optimum solution. Future work is necessary to test the performance of our parameter estimation procedure in the three-phase flow case.

Acknowledgments. One of the authors (G.M.M.) thanks CONACYT for partial funding of his research work at Texas A&M University.

REFERENCES

Abriola, L. M., *Multiphase Migration of Organic Compounds in a Porous Medium*, Springer, 1984.

Aziz, K., and A. Settari, *Petroleum Reservoir Simulation*, Applied Science, 1979.

Collins, R. E., *Flow of Fluids Through Porous Materials*, The Petroleum Publishing Company, 1976.

de Boor, C., *A Practical Guide to Splines*, Springer-Verlag, 1978.

Donaldson, E. L., and G. W. Dean, Two- and Three-Phase Relative Permeability Studies, *RI 6826*, U.S. Bureau of Mines, 1966.

Kool, J. B., J. C. Parker, and M. T. van Genuchten, Parameter estimation for unsaturated flow and transport models: A Review, *J. Hydrology*, **91**, 255–293, 1987.

Luckner, L., and W. Schestakow, *Migration Processes in the Soil and Groundwater Zone*, Lewis Publishers, 1991.

Mejia-Velazquez, G. M., A method for estimating three-phase flow functions, Ph.D. thesis, Texas A&M University, College Station, TX, 1992.

Mejia, G. M., K. K. Mohanty, and A. T. Watson, Use of in situ saturation data in estimation of two-phase flow functions in porous media, *J. Petroleum Sci. Eng.*, **12**, 233–245, 1995.

Mohanty, K. K., and A. E. Miller, Factors influencing unsteady relative permeability of a mixed-wet reservoir rock, *SPE Form. Eval.*, **6**, 349–358, 1991.

Nordtvedt, J. E., G. Mejia, P. Yang, and A. T. Watson, Estimation of capillary pressure and relative permeability functions from centrifuge experiments, *SPE Reservoir Eng.*, **8**, 292–298, 1993.

Oak, M. J., L. E. Baker, and D. C. Thomas, Three-phase relative permeability of Berea sandstone, *J. Pet. Technol.*, August, 1054–1061, 1990.

Parker, J. C., R. J. Lenhard, and T. Kuppusamy, Physics of immiscible flow in porous media (Project Summary), *U.S. Environmental Protection Agency, EPA/600/S2-87/101*, Ada, OK, 1988.

Richmond, P. C., Estimating multiphase flow functions in porous media from dynamic displacement experiments, Ph.D. thesis, Texas A&M University, College Station, TX, 1988.

Richmond, P. C. and A. T. Watson, Estimation of multiphase flow functions from displacement experiments, *SPE Reservoir Eng.*, **5**, 121–127, 1990.

Saraf, D. N., J. P. Batycky, C. H. Jackson, and D. B. Fisher, An experimental investigation of three-phase flow to water/oil/gas mixtures through water-wet sandstones, paper SPE 10761 presented at the SPE California Regional Meeting, San Francisco, CA, March 24–26, 1982.

Sarem, A. M., Three-phase relative permeability measurements by unsteady-state method, *Soc. Pet. Eng. J.*, September, 199–205, 1966.

Watson, A. T., P. C. Richmond, P. D. Kerig, and T. M. Tao, A regression-based method for estimating relative permeabilities from displacement experiments, *SPE Reservoir Eng.*, **3**, 953–958, 1988.

Weisberg, S., *Applied Linear Regression*, John Wiley & Sons, 1985.